Wide-Bandgap Device Application: Devices, Circuits, and Drivers

Wide-Bandgap Device Application: Devices, Circuits, and Drivers

Guest Editors

Yimeng Zhang
Lejia Sun
Yuming Zhang

MDPI

Basel • Beijing • Wuhan • Barcelona • Belgrade • Novi Sad • Cluj • Manchester

Guest Editors
Yimeng Zhang
School of Microelectronics
Xidian University
Xi'an
China

Lejia Sun
School of Microelectronics
Xidian University
Xi'an
China

Yuming Zhang
School of Microelectronics
Xidian University
Xi'an
China

Editorial Office
MDPI AG
Grosspeteranlage 5
4052 Basel, Switzerland

This is a reprint of the Special Issue, published open access by the journal *Electronics* (ISSN 2079-9292), freely accessible at: www.mdpi.com/journal/electronics/special_issues/LZ765U6R4U.

For citation purposes, cite each article independently as indicated on the article page online and using the guide below:

Lastname, A.A.; Lastname, B.B. Article Title. *Journal Name* **Year**, *Volume Number*, Page Range.

ISBN 978-3-7258-3216-3 (Hbk)
ISBN 978-3-7258-3215-6 (PDF)
https://doi.org/10.3390/books978-3-7258-3215-6

Contents

electronics

Communication

Enhanced Short-Circuit Robustness of 1.2 kV Split Gate Silicon Carbide Metal Oxide Semiconductor Field-Effect Transistors for High-Frequency Applications

Kanghee Shin, Dongkyun Kim, Minu Kim, Junho Park and Changho Han *

School of Electronic Engineering, Kumoh National Institute of Technology, Gumi 39177, Gyeongbuk, Republic of Korea; khshin@kumoh.ac.kr (K.S.); dk1217@kumoh.ac.kr (D.K.); gmu0054@kumoh.ac.kr (M.K.); pjh1106@kumoh.ac.kr (J.P.)
* Correspondence: changho.han@kumoh.ac.kr

Abstract: Split Gate SiC MOSFETs (SG-MOSFETs) have been demonstrated to exhibit excellent power dissipation at high operating frequencies due to their low specific reverse transfer capacitance ($C_{rss,sp}$); however, there are several reliability issues of SG-MOSFETs, including electric field crowding at the gate oxide and insufficient short-circuit (SC) robustness. In this paper, we propose a device structure to enhance the short-circuit withstand time ($SCWT$) of 1.2 kV SG-MOSFETs. The proposed P-shielded SG-MOSFETs (PSG-MOSFETs) feature a P-shielding region that expands the depletion region within the JFET region under both blocking mode and SC conditions. Compared to the conventional structure, this reduces the maximum electric field in the gate oxide, enabling a higher doping concentration in the JFET region, which can reduce the specific on-resistance ($R_{on,sp}$) to minimize power dissipation during device operation. The SC robustness of PSG-MOSFETs, with an $R_{on,sp}$ identical to those of SG-MOSFETs, was investigated by adjusting the width of the P-shielding region (W_P). Furthermore, the $C_{rss,sp}$ of PSG-MOSFETs was compared with that of SG-MOSFETs to analyze the relationship between the W_P and high-frequency figure of merit (HF-FOM), defined as $R_{on,sp} \times C_{rss,sp}$. These results demonstrate that the PSG-MOSFET achieved an enhanced SC robustness and HF-FOM in comparison to the SG-MOSFET. Thus, the proposed PSG-MOSFET is a highly suitable candidate for high frequency and reliable applications.

Keywords: silicon carbide; metal oxide semiconductor field-effect transistors; split gate; short-circuit; high-frequency figure-of-merit

Academic Editors: Yimeng Zhang, Lejia Sun and Yuming Zhang

Received: 1 December 2024
Revised: 27 December 2024
Accepted: 30 December 2024
Published: 3 January 2025

Citation: Shin, K.; Kim, D.; Kim, M.; Park, J.; Han, C. Enhanced Short-Circuit Robustness of 1.2 kV Split Gate Silicon Carbide Metal Oxide Semiconductor Field-Effect Transistors for High-Frequency Applications. *Electronics* **2025**, *14*, 163. https://doi.org/10.3390/electronics14010163

1. Introduction

Silicon carbide (SiC) devices, including Schottky barrier diodes (SBDs) and metal oxide semiconductor field-effect transistors (MOSFETs), have become key components in power electronic systems [1–3]. SiC devices offer a low specific on-resistance ($R_{on,sp}$), superior blocking capability, and excellent thermal conductivity compared to silicon (Si) insulated gate bipolar transistors (IGBTs) [4–6]. Furthermore, the fast switching speed of SiC devices, operating at frequencies exceeding 100 kHz, can achieve the advantage of reducing the volume of passive components in power electronic systems [7]. In particular, the utilization of SiC MOSFETs as switching components is becoming increasingly prevalent in DC–DC converters, on-board chargers, and inverters for electric vehicles due to their superior performance [8–10].

Reducing the $R_{on,sp}$ and specific reverse transfer capacitance ($C_{rss,sp}$) is crucial for minimizing lower power dissipation, as a significant portion of the power dissipation occurs during the switching operation [11,12]. In particular, during the charging of the $C_{rss,sp}$, a simultaneous application of high voltage and current occurs, which can result in a significant power dissipation in SiC MOSFETs. Thus, it is crucial to achieve a lower high-frequency figure of merit (HF-FOM), defined as $R_{on,sp} \times C_{rss,sp}$ [13,14]. It has been demonstrated that the superior HF-FOM of 1.2 kV Split Gate SiC MOSFETs (SG-MOSFETs) can be achieved without an increase in $R_{on,sp}$ [15–20]. However, electric field crowding occurs at the exposed edge of the poly-Si gate of SG-MOSFETs in the blocking mode, which may cause potential issues with the gate oxide [20–22].

Ruggedness and reliability, particularly short-circuit (SC) robustness, are also vital considerations for SiC MOSFETs. SiC MOSFETs exhibit a shorter short-circuit withstand time ($SCWT$) due to the higher electric field and current density compared to Si IGBTs [23,24]. Under the SC conditions, the lattice temperature of SiC MOSFETs rises rapidly, leading to thermal runaway and surges in drain current (I_D). These phenomena have the significant potential to result in device failure [24,25]. Several approaches for improving the SC robustness of SiC MOSFETs have been reported, but they should entail an increase in $R_{on,sp}$ [25–30]. Additionally, the electric field crowding at the gate oxide of SG-MOSFETs results in a shorter $SCWT$ than conventional MOSFETs (C-MOSFETs) [31]. Nevertheless, limited research has focused on enhancing the SC robustness of SG-MOSFETs.

In this paper, we proposed an approach to enhance the trade-off between the SC robustness and $R_{on,sp}$ of 1.2 kV Split Gate SiC MOSFETs (SG-MOSFETs). The proposed P-shielded SG-MOSFETs (PSG-MOSFETs) feature a P-shielding region within the JFET region, which serves to expand the depletion region under SC conditions and in blocking mode. The depletion region induced by the P-shielding region can narrow the current path in the JFET region, which results in the suppression of the surge current under SC conditions. In blocking mode, the expanded depletion region of PSG-MOSFETs can serve to protect the gate oxide from the electric field and reduce the reverse transfer capacitance (C_{rss}).

2. Device Structures and Simulation Methods

Sentaurus TCAD simulations were employed to design and analyze 1.2 kV SiC MOSFETs based on a half-cell size. Figure 1a–c show the 2-D schematic cross-sectional views of the 1.2 kV SiC C-MOSFET, SG-MOSFET, and PSG-MOSFET, respectively. A SiC N-drift region with a background doping concentration of 8×10^{15} cm^{-3} and a thickness of 10 μm on a SiC N^{+} substrate was implemented. All devices were designed with an inversion channel, an identical channel length of 0.5 μm and a gate oxide thickness of 50 nm. In order to ensure the reliability of the gate oxide, the SG-MOSFET and PSG-MOSFET were designed with an identical length of the overlap region of 0.3 μm between the poly-Si gate and JFET region. The JFET region was designed with a width of 1.5 μm, and a depth of 0.8 μm. Additionally, the JFET region of the C-MOSFET and SG-MOSFET was designed with a doping concentration (N_{JFET}) of 3×10^{16} cm^{-3}. Note that PSG-MOSFETs were designed with a higher N_{JFET} than C-MOSFETs and SG-MOSFETs due to their reduced $E_{ox,max}$. This can reduce the $R_{on,sp}$, which can minimize power dissipation during device operation.

All devices were created through following process simulations, and voltage was applied to each terminal of the devices to analyze its electrical characteristics through device simulations. Figure 2 shows the process simulation flow for the PSG-MOSFET. The PSG-MOSFET followed the same process flow as the C-MOSFET and SG-MOSFET. Note that all ion implantation processes used the Monte Carlo method for precision of simulations. The P-shielding region within the JFET region was formed through ion implantation for the P-source without any additional patterning steps. Additionally, the

P-shielding region was designed with a depth of 0.5 μm and a width (W_P) ranging from 0.3 μm to 0.9 μm. Furthermore, it should be noted that the P-shielding region is electrically connected to the source, which effectively improves the electrical characteristics of PSG-MOSFETs. In order to achieve this connection, a contact placement for simulation was created on the surface of the P-shielding region and subsequently merged with the source contact placement. Table 1 lists all device parameters used in TCAD simulations. Note that all device simulations were conducted with the same device area.

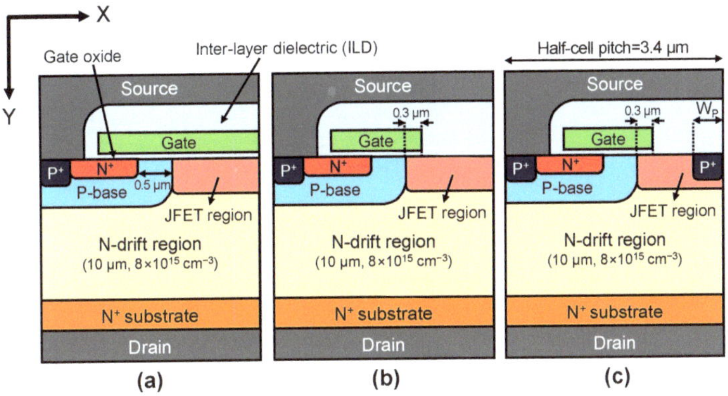

Figure 1. 2-D Schematic cross-sectional views of the (**a**) C-MOSFET, (**b**) SG-MOSFET and (**c**) PSG-MOSFET.

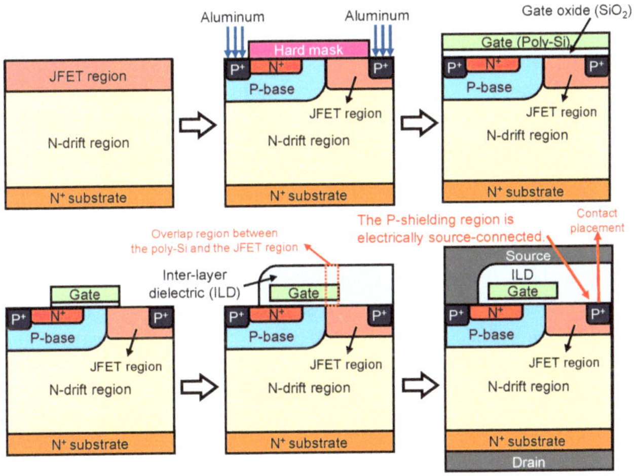

Figure 2. Process simulation flow for the PSG-MOSFET.

In order to enhance the precision of TCAD simulations, the following models were employed: the Shockley–Read–Hall recombination model and the Auger recombination model were used to predict the electrical characteristics of devices under different temperature conditions [32,33]; the Okuto–Crowell avalanche model was used to accurately predict the reliability and breakdown mechanisms of high-voltage devices [34,35]; and finally, the incomplete ionization model was used to accurately predict and design the conductivity of devices with high doping concentration [36]. The breakdown voltage (BV) was defined as the drain-source voltage (V_{DS}) when the maximum electric field in SiC, $E_{ox,max}$ or I_D exceeded 3 MV/cm, 5 MV/cm or 10 μA, respectively. Furthermore, the $R_{on,sp}$ was defined as the inverse of the slope at a V_{DS} of 2.3 V from the output curve under a gate-source voltage (V_{GS}) of 18 V, and the threshold voltage (V_{th}) was defined as the gate voltage at which the I_D reached 1 mA, following the extraction of the transfer curve under a V_{DS} of 5 V. Finally, the *SCWT* was defined as the gate pulse duration at which the drain current surges and thermal runaway occurs under SC conditions [24–31].

Table 1. Device parameters used in TCAD simulations.

Parameters [Unit]	Values
Channel length [μm]	0.5
Gate oxide thickness [nm]	50
Half-cell pitch [μm]	3.4
Width of the JFET region [μm]	1.5
Depth of the JFET region [μm]	0.8
Thickness of the N-drift region [μm]	10
Length of the overlap region between the poly-Si gate and the JFET region [μm]	0.3
Width of the P-shielding region (W_P) [μm]	0.3–0.9
Depth of the P-base [μm]	1.0
Doping concentration of the N-drift region [cm^{-3}]	8×10^{15}
Doping concentration of the JFET region (N_{JFET}) [cm^{-3}]	3×10^{16}–1×10^{17}
Doping concentration of the P-shielding region [cm^{-3}]	1.2×10^{20}

3. Results and Discussion

Figure 3 shows the electric field distribution of the SG-MOSFET in blocking mode with a V_{GS} of 0 V and a drain-source voltage (V_{DS}) of 1.2 kV. The SG-MOSFET exhibited an $E_{ox,max}$ of 4.79 MV/cm because electric field crowding occurs at the exposed edge of the poly-Si gate. The critical electric field of SiO_2 in SiC MOSFETs is known to be 8–9 MV/cm [37]; however, the $E_{ox,max}$ should be to 3 MV/cm or less to ensure the reliability of the gate oxide in the device [38,39]. In contrast, the PSG-MOSFET with a W_P of 0.3 μm achieved an $E_{ox,max}$ of 2.95 MV/cm, representing a 38.28% reduction compared to the SG-MOSFET. This reduced $E_{ox,max}$ of the PSG-MOSFET is due to the dispersion of the electric field from the gate oxide into the P-shielding region, which results in a maximum electric field in the P-shielding region of 2.32 MV/cm. In other words, the high electric field in the gate oxide is dispersed by the on-sided depletion region that occurs at the PN junction between the additional P-shielding region and the JFET region. Accordingly, this results in the suppression of the electric field crowding at the gate oxide of the PSG-MOSFET.

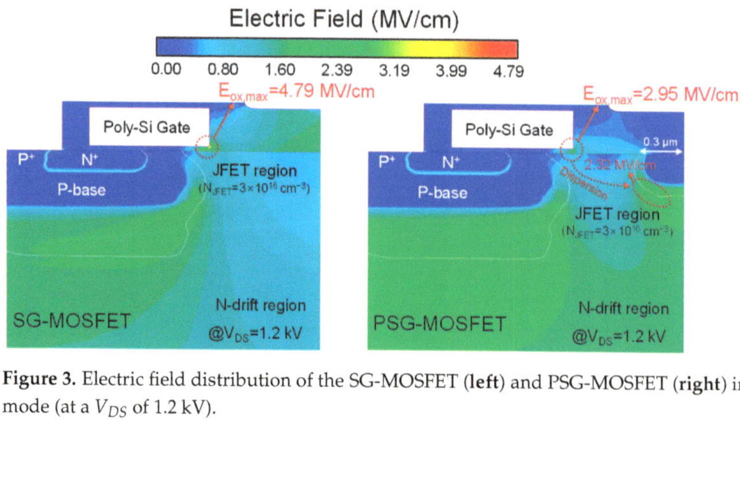

Figure 3. Electric field distribution of the SG-MOSFET (**left**) and PSG-MOSFET (**right**) in blocking mode (at a V_{DS} of 1.2 kV).

Figure 4 shows the $E_{ox,max}$ of PSG-MOSFETs according to the W_P and N_{JFET}. As the N_{JFET} increases, the $E_{ox,max}$ also increases due to a reduction in the width of the depletion region of the P-shielding region. Conversely, as the W_P increases, the $E_{ox,max}$ significantly decreases. Since an increase in $R_{on,sp}$ denotes an increase in conduction loss during device operation, optimization of N_{JFET} and W_P is crucial. These results indicate that the P-shielding region can effectively suppress electric field crowding at the gate oxide, despite the higher N_{JFET} of PSG-MOSFETs compared to the SG-MOSFET.

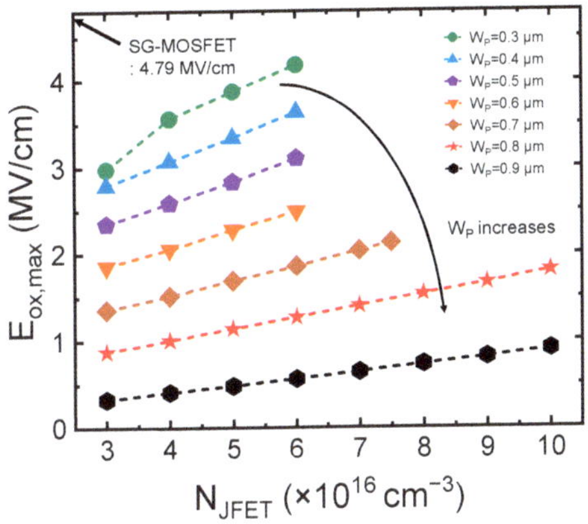

Figure 4. $E_{ox,max}$ of PSG-MOSFETs according to the W_P and N_{JFET}.

Figure 5 shows the $R_{on,sp}$ of PSG-MOSFETs according to the W_P and N_{JFET}. As the N_{JFET} increases, the $R_{on,sp}$ decreases due to the expansion of the current path of the device in the conduction mode. Conversely, as the W_P increases, the $R_{on,sp}$ significantly elevates. In particular, the PSG-MOSFET (W_P = 0.9 μm and $N_{JFET} = 1 \times 10^{17}$ cm^{-3}) exhibits a $R_{on,sp}$ of 11.41 mΩ·cm^2. In order to achieve an identical $R_{on,sp}$ as the SG-MOSFET, it is required to design the N_{JFET} of the PSG-MOSFET with a W_P of 0.9 μm to be much higher than 1×10^{17} cm^{-3}. However, the implementation of a significantly higher N_{JFET} results in a reduction in the V_{th} of the device. For the design of the inversion channel, the W_P with a range of 0.3 μm to 0.8 μm was employed, and devices with an identical $R_{on,sp}$ as the SG-MOSFET were selected, as shown in Table 2.

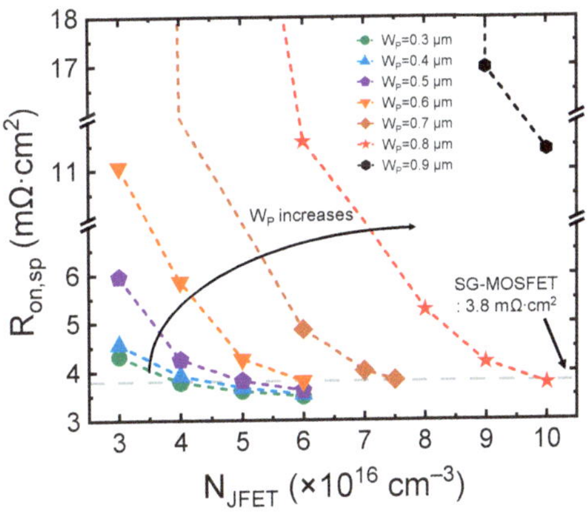

Figure 5. $R_{on,sp}$ of PSG-MOSFETs according to the W_P and N_{JFET}.

Table 2. Device structures with identical $R_{on,sp}$ as SG-MOSFETs.

Descriptions	Conditions [Unit]		$R_{on,sp}$ [mΩ·cm^2]	V_{th} [V]
	W_P [μm]	N_{JFET} [cm^{-3}]	(@V_{GS} = 18 V and V_{DS} = 2.3 V)	(@I_D = 1 mA)
Device #A	0.3	4×10^{16}	3.78	3.84
Device #B	0.4	5×10^{16}	3.67	3.67
Device #C	0.5	5×10^{16}	3.81	3.67
Device #D	0.6	6×10^{16}	3.80	3.52
Device #E	0.7	7.5×10^{16}	3.82	3.37
Device #F	0.8	1×10^{17}	3.75	2.96

TCAD mixed-mode simulations were conducted to analyze the SC robustness of devices. Figure 6 shows the schematic implemented in the SC test. The gate voltage (V_G) was pulsed from 0 V to 20 V in 0.1 μs. Additionally, an 800 V voltage source (V_{DD}) was employed at the drain stage. The external gate resistance (R_G) was set to 20 Ω, and the source/drain resistance (R_S and R_D) and inductance (L_S and L_D) were set to 1 mΩ and 1 nH, respectively.

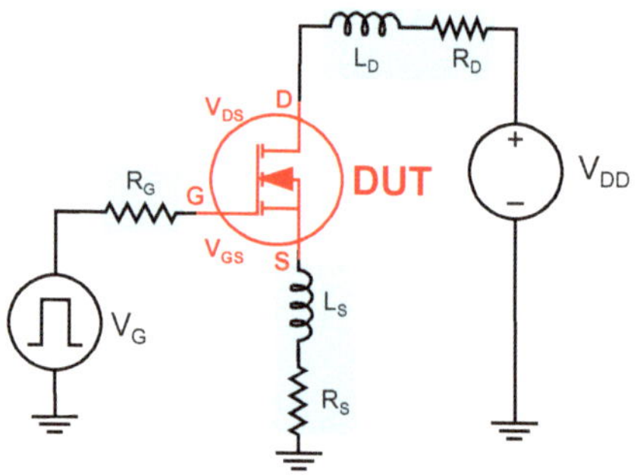

Figure 6. Schematic of SC test implemented in the TCAD mixed-mode simulation.

Figure 7 compares the *SCWT* and $R_{on,sp}$ of all the devices. The C-MOSFET has a $R_{on,sp}$ of 3.65 mΩ·cm^2 and *SCWT* of 5.8 μs. By comparison, the SG-MOSFET exhibits a 4.11% increase in $R_{on,sp}$ (3.8 mΩ·cm^2) and an 8.62% decrease in *SCWT* (5.3 μs), indicating a less favorable trade-off between *SCWT* and $R_{on,sp}$ than the C-MOSFET. In contrast, the PSG-MOSFET achieves a superior trade-off between *SCWT* and $R_{on,sp}$ as the W_P increases. Accordingly, it is observed that Device #F ($N_{JFET} = 1 \times 10^{17}$ cm^{-3} and W_P = 0.8 μm) exhibits the best trade-off between *SCWT* and $R_{on,sp}$. Note that PSG-MOSFETs did not make any sacrifices of static characteristics, including the $R_{on,sp}$, to achieve enhanced SC robustness.

Figure 8 shows the SC characteristics of all the devices. Additionally, the solid line indicates the I_D, while the dotted line denotes the maximum lattice temperature. Note that all the devices exhibited thermal runaway and drain surge current at the gate pulse duration of the *SCWT*. PSG-MOSFETs demonstrate enhanced SC robustness compared to the SG-MOSFET with identical $R_{on,sp}$. It is important to note that the proposed device reduces the probability of device failure due to thermal runaway under SC conditions, and does not require the sacrifice of static characteristics such as the $R_{on,sp}$ to achieve this. Meanwhile, as W_P increases, PSG-MOSFETs exhibit a decrease in the peak drain current under SC conditions (I_{SC}) and peak value in maximum lattice temperature (T_{peak}).

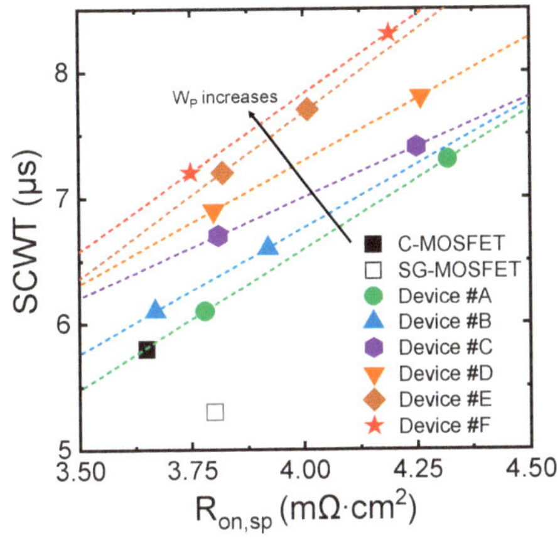

Figure 7. Comparison of *SCWT* and $R_{on,sp}$ of C-MOSFET, SG-MOSFET, and Devices #A–#F.

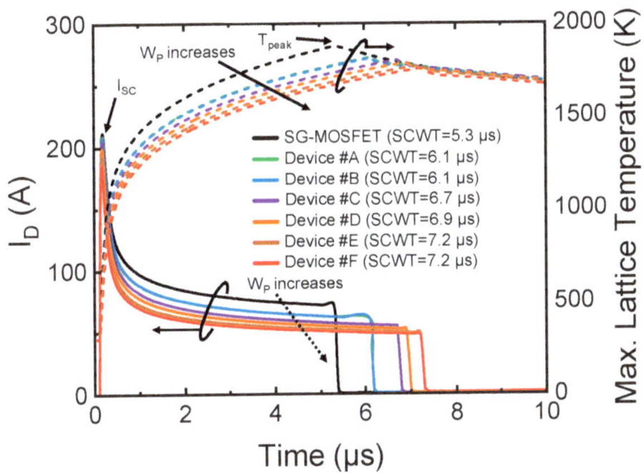

Figure 8. SC characteristics of the C-MOSFET, SG-MOSFET, and Devices #A–#F.

Figure 9 shows the I_{SC} of the SG-MOSFET and PSG-MOSFETs under SC conditions. It should be noted that it is a comparison of the I_{SC} at the *SCWT* duration for each device. Device #F, which shows the best trade-off between *SCWT* and $R_{on,sp}$, exhibits an I_{SC} of 186 A, representing a 12.62% reduction relative to the SG-MOSFET. This reduction in I_{SC} can contribute to a lower rate of increase in lattice temperature (R_T) under SC conditions, effectively suppressing the positive feedback loop of the lattice temperature, which can enhance *SCWT*. Accordingly, Device #F exhibits an *SCWT* of 7.2 μs, representing a 35.85% enhancement in comparison to the SG-MOSFET.

Figure 10 shows the output characteristics of the SG-MOSFET and PSG-MOSFETs. Due to the comparable slope of devices at low V_{DS}, PSG-MOSFETs can achieve a $R_{on,sp}$ similar to that of the SG-MOSFET. In contrast, it is observed that PSG-MOSFETs rapidly saturate the drain current at higher V_{DS} compared to the SG-MOSFET. Furthermore, this tendency appears to intensify as the W_P increases. This indicates that the I_{SC} is proportional to the drain saturation current ($I_{D,sat}$). Figure 11 shows the total current density of the SG-MOSFET (on the left) and Device #F (on the right) in conduction mode (at a V_{DS} of 20 V) and under SC conditions (at a V_{DS} of 800 V). Device #F exhibits a wider depletion region within the JFET region compared to the SG-MOSFET in conduction mode and under SC conditions. The narrower current path facilitated by the P-shielding region can effectively suppress the $I_{D,sat}$ under SC conditions. According to Equation (1), the $I_{D,sat}$ is inversely proportional to $SCWT$ [40]. Thus, the reduction in $I_{D,sat}$ can enhance the SC robustness of PSG-MOSFETs.

$$SCWT = A_{chip} \times \frac{(T_{critical} - T_{initial}) \times W_{SiC} \times C_V}{I_{D,sat} \times V_{DS}} \quad (1)$$

where $T_{critical}$ is the critical temperature, $T_{initial}$ is the initial temperature, A_{chip} is the chip area, W_{SiC} is the thickness of the chip, C_v is the volumetric specific heat capacity, $I_{D,sat}$ is the drain saturation current, and V_{DS} is the drain-source voltage [40].

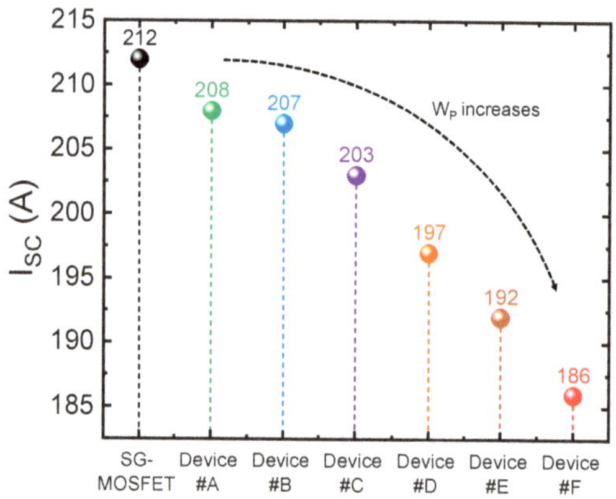

Figure 9. I_{SC} of the SG-MOSFET and Devices #A–#F under SC conditions.

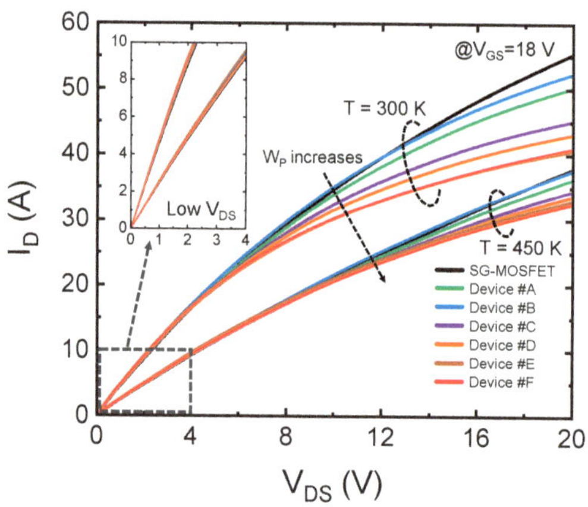

Figure 10. Output characteristics of the SG-MOSFET and Devices #A–#F.

Figure 12 shows the lattice temperature distribution of the SG-MOSFET and PSG-MOSFETs under SC conditions. It should be noted that it is a comparison of the lattice temperature distribution at the $SCWT$ duration for each device. As the W_P increases, the magnitude and area of lattice temperature of SiC is observed to decrease. In particular, Device #F exhibits a 6.21% reduction in T_{peak} compared to the SG-MOSFET. According to Equation (2), the $I_{D,sat}$ is proportional to the R_T [40]. Due to the reduction in $I_{D,sat}$, the positive feedback of the lattice temperature is effectively suppressed, which results in the enhanced SC robustness of the PSG-MOSFET.

$$R_T = \frac{dT}{dt} = \frac{K_T \times I_{D,sat} \times V_d}{A_{chip} \times W_{SiC} \times C_V} \quad (2)$$

where K_T is the compensation factor for the non-uniform temperature distribution through the wafer thickness [40].

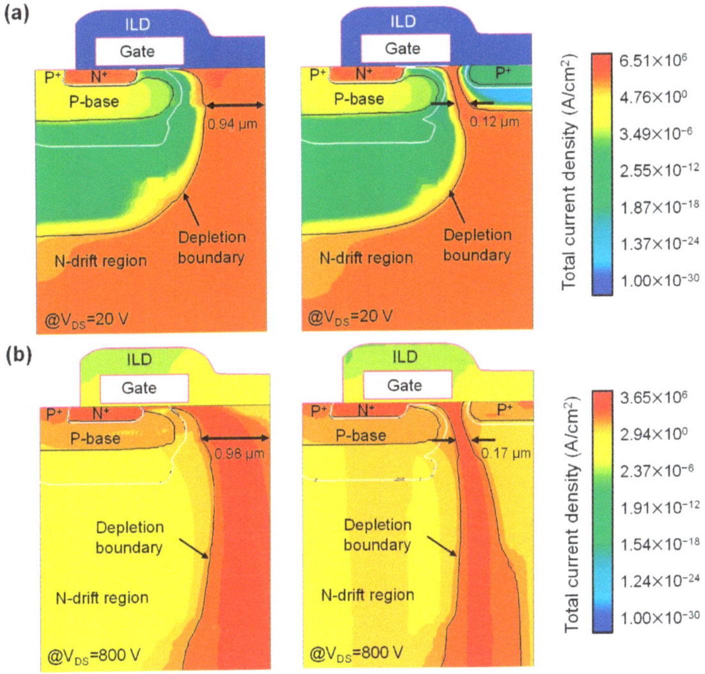

Figure 11. Total current density of the SG-MOSFET (**left**) and Device #F (**right**) in (**a**) conduction mode (at a V_{DS} of 20 V) and (**b**) under SC conditions (at a V_{DS} of 800 V).

The $C_{rss,sp}$ characteristics of all the devices were obtained at 1 MHz through TCAD mixed-mode simulation, as shown in Figure 13. Additionally, the $C_{rss,sp}$ is normalized as the device area. The SG-MOSFET exhibits lower $C_{rss,sp}$ than the C-MOSFET, which is a result of the split gate structure. However, the P-shielding region of PSG-MOSFETs has the effect of expanding the depletion region, which results in a reduction in depletion capacitance (C_{dep}). As a result, PSG-MOSFETs exhibit significantly lower $C_{rss,sp}$ than the SG-MOSFET, according to Equation (3). Meanwhile, as the W_P increases, the $C_{rss,sp}$ of PSG-MOSFETs is effectively further reduced.

$$C_{rss,sp} = \frac{1}{A_{active}} \times \frac{C_{ox} \times C_{dep}}{C_{ox} + C_{dep}} \quad (3)$$

where A_{active} is the area of the active cell, C_{ox} is the oxide capacitance, and C_{dep} is the depletion capacitance.

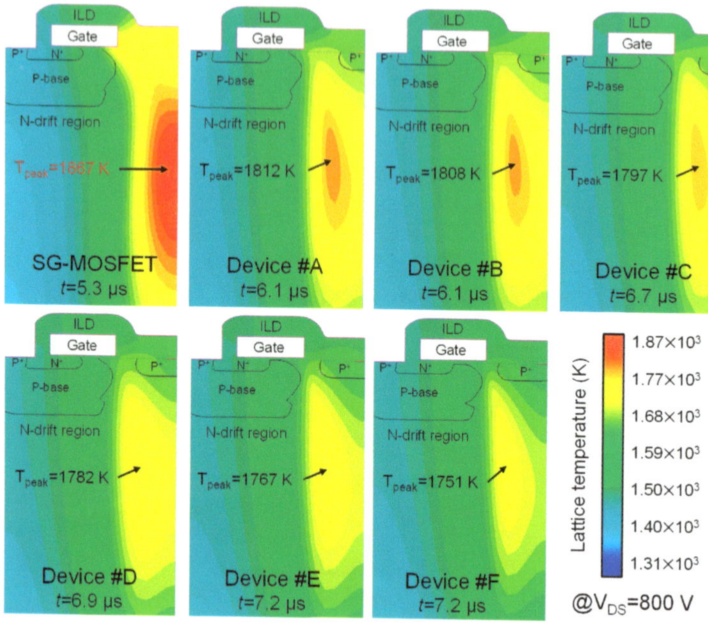

Figure 12. Lattice temperature distribution of the SG-MOSFET and Devices #A–#F under SC conditions.

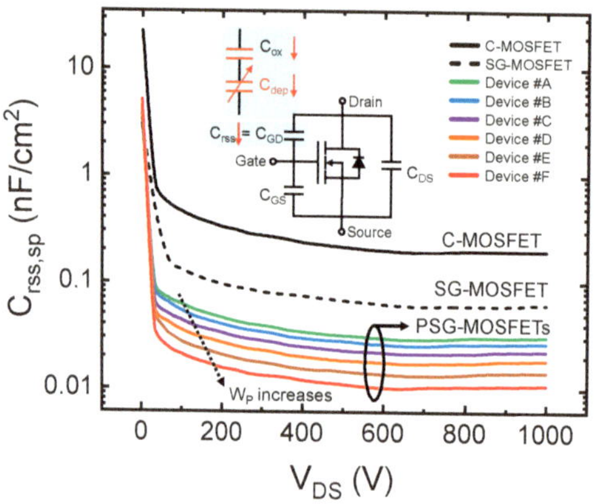

Figure 13. Specific reverse transfer capacitance characteristics of the C-MOSFET, SG-MOSFET, and Devices #A–#F.

Figure 14 compares the $R_{on,sp}$, $C_{rss,sp}$ and HF-FOM [$R_{on,sp} \times C_{rss,sp}$] across all the devices. As W_P increases, the HF-FOM of PSG-MOSFETs improves significantly. Accordingly, Device #F represents the optimized PSG-MOSFET, achieving the best HF-FOM and enhanced SC robustness compared to the C-MOSFET and SG-MOSFET.

Table 3 summarizes the TCAD simulation results in this work. The $E_{ox,max}$, $R_{on,sp}$, $SCWT$, $C_{rss,sp}$, and HF-FOM of Device #F were 1.81 MV/cm, 3.75 mΩ·cm², 7.2 μs, 10.87 pF/cm², and 0.041 pΩ·F, respectively. Compared to the SG-MOSFET, Device #F achieved a 62.21% decrease in $E_{ox,max}$, a 35.85% increase in $SCWT$, an 82.48% decrease in C_{rss}, and an 82.92% decrease in HF-FOM. This demonstrates that the P-shielding region of the PSG-MOSFET effectively enhances SC robustness and HF-FOM, without increasing the $R_{on,sp}$.

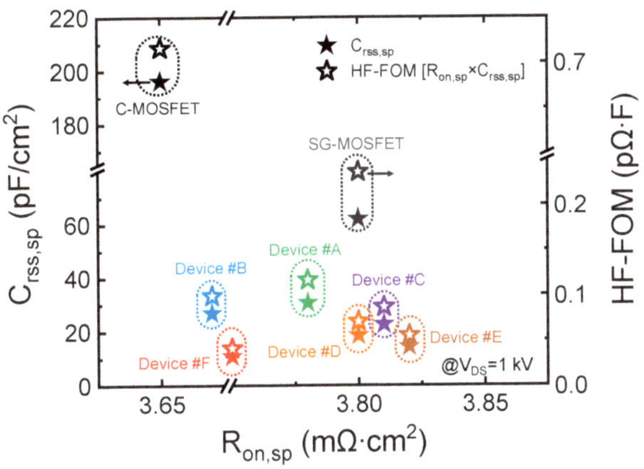

Figure 14. Comparison of the $R_{on,sp}$, $C_{rss,sp}$, and HF-FOM in the C-MOSFET, SG-MOSFET, and Devices #A–#F.

Table 3. Summary of TCAD simulation results.

Electrical Characteristics [Unit]	C-MOSFET	SG-MOSFET	Device #A	Device #B	Device #C	Device #D	Device #E	Device #F
$R_{on,sp}$ [mΩ·cm^2]	3.65	3.80	3.78	3.67	3.81	3.80	3.82	3.75
V_{th} [V]	4.0	4.0	3.84	3.67	3.67	3.52	3.37	2.96
BV [V]	1536	812	1649	1642	1752	1883	1817	1812
$E_{ox,max}$ [MV/cm]	4.08	4.79	3.56	3.35	2.84	2.49	2.13	1.81
$SCWT$ [μs]	5.8	5.3	6.1	6.1	6.7	6.9	7.2	7.2
$C_{rss,sp}$ [pF/cm^2] (@V_{DS} = 1 kV)	196.19	62.04	30.71	26.89	22.59	18.47	14.33	10.87
$R_{on,sp} \times C_{rss,sp}$ [pΩ·F] (@V_{DS} = 1 kV)	0.72	0.24	0.12	0.099	0.086	0.07	0.054	0.041

4. Conclusions

In this paper, 1.2 kV P-shielded Split Gate SiC MOSFETs were designed to enhance the SC robustness compared to SG-MOSFETs. In order to achieve this improvement, PSG-MOSFETs feature a P-shielding region within the JFET region, which can be formed without requiring additional patterning steps. The analysis of the $R_{on,sp}$, $E_{ox,max}$, and $SCWT$ of PSG-MOSFETs in relation to the N_{JFET} and W_P demonstrated that the optimal trade-off between SC robustness and $R_{on,sp}$ occurs at an N_{JFET} of 1×10^{17} cm^{-3} and a W_P of 0.8 μm. Under these conditions, the PSG-MOSFET achieved an $E_{ox,max}$ reduction to 0.38×, an $SCWT$ increase to 1.36×, and a HF-FOM [$R_{on,sp} \times C_{rss,sp}$] reduction to 0.17×, while maintaining a $R_{on,sp}$ identical to that of the SG-MOSFET. These results indicate that the proposed PSG-MOSFET is a highly suitable candidate for high-frequency and reliable applications.

Author Contributions: Conceptualization, K.S., D.K., M.K., J.P. and C.H.; data curation, K.S.; formal analysis, K.S. and D.K.; investigation, C.H.; project administration, C.H.; writing—original draft preparation, K.S.; writing—review and editing, K.S. and C.H. All authors have read and agreed to the published version of the manuscript.

Funding: This research was supported by the Korea Institute for Advancement of Technology (KIAT) grant funded by the Korea Government. (Ministry of Education-Ministry of Trade, Industry and Energy) (P0022161, Semiconductor Major Track).

Data Availability Statement: All data are presented in this paper in the form of figures.

Conflicts of Interest: The authors declare no conflicts of interest.

References

1. Millán, J.; Godignon, P.; Perpiñà, X.; Pérez-Tomás, A.; Rebollo, J. A Survey of Wide Bandgap Power Semiconductor Devices. *IEEE Trans. Power Electron.* **2014**, *29*, 2155–2163. [CrossRef]
2. She, X.; Huang, A.Q.; Lucía, Ó.; Ozpineci, B. Review of Silicon Carbide Power Devices and Their Applications. *IEEE Trans. Ind. Electron.* **2017**, *64*, 8193–8205. [CrossRef]
3. Kimoto, T. Material science and device physics in SiC technology for high-voltage power devices. *Jpn. J. Appl. Phys.* **2015**, *54*, 040103. [CrossRef]
4. Östling, M.; Ghandi, R.; Zetterling, C.-M. SiC power devices—Present status, applications and future perspective. In Proceedings of the 23rd IEEE International Symposium on Power Semiconductor Devices and IC's (ISPSD), San Diego, CA, USA, 23–26 May 2011.
5. Kimoto, T. Fundamentals, Commercialization, and Future Challenges of SiC Power Devices. In Proceedings of the 2023 IEEE International Meeting for Future of Electron Devices (IMFEDK), Kyoto, Japan, 16–17 November 2023.
6. Baliga, B.J. Trends in power semiconductor devices. *IEEE Trans. Electron Devices* **1996**, *43*, 1717–1731. [CrossRef]
7. Chen, Z.; Huang, A.Q. High Performance SiC Power Module Based on Repackaging of Discrete SiC Devices. *IEEE Trans. Power Electron.* **2023**, *38*, 9306–9310. [CrossRef]
8. Buffalo, M.; Favero, D.; Marcuzzi, A.; De Santi, C.; Meneghesso, G.; Zanoni, E.; Meneghini, M. Review and Outlook on GaN and SiC Power Devices: Industrial State-of-the-Art, Applications, and Perspectives. *IEEE Trans. Electron Devices* **2024**, *71*, 1344–1355. [CrossRef]
9. Losse, P.A.; Bolotnikov, A.; Yu, L.C.; Dunne, G.; Esler, D.; Erlbaum, J.; Rowden, B.; Gowda, A.; Halverson, A.; Ghandi, R.; et al. SiC MOSFET design considerations for reliable high voltage operation. In Proceedings of the 2017 IEEE International Reliability Physics Symposium (IRPS), Monterey, CA, USA, 2–6 April 2017.
10. Baliga, B.J. An Academic's Perspective on SiC Power Devices: Retrospection and Prognostication. In Proceedings of the 2022 IEEE Workshop Wide Bandgap Power Devices Applications (WIPDA), Redondo Beach, CA, USA, 7–9 November 2022.
11. Bernstein, J.B.; Bahl, S.; Schlecht, M.F. A low capacitance power MOSFET with an integral gate driver. In Proceedings of the 1987 IEEE Power Electronics Specialists Conference (PESC), Blacksburg, VA, USA, 21–26 June 1987.
12. Lorenz, L. Study of the switching performance of a power MOSFET circuit. In Proceedings of the 1984 IEEE Power Electronics Specialists Conference (PESC), Gaithersburg, MD, USA, 18–21 June 1984.
13. Han, K.; Baliga, B.J. The 1.2-kV 4H-SiC OCTFET: A New Cell Topology With Improved High-Frequency Figures-of-Merit. *IEEE Electron Device Lett.* **2018**, *40*, 299–302. [CrossRef]
14. Sung, W.; Han, K.; Baliga, B.J. Optimization of the JFET region of 1.2 kV SiC MOSFETs for improved high frequency figure of merit (HF-FOM). In Proceedings of the 2017 IEEE Workshop Wide Bandgap Power Devices Applications (WIPDA), Albuquerque, NM, USA, 30 October–1 November 2017.
15. Han, K.; Baliga, B.J.; Sung, W. Split-Gate 1.2-kV 4H-SiC MOSFET: Analysis and Experimental Validation. *IEEE Electron Device Lett.* **2017**, *38*, 1437–1440. [CrossRef]
16. Xu, H.-Y.; Wang, Y.; Bao, M.-T.; Cao, F. Low Switching Loss Split-Gate 4H-SiC MOSFET With Integrated Heterojunction Diode. *IEEE J. Electron Devices Soc.* **2022**, *10*, 554–561. [CrossRef]
17. Han, K.; Baliga, B.J. Operation of 1.2-kV 4H-SiC Accumulation and Inversion Channel Split-Gate (SG) MOSFETs at Elevated Temperatures. *IEEE Trans. Electron Devices* **2018**, *65*, 3333–3338. [CrossRef]
18. Zhou, X.; Yue, R.; Zhang, J.; Dai, G.; Li, J.; Wang, Y. 4H-SiC Trench MOSFET with Floating/Grounded Junction Barrier-controlled Gate Structure. *IEEE Trans. Electron Devices* **2017**, *11*, 4568–4574. [CrossRef]
19. Han, K.; Baliga, B.J. Analysis and Experimental Quantification of 1.2-kV 4H-SiC Split-Gate Octagonal MOSFET. *IEEE Electron Devices Lett.* **2019**, *40*, 1163–1166. [CrossRef]
20. Yu, H.; Wang, J.; Liang, S.; Deng, G.; Liu, H.; Ji, B.; Shen, Z.J. 1.2-kV silicon carbide planar split-gate MOSFET with source field plate for superior figure-of-merits. *IET Power Electron.* **2022**, *15*, 1502–1510. [CrossRef]
21. Vudumula, P.; Kotamraju, S. Design and Optimization of 1.2-kV SiC Planar Inversion MOSFET Using Split Dummy Gate Concept for High-Frequency Applications. *IEEE Trans. Electron Devices* **2019**, *66*, 5266–5271. [CrossRef]
22. Yu, H.; Wang, J.; Zhang, J.; Liang, S.; Shen, Z.J. Theoretical Analysis and Experimental Characterization of 1.2-kV 4H-SiC Planar Split-Gate MOSFET With Source Field Plate. *IEEE Trans. Electron Devices* **2024**, *71*, 1508–1512. [CrossRef]
23. Kim, D.; Sung, W. Improved Short-Circuit Ruggedness for 1.2kV 4H-SiC MOSFET Using a Deep P-Well Implemented by Channeling Implantation. *IEEE Electron Device Lett.* **2021**, *12*, 1822–1825. [CrossRef]
24. Diao, S.; Sun, J.; Zhou, Z.; Zhang, Z.; Schöner, A.; Zheng, Z.; He, W. Determination of failure degree of 1.2 kV SiC MOSFETs after short-circuit test using an improved test setup. *Nanotechnol. Precis. Eng.* **2020**, *3*, 235–240. [CrossRef]
25. Lin, C.; Ren, N.; Xu, H.; Liu, L.; Zhu, Z.; Sheng, K. 1.2-kV Planar SiC MOSFETs With Improved Short-Circuit Capability by Adding Plasma Spreading Layer. *IEEE Trans. Electron Devices* **2023**, *70*, 4730–4736. [CrossRef]

26. Hatta, H.; Tominaga, T.; Hino, S.; Miura, N.; Tomohisa, S.; Yamakawa, S. Suppression of Short-Circuit Current with Embedded Source Resistance in SiC-MOSFET. *Mater. Sci. Forum* **2018**, *924*, 727–730. [CrossRef]
27. Unger, C.; Pfost, M. Particularities of the Short-Circuit Operation and Failure Modes of SiC-MOSFETs. *IEEE J. Emerg. Sel. Topics Power Electron.* **2021**, *9*, 6432–6440. [CrossRef]
28. Kim, J.J.; Park, J.-H.; Sabri, S.; Fetzer, B.; Hull, B.; Ryu, S.-H. Investigation into Relationship of the Switching Performance and Short-Circuit Withstand Time on 1.2 kV 4H-SiC Power MOSFETs. In Proceedings of the IEEE International Symposium on Power Semiconductor Devices and IC's (ISPSD), Bremen, Germany, 2–6 June 2024.
29. Kim, D.; Yun, N.; Morgan, A.J.; Sung, W. The Effect of Deep JFET and P-Well Implant of 1.2 kV 4H-SiC MOSFETs. *IEEE J. Electron Devices Soc.* **2022**, *10*, 989–995. [CrossRef]
30. Yang, T.; Wang, Y.; Yue, R. SiC Trench MOSFET With Reduced Switching Loss and Increased Short-Circuit Capability. *IEEE Trans. Electron Devices* **2020**, *67*, 3685–3690. [CrossRef]
31. Liu, H.; Wei, J.; Wei, Z.; Liu, S.; Shi, L. Experimental Comparison of a New 1.2 kV 4H-SiC Split-Gate MOSFET with Conventional SiC MOSFETs in Terms of Reliability Robustness. *Electronics* **2023**, *12*, 2551. [CrossRef]
32. Schenk, A. A model for the field and temperature dependence of Schockley-Read-Hall lifetimes in silicon. *Solid-State Electron.* **1992**, *35*, 1585–1596. [CrossRef]
33. Klaassen, D.B.M. A unified mobility model for device simulation—II. Temperature dependence of carrier mobility and lifetime. *Solid-State Electron.* **1992**, *35*, 961–967. [CrossRef]
34. Okuto, Y.; Crowell, C.R. Energy-Conservation Considerations in the Characterization of Impact Ionization in Semiconductors. *Phys. Rev. B.* **1972**, *6*, 3076. [CrossRef]
35. Okuto, Y.; Crowell, C.R. Threshold energy effect on avalanche breakdown voltage in semiconductor junctions. *Solid-State Electron.* **1975**, *18*, 161–168. [CrossRef]
36. Scaburri, R.; Desalvo, A.; Nipoti, R. Incomplete Ionization of the n-Type Dopant Phosphorus in 4H-SiC, Including Screening by Free Carriers. *Mater. Sci. Forum* **2011**, *679–680*, 397–400. [CrossRef]
37. Shi, L.; Qian, J.; Jin, M.; Bhattacharya, M.; Yu, H.; Shimbori, A.; White, M.H.; Agarwal, A.K. Investigation on gate oxide reliability under gate bias screening for commercial SiC planar and trench MOSFETs. *Mater. Sci. Semicond. Process* **2024**, *174*, 108194. [CrossRef]
38. Yu, L.C.; Dunne, G.T.; Matocha, K.S.; Cheung, K.P.; Suehle, J.S.; Sheng, K. Reliability Issues of SiC MOSFETs: A Technology for High-Temperature Environments. *IEEE Trans. Device Mater. Reliab* **2010**, *10*, 418–426.
39. Yu, L.; Cheung, K.P.; Campbell, J.; Suehle, J.S.; Sheng, K. Oxide Reliability of SiC MOS Devices. In Proceedings of the IEEE International Integrated Reliability Workshop Final Report (IIRW), South Lake Tahoe, CA, USA, 12–16 October 2008.
40. Han, K.; Kanale, A.; Baliga, B.J.; Ballard, B.; Morgan, A.; Hopkins, D.C. New Short Circuit Failure Mechanism for 1.2 kV 4H-SiC MOSFETs and JBSFETs. In Proceedings of the 2018 IEEE Workshop Wide Bandgap Power Devices Applications (WIPDA), Atlanta, GA, USA, 13 October–2 November 2018.

electronics

Article

A SiC Photo-Conductive Switch-Based Pulse Generator with Nanoseconds and High Voltage for Liver Cancer Cells Ablation Therapy

Haocheng Yin, Zeyu Zhang, Yapeng Liu, Yutian Wang *, Hui Guo and Yuming Zhang

Key Laboratory of Wide Band-Gap Semiconductor Materials and Devices, School of Microelectronics, Xidian University, Xi'an 710071, China; hcyin@xidian.edu.cn (H.Y.); 22111213866@stu.xidian.edu.cn (Z.Z.)
* Correspondence: ytwang@xidian.edu.cn

Abstract: Electroporation ablation, as an innovative cancer treatment, not only preserves the structure and function of affected organs but also significantly reduces surgical risks, offers patients a safer and more effective therapeutic option, and demonstrates immense potential in the field of oncology. This paper presents the innovative design of a high-voltage nanosecond pulse generator triggered by a silicon carbide (SiC) photoconductive switch. The generator is capable of stably outputting adjustable voltages ranging from 10 kV to 15 kV, with pulse widths precisely controlled between 10 and 15 nanoseconds, and an operating frequency adjustable from 1 Hz to 10 Hz. This device enables instant activation and deactivation of the pulse generator during ablation, enhancing the efficiency of strong electric field applications and preventing overtreatment due to delayed shutdown. This paper introduces the structure and basic principles of this novel SiC photoconductive switch-triggered pulse device and reports on the impact of device-related pulse parameters on the ablation effect of hepatocellular carcinoma cells through cell experiments. Under optimal ablation parameters, the CCK8 results show that the number of viable cells is only 0.7% of that in the untreated control group after 12 h of subculture following ablation. These findings hold significant importance for expanding the application areas of SiC devices.

Keywords: pulse power generator (PPG); high voltage; nanoseconds; electroporation; tumor cell ablation

check for updates

Citation: Yin, H.; Zhang, Z.; Liu, Y.; Wang, Y.; Guo, H.; Zhang, Y. A SiC Photo-Conductive Switch-Based Pulse Generator with Nanoseconds and High Voltage for Liver Cancer Cells Ablation Therapy. *Electronics* **2024**, *13*, 4816. https://doi.org/10.3390/electronics13234816

Academic Editor: Pedro J. Villegas

Received: 13 October 2024
Revised: 18 November 2024
Accepted: 29 November 2024
Published: 6 December 2024

1. Introduction

Cancer, medically defined as a malignant tumor, originates from abnormal changes in cellular genetic material. These changes drive cells to divide and proliferate uncontrollably and irregularly, ultimately forming masses that invade adjacent tissues and may even spread to other parts of the body through the bloodstream or lymphatic system. Currently, the primary treatments for cancer include surgical therapy [1], radiotherapy [2], chemotherapy [3], immunotherapy [4], and targeted therapy [5], among others [6]. While these traditional therapies have shown high effectiveness in eliminating cancer cells, they also cause significant functional damage to the human body. Surgical therapy, often regarded as a direct and effective means, is costly and carries certain risks such as infection and bleeding, without guaranteeing a completely trauma-free procedure. Furthermore, patients face risks of complications such as infection, bleeding, and pulmonary embolism after surgery, all of which increase patients' suffering and treatment burden. Moreover, surgery may sometimes fail to remove the tumor completely due to its location or size, leaving the possibility for recurrence or regrowth [7–9]. Chemotherapy, as another common treatment, also has side effects that cannot be ignored. While chemotherapy drugs kill cancer cells, they also damage normal cells, triggering a series of side effects such as nausea, vomiting, hair loss, and decreased white blood cell count, all of which severely affect the quality of patients' lives. In more severe cases, excessive chemotherapy may even shorten the patient's lifespan [10,11]. Additionally, since certain types of tumors are not sensitive to chemotherapy drugs, forcibly administering chemotherapy may yield minimal results [12]. Radiotherapy differs from surgery and chemotherapy. Although modern radiotherapy techniques have achieved considerable precision, it is still difficult to completely avoid radiation damage to surrounding healthy tissues and lead to a series of side effects such as skin inflammation, nausea, vomiting, and fatigue. Furthermore, radiotherapy has a relatively long treatment cycle, slow onset of efficacy, and a high recurrence rate [13,14].

With advancements in technology, ablation therapy has gradually attracted the attention of researchers. Ablation methods, primarily including radiofrequency ablation (RFA), cryoablation, microwave ablation, laser ablation, and ultrasound ablation, significantly reduce patients' suffering during cancer treatment by directly destroying local cancer cells or tissues, demonstrating considerable application potential [15]. However, these traditional ablation methods still face numerous challenges in clinical applications. RFA, though effective, involves a relatively lengthy ablation process, and the expansion effect during tumor necrosis may lead to residual cancer cells. Meanwhile, in some cases, increased vascular endothelial growth factor (VEGF) can inhibit drug efficacy, thereby increasing the risk of tumor recurrence [16,17]. Cryoablation is limited by its high cost and time-consuming procedure, and multi-needle ablation may lead to a significant reduction in platelets, increasing the risk of bleeding [18,19]. Microwave ablation currently faces issues with imperfect equipment, resulting in unstable treatment effects [20]. Laser ablation may cause local temperatures to rise excessively during treatment, damaging surrounding healthy tissues and nerves, leading to complications. Furthermore, the effectiveness of laser ablation is closely related to needle placement, making it difficult to operate [21,22]. As for ultrasound ablation, accurate lesion localization can sometimes be challenged due to factors such as echo, bone obstruction, and lung air, resulting in a longer treatment duration. Additionally, there is a possibility of missing the target when treating irregular tumor tissues [23].

With the continuous progress of science and technology, researchers have gained a deeper understanding of the electromagnetic effects on biological cells, leading to the development of an innovative ablation technique—pulsed electric field ablation (PEFA). This technique widely utilizes high-voltage electric pulses to act on the phospholipid bilayer of cell membranes, generating transmembrane potentials and forming unstable potential differences, thereby inducing irreversible permeabilizing damage to the cell membrane and ultimately leading to cell apoptosis [24]. Specifically, this ablation method relies mainly on high-voltage direct current pulses with pulse widths precisely controlled between 5 and 100 microseconds. PEFA technology has been successfully applied in clinical practice for the ablation of liver and pancreatic tumors, and it has also shown great potential in the treatment of arrhythmias such as atrial fibrillation [25,26]. Compared with traditional thermal and cold ablation techniques, PEFA exhibits higher targeting precision, more precise action on target tissues, significantly reduced complications, and a substantially shortened ablation time [27]. With continuous technological iterations and upgrades, PEFA has evolved from millisecond pulses to microsecond pulses and further to nanosecond pulses, bringing about broader application prospects and more possibilities to this technology [28,29]. Looking ahead, PEFA is expected to play a significant role in various medical fields such as tumor ablation and arrhythmia treatment, providing patients with more and better treatment options.

Compared with cold and thermal ablation, PEFA has several notable advantages. Firstly, in terms of safety, as a non-thermal ablation method, electric pulse ablation almost eliminates the risk of thermal-related injuries. It demonstrates high selectivity for surrounding tissues such as the phrenic nerve and esophagus, avoiding damage to these normal tissues and significantly reducing the risk of damaging normal tissues during surgery [16,30,31]. Secondly, PEFA also excels in ablation efficiency. By using a circular multi-electrode ablation catheter, this technology simplifies the surgical process, enabling the entire procedure to be completed through a single catheter [32]. Its multi-point discharge feature allows each discharge to occur in just 2–4 s, significantly shortening the ablation time and improving surgical efficiency compared to methods like RFA and cryoablation. Finally, in terms of applicability, PEFA shows broad potential. Since its mechanism of action does not depend on specific tumor types, it may be applicable to various types of tumors, providing a wider range of options for tumor treatment [33]. In summary, PEFA has clear advantages over cold and thermal ablation in terms of safety, efficiency, and applicability; it is expected to become one of the crucial means for future tumor treatment.

Currently, PEFA devices commonly use traditional semiconductor power devices as pulse switches to emit treatment pulses. However, these devices typically operate at kilovolt-level pulse source voltages during treatment, and semiconductor power devices struggle to achieve nanosecond-level rapid shut-off at such high voltages [34]. This limitation can easily lead to ablation beyond the expected range or overtreatment during high-precision surgeries, particularly in areas where thermal ablation is contraindicated, such as the brain, liver, and blood vessels, where overtreatment may cause unnecessary organ damage to patients [35]. Additionally, the rising edge slope (dv/dt) of the treatment pulses output by traditional semiconductor power devices such as pulse switches is relatively gentle, approximately between 100 V/ns and 300 V/ns, which is insufficient to generate instantaneous strong electric fields within cells to achieve cell perforation. Although increasing the rising edge slope by paralleling multiple semiconductor power devices is possible, it also increases the width of the treatment pulses, leading to additional thermal accumulation during treatment and increasing the risk of thermal damage to healthy tissues. To overcome this challenge, this paper reports on a novel high-power pulse ablation device with a maximum voltage of 15 kV, a frequency of 10 Hz, and a pulse width of only 5 ns. This pulse source uses a SiC photoconductive switch for control and has been tested for pulse ablation on A549 cells to study the ablation effect of the high-voltage pulse source on liver cancer cells. This innovation provides strong support for the further development of PEFA technology.

2. Equipment Set Up and Material

2.1. Pulse Equipment Set Up

Unlike the square wave pulses used in traditional electro-pulse ablation, the pulse waveform employed in our study, as shown in Figure 1c, is presented in the form of a steep pulse. Traditional square wave pulses switch rapidly between high and low voltage levels, with a stable waveform characterized by clear rising and falling edges, as well as constant pulse width and frequency. The stable waveform of square wave pulses ensures uniform electric field distribution and a consistent duration of action, thus offering advantages such as good stability, high controllability, and consistent ablation effects. In contrast to square waves, steep pulses have a sharp waveform, reaching a high peak voltage in a short time and quickly falling back. They are characterized by concentrated energy and a short duration of action, resulting in a more significant electroporation effect on cell membranes compared to square waves. At the same time, steep pulse ablation offers advantages such as high efficiency, precision, and safety. Its rapid rising and falling edges ensure the quick action and disappearance of the electric field, reducing damage to normal tissues. However, there is relatively little research on ablation using steep pulses currently, which also makes our research meaningful.

To generate nanosecond high-voltage pulses, a photoconductive pulse-triggering circuit based on SiC material is used in this experiment. 4H-SiC-doped V as the material for PCSS in cancer therapy has several advantages. Firstly, SiC does not contain any harmful components like GaAs. Secondly, no avalanche effect has been found in SiC so far. Therefore, its photoelectric linear characteristics are easily triggered by laser control. Finally, due to the ability of a single SiC photoconductive switch to operate at 15 KV for a long time, the trigger circuit is simple. The circuit consists of a pulsed high-voltage power supply (Wiswan DE70P6), a high-voltage capacitor, a SiC photoconductive switch, and a load module, as shown in Figure 1.

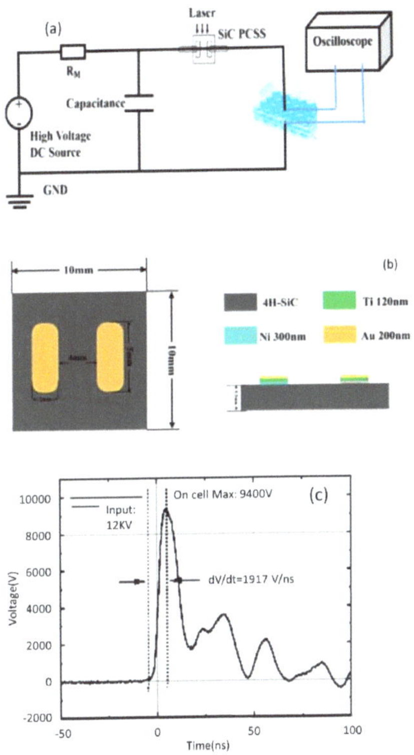

Figure 1. (**a**) Schematic diagram of a pulse system based on photoconductive switches for cancer treatment. (**b**) The photoconductive switch used in this experiment. (**c**) The load waveform collected on the cell under 12 KV input voltage.

The entire pulse system consists of two circuits. Firstly, the charging circuit is composed of a high-voltage power supply, a charging resistor, and a charging capacitor. The charging circuit is composed of the same capacitor PCSS and a cell plate as the discharge circuit, where the cell plate can be equivalently regarded as a load resistor. Before applying the pulse voltage, we tested the resistance value of the cells with a multimeter, which was approximately at the level of 100K ohms.

The VCSI 4H-SiC substrates used for fabricated PCSS in this experiment were provided by SICC Company Ltd.1. (Jinan, China), with a thickness of 500 μm and a size of 10 × 10 mm. We first cleaned the SiC material using the microelectronics standard cleaning process and then deposited a layer of nickel metal according to the electrode shape in Figure 1, with a thickness of 300 nm. Then, it is annealed at temperature 1100 °C for 60 s under nitrogen conditions. Finally, we covered the nickel metal with titanium and gold, with thicknesses of 120 nm and 200 nm, respectively.

When there is no strong laser triggering PCSS, the resistance of PCSS is around 10^{10} ohms, so the discharge circuit is not connected. During operation, the charging circuit charges the capacitor through a high-voltage source and charging resistor and then uses a 532 nm laser to touch the PCSS. Due to the generation of the 10^{17} cm^{-3} concentration of photo-generated carriers in PCSS under the action of laser pulses, the switching resistance decreases from 10^{10} ohms to 30 ohms within 1–3 ns, and the capacitor discharges instantaneously to generate high-voltage pulses. High-voltage pulses are applied to the tip of the electrode needle by linking it to the cell plate wall, generating an electric field strength of 12 kV/cm. The electrodes space 4 mm, ensuring that the switching material can withstand high voltages of up to 20 KV to deliver a stronger nanosecond pulse. A Q-switched Nd: YAG laser form is used as the optical trigger source, whose full width at half maximum is 10 ns at a wavelength of 532 nm. The load side is a cell plate filled with cells for this experiment.

During operation, the high voltage originally charged the high-voltage capacitor. After the capacitor was full, the pulsed laser triggered the photoconductive switch to instantaneously discharge the capacitor. For a standard 50 ohm capacitor, a quasi-Gaussian pulse with a pulse width of about 15 nanoseconds was generated from 1 kV to 12 kV. Meanwhile, for the subsequent study of the equivalent resistance model of the cell, we connected a 2G bandwidth oscilloscope (Siglent SDS6204 H10 Pro, Shenzhen, China) through a high-voltage probe (P6039A of Pintech, Guangzhou, China) and connected it to both ends of the cell plate to record the pulse waveforms of the cell plate under the high-voltage electricity.

2.2. Human Pulmonary Carcinoma Cell Preparation

The procedure for preparing hepatocellular carcinoma cells for the experiments follows the standard protocols [28] and has been adaptively adjusted according to the laboratory conditions and actual needs, as detailed below. For cell cryopreservation, a combination of Dulbecco's Modified Eagle Medium (DMSO, sourced from ThermoFisher Scientific, Waltham, MA, USA) and a serum-free cell cryopreservation solution (FBS, obtained from Merck KGaA, Darmstadt, Germany) is utilized. To formulate the cell culture medium, Dulbecco's Modified Eagle Medium (DMEM, also from ThermoFisher Scientific, USA), FBS, and a penicillin/streptomycin solution (1000 U/mL, ThermoFisher Scientific, USA) are employed. The human pulmonary carcinoma cells (A549 cells) are maintained in the State Key Laboratory of Oncology at the Air Force Medical University. Preparation of the cell culture medium entails mixing DMEM, adjusting its pH to a range of 7.2 to 7.4, with 10% fetal bovine serum (FBS) and 1% penicillin/streptomycin solution. The cells are cultured in a Midi 40 carbon dioxide incubator (Thermo, USA), maintained at 37 °C, with a 5% CO_2 concentration and 80–90% humidity. Every two days, the medium is replaced to promote optimal cell growth. Subculturing is performed when cell density reaches 80%-90%, involving cell detachment with a 0.25% trypsin solution, followed by the addition of an appropriate volume of fresh culture medium to achieve uniform cell dispersion for subculturing. Upon reaching 80–90% cell density again, viable cells are chosen for cryopreservation. All cell preparations follow the standard procedure for cell suspension preparation [36]. These cells are mixed with a solution comprising 10% DMSO and 90% FBS, transferred to cryopreservation tubes, securely sealed, and stored in liquid nitrogen for future applications. Prior to use, the cells are thawed in a 37 °C water bath.

2.3. Cell Experiment Set Up

In the cell experiments, a specific volume (400 microliters) of cell suspension containing approximately 2.5×10^5 cells, which was selected based on the standard references [37,38], was dispensed into each well of a 48-well plate. Sterile needle electrodes, which are coated with an insulating material except at the tip, are then obtained. These electrodes are inserted into the wells of the cell culture plate with a predetermined spacing, as dictated by the experimental protocol. In the cell experiment, we initially used the electrode cups recommended in the relevant literature. However, we observed that the pulse output waveform was altered when using these electrode cups. This phenomenon could be attributed to the fact that the electrode cups themselves act as capacitors. Due to the charging and discharging characteristics of capacitors, the rising and falling edges of the pulse become more gradual, leading to the distortion of the pulse waveform. After consulting the literature [39], we decided to adopt a method of directly inserting the electrode needles into the culture dishes. Prior to the experiment, the workbench and its surroundings were thoroughly disinfected to preclude any contamination that might skew the experimental outcomes. Following pulse ablation, the cells' physical attributes were examined using a transmission microscope (MSD530, Murzider, Dongguan, China). For all cell experiments, electrode cups with a 1 mm gap were utilized, sourced from ThermoFisher Scientific (USA). The CCK8 dye employed for the cell proliferation assessment was procured from Selleck (Houston, TX, USA). In the CCK8 assay, the cells are permitted to proliferate for a duration of 12 h before being analyzed using a Varioskan LUX microplate reader, Shanghai, China. After this period, a precise volume (400 μL) of CCK-8 reagent is added to each well, with meticulous care taken to prevent bubble formation. The culture plate is then gently agitated to ensure homogeneous mixing of the solution. The absorbance value of each well is subsequently measured at a wavelength of 450 nm using a microplate reader. Statistical analysis is conducted on the recorded absorbance values to discern differences among various treatment groups and to compute cell viability or survival rates. In the experimental process, the final experimental results are obtained by calculating the average value. Specifically, 30 groups of cells are selected for each parameter ablation experiment, with each group originating from the same batch of cultured liver cancer cells. After the experiment, when calculating the cell survival rate, the three groups with the highest values and the three groups with the lowest values were excluded. The average survival rate of the remaining 24 groups of cells was then calculated, which is considered to be the final cell survival rate after ablation for that parameter. The same calculation method was applied to the control group.

3. Results and Analysis

3.1. Voltage Effect

Studies have shown that enhancing the electric field intensity can significantly improve the ablation effect on cancer cells; when a specific threshold is reached, it can ensure the complete ablation of cells within the targeted ablation area. However, this does not imply that the electric field intensity can be increased indefinitely. When the electric field intensity reaches a saturation level, further increases not only fail to enhance the ablation effect but may also trigger a series of adverse effects, such as expanding the scope of tissue damage and inducing cardiac arrhythmias. Therefore, during the administration of electroporation ablation therapy, precise control of the electric field intensity is essential to ensure both effective ablation and the avoidance of unnecessary risks and side effects due to exceeding the saturation value. Given this, it is necessary to further verify experiments whose voltage settings can achieve the threshold voltage required for effective ablation when using our self-developed pulse generator.

In the test of electric field intensity on ablation effect, we maintained constant conditions of a frequency of 10 Hz, 300 pulses, and an electrode spacing of 1 cm. By adjusting the voltage across the electrodes, we varied the electric field intensity from 10 kV/cm to 12 kV/cm for ablation testing. During the experiment, we used a temperature gun for real-time monitoring and recorded an indoor temperature of 25.2 °C and an initial cell sample temperature of 24.8 °C. The waveform of the applied voltage on the cell under 12 KV is referred to in Figure 1c.

After the ablation treatment, we first observed the morphological changes in the cells using an optical microscope, and the related images are displayed in Figure 2. Subsequently, 0.1 mL of DMEM culture medium was added to each well, and the 48-well cell culture plate was placed in a CO_2 incubator. After 12 h of incubation, we conducted a CCK8 cell viability assay, and the experimental results are shown in Figure 3b.

Throughout the experiment, we focused our observations using an optical microscope on the central region between the electrodes. When a voltage of 10 kV was applied, no significant cell shrinkage was observed. As the ablation electric field intensity gradually increased from 10 kV/cm to 12 kV/cm, the number of shrunken cells around the electrodes significantly increased, reaching 68%. This phenomenon is clearly demonstrated in the red-boxed area in Figure 2b. Meanwhile, during the ablation process, the temperature remained relatively stable; compared to 10 kV/cm, the temperature after ablation only rose by 0.6 °C even at a field intensity of 12 kV/cm. The average temperature change per second for each group during the experiment is shown in Figure 3a. Confirming the fact that minimal temperature increase has almost no impact on cell viability, the entire ablation process is not dominated by thermal effects, and the final ablation effect should be mainly attributed to the action of the pulsed electric field.

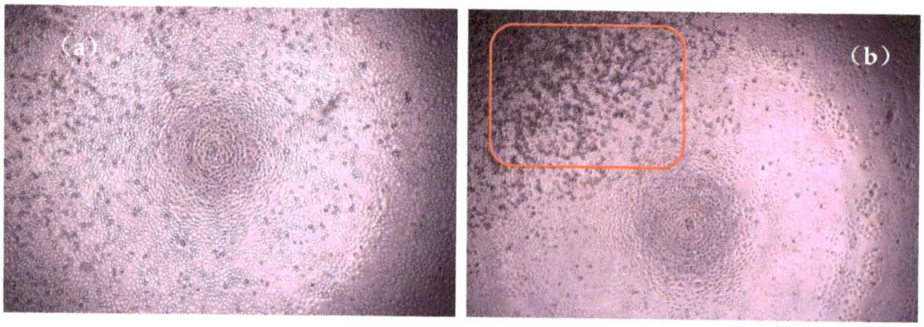

Figure 2. Optical microscope images after ablation at different pulsed electric field intensities: (**a**) field intensity of 10 kV/cm, 10 Hz, 30 s; (**b**) field intensity of 12 kV/cm, 10 Hz, 30 s.

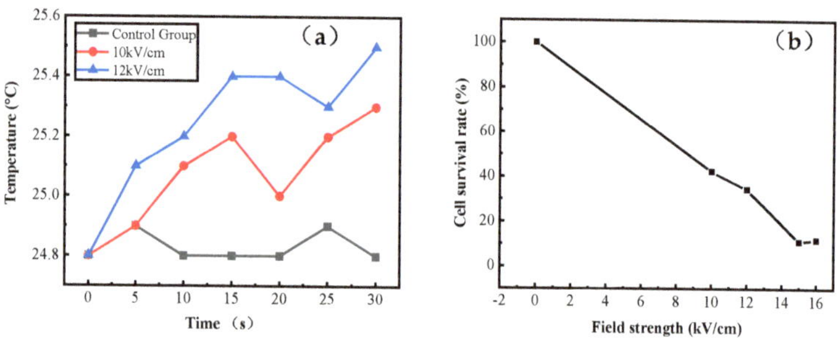

Figure 3. Graphs show temperature change over time and cell survival rate after ablation experiment. (**a**) Black squares represent the control group, red represents 10 kV/cm, and blue represents 12 kV/cm; (**b**) CCK8 results of cells after ablation at different electric field intensities and subsequent subculture for 12 h.

Although optical microscope images show that cells did not exhibit extensive shrinkage at a field intensity of 10 kV, this does not imply that the ablation effect is ineffective. The fundamental reason for this is that pulsed electric fields can induce permanent pores in the cell membrane, thereby enhancing its permeability and causing intracellular substances to leak out through osmosis. This process ultimately leads to cellular energy depletion, disrupts normal cellular physiological functions, and triggers cell death. It is noteworthy that even after experiencing permanent electroporation, cells may still maintain a certain degree of structural integrity, making it difficult to accurately determine cell death solely through an optical microscope observation. To more precisely quantify the ablation effect on cells more precisely, we employed the CCK8 cell viability assay. In the CCK8 test, we maintained consistent experimental parameters and performed the test in electrode cups. After the ablation procedure, we promptly disinfected the surface of the electrode cups in a biosafety laboratory without affecting the samples and then transferred the cells to a 48-well plate and added an appropriate amount of culture medium. After 12 h of incubation, we introduced the CCK8 reagent and used the Varioskan LUX microplate reader to conduct detailed analysis and detection of the CCK8 values. The survival rates of the cells in each group were rigorously calculated based on Equation (1), as follows:

$$R = \frac{Z_{CG} - Z_n}{Z_{CG}} \tag{1}$$

where Z_{CG} represents the CCK8 value of the control group after 12 h; Z_n represents the CCK8 value of the test group after 12 h. The CCK8 experimental results are depicted in Figure 3. As the voltage was increased from 10 kV to 12 kV, the survival rate of the cells cultured in Petri dishes for 12 h significantly decreased from 42.5% to 24.6% compared to the control group. Previously, after ablation at a field intensity of 12 kV, the proportion of cell shrinkage observed through an optical microscope was approximately 68%. Assuming that shrunken cells are dead cells, the proportion of cells maintaining normal morphology after ablation should be around 30%. This proportion is highly consistent with the 34.6% survival rate observed 12 h after ablation at 12 kV/cm. We speculate that the reason for this phenomenon may be the reduction in the number of viable cells after ablation, which indirectly affects the number of viable cells after 12 h of culture. Furthermore, the CCK8 results also reveal that during the ablation process, there is a positive correlation between the electric field intensity and the number of liver cancer cell deaths. However, this relationship is not linear between 10 kV/cm and 15 kV/cm. When the applied frequency and the number of pulses are kept constant at 10 Hz and 300, respectively, a significant enhancement in the ablation effect can be achieved even with a slight increase in electric field intensity once it exceeds 10 kV/cm. This finding provides an important scientific basis for optimizing the parameters of electric field ablation therapy.

Existing research indicates that cells predominantly die through cellular apoptosis under nanosecond pulses [40,41]. When using our pulse generator, cells shrink and die within just a few seconds of the ablation process, a phenomenon that can even be directly observed under a microscope. This phenomenon is caused by strong electric field stimulation. When we reduce the electric field intensity to below 10 kV/cm, the ablation effect diminishes, and the cell shrinkage phenomenon almost disappears. However, the ablation effect is still lower than that achieved with square wave pulses. This phenomenon may be related to the differences between square wave pulses and steep pulses. When using high-voltage steep pulses, the electric field reaches its peak rapidly and then quickly falls back. Higher voltages result in stronger penetration effects, and steep pulse electric fields are more likely to penetrate deep into cells and affect mitochondria, which aligns with existing reports that the ablation effect is dependent on mitochondria [42]. When the electric field intensity is lower, the duration for which the pulse electric field reaches the threshold on cancer cells becomes the dominant factor. At this point, the ablation effect is primarily influenced by the electroporation of the cell membrane. Therefore, when using our pulse device, a higher electric field intensity is required to achieve the desired effect. However, high electric field intensities pose a higher risk to device stability. Thanks to the advantages of SiC photoconductive switches, our device excels in terms of both safety and precise control compared to other high-voltage steep pulse devices. This is also a significant advantage of our self-developed pulse power device.

To further explore the threshold of electric field intensity's impact on ablation efficiency, we gradually increased the electric field intensity to 15 kV/cm and 16 kV/cm. As other experimental parameters remain unchanged, when the electric field intensity was increased to 15 kV/cm, the cell survival rate decreased to 11.3% after 12 h of culture. However, when the electric field intensity increased to 16 kV/cm, the cell survival rate decreased to 12.1%. Considering that inherent differences between samples in biological experiments may introduce experimental errors, we can reasonably infer that when the electric field intensity reaches 15 kV/cm, the ablation effect begins to saturate. In other words, the effect of pulsed electric fields on cells reaches a plateau at 15 kV/cm, and thereafter, even if the electric field intensity continues to increase, the ablation efficiency is no longer significantly enhanced. This discovery holds significant practical significance, as continuously and indefinitely increasing the electric field intensity typically brings unnecessary risks and side effects, such as tissue damage. Therefore, our research results indicate that under specific experimental conditions, there is an optimal range of electric field intensities that can ensure ablation efficiency while minimizing potential risks and side effects. This finding provides an important scientific basis and reference for the clinical application of electric field ablation technology.

3.2. Frequency Effect

Traditional high-frequency pulse generators face numerous challenges when pursuing a high voltage output. A reduction in voltage often necessitates a narrowing of the treatment area to compensate for insufficient electric field intensity. Furthermore, when voltage is significantly increased, ensuring precise shut-off of the pulse generator power supply becomes another technical bottleneck, which is crucial for guaranteeing treatment safety. By addressing these challenges, we have independently developed a novel pulse generator that not only stably provides voltages up to 15 kV, ensuring a wide treatment range, but also integrates photoconductive switching technology to achieve fine control over the pulse generator's on–off operation, effectively avoiding the risk of overtreatment due to shut-off delay. Additionally, pulse devices typically face limitations in their stable operating frequency range when outputting high voltages. Even if attempts are made to increase the frequency, maintaining pulse stability and consistency become difficult. Particularly in the low-frequency region (commonly defined as less than 100 Hz), research on the ablation effects of electrical pulses is insufficient, yet this is precisely the operating frequency band of our novel pulse generator. Therefore, there is an urgent need to delve deeper into the ablation mechanisms of low-frequency electrical pulses.

To systematically assess the impact of frequency on ablation effectiveness, we designed a series of experiments where the number of pulses and electric field intensity were kept constant at 300 pulses and 10 kV/cm, respectively, while the frequency was adjusted to observe changes in ablation effectiveness. The experimental design rigorously followed the framework of previous studies on voltage impact to ensure the comparability and accuracy of the results. The trend of temperature changes during ablation is shown in Figure 4a, providing important data for analyzing ablation efficiency. After ablation, the CCK8 assay was used to evaluate cell survival after 12 h of culture, with the results presented in Figure 4b. These data directly reflect the biological effects of pulse ablation at different frequencies. In this study, we used the ratio of viable cells after 12 h of culture post-ablation as a measure of pulse ablation efficiency, and through scientific analysis of this key indicator, we aim to deeply reveal the underlying mechanisms of low-frequency electrical pulse ablation, providing a solid theoretical foundation for optimizing treatment protocols and expanding clinical applications.

The experimental results indicate that when the frequency is increased from 1 Hz to 5 Hz, the cell survival rate after 12 h of ablation improves from 82.3% to 72.9%. As the frequency continues to rise to 10 Hz, while other electrical pulse parameters remain unchanged, the ablation effectiveness increases to 57.5%. Comparing these results with previous studies on voltage impact, it is found that under the same parameters, the ablation rate is 58.6%, which is essentially consistent with the 57.5% ablation rate at 10 Hz. When the frequency is doubled from 5 Hz to 10 Hz, the ablation efficiency increases from 27.1% to 57.5%, representing a nearly 50% improvement. In contrast, in previous experiments, when the voltage increased from 10 kV/cm to 12 kV/cm, the ablation efficiency rose from 56.5% to 75.4%, an increase of approximately 30%. These findings suggest that when using our self-developed pulse generator for cell ablation, as the applied frequency falls within the low-frequency range, the ablation effect correspondingly improves with increasing frequency. However, the ablation effect at 10 Hz is still lower than that achieved with an electric field of 12.5 kV/cm. From this perspective, when the frequency and voltage are increased by the same ratio, electric field intensity has a more promotive effect on cell ablation than frequency.

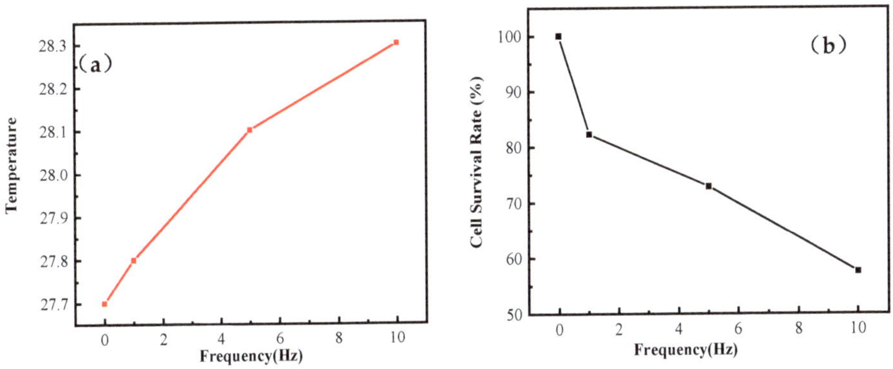

Figure 4. Temperature variation over time during the experiment and cell viability after 12 h. (**a**) Final temperature after ablation at different frequencies; (**b**) CCK8 results of cells after 12 h of subculture following ablation at different frequencies.

Previous studies have shown that pulse electric field intensity has a greater impact on cell ablation than pulse frequency [28,37,38]. The magnitude of electric field intensity directly determines the electric field's ability to penetrate cell membranes and its driving force on charges within tissues. During ablation, the level of voltage directly influences the degree of cell damage and the ablation range caused by the electric field. Frequency, on the other hand, primarily affects the changing rate of the electric field during ablation. Although frequency also has an impact, its influence is relatively smaller within the low-frequency range. The experimental data reveal that as the frequency gradually increased from 1 Hz to 5 Hz, the cell survival rate after 12 h of ablation significantly decreased, dropping from 82.3% to 72.9%. Further increasing the frequency to 10 Hz resulted in a substantial decrease in the survival rate to 41.7%, while keeping other electrical pulse parameters constant. By comparing these results with previous studies on voltage impact, it is found that under the same parameter settings, the cell survival rate is 41.7%, which is almost identical to the 42.5% survival rate at 10 Hz in previous studies. Notably, when the frequency is doubled from 5 Hz to 10 Hz, the survival rate decreases by nearly 50% (from 72.9% to 41.7%), which is a much larger improvement than the approximate 30% decrease in the survival rate (from 42.5% to 34.6%) observed in previous experiments when the voltage was increased from 10 kV/cm to 12 kV/cm. These results strongly demonstrate that using our self-developed low-frequency pulse generator for cell ablation shows a positive correlation with increasing frequency within the low-frequency range. However, it should be noted that even at 10 Hz, the ablation effect is slightly weaker than that achieved with an electric field intensity of 12.5 kV/cm. This indicates that, when the frequency and voltage are increased by the same ratio, electric field intensity (directly determined by voltage) has a more significant promotive effect on cell ablation than frequency.

Electric field intensity, as a direct reflection of voltage, directly determines the electric field's ability to penetrate cell membranes and drive charges within tissues. During ablation, a higher voltage means that the electric field can disrupt cell structures and expand the ablation range more effectively. In contrast, frequency mainly affects the changing rate of the electric field. Within the low-frequency range, although frequency has a certain impact on ablation effectiveness, its role is relatively weaker. Meanwhile, the potential optimization space of frequency should not be overlooked, especially when exploring possible nonlinear effects or synergistic interactions at specific frequencies. In summary, by fine-tuning the combination of voltage and frequency, it is possible to achieve more efficient and precise cell ablation effects. On the other hand, during the experiment, the temperature showed an increasing trend. At frequencies of 1 Hz, 5 Hz, and 10 Hz, the temperature rises by 0.1 °C, 0.4 °C, and 0.6 °C, respectively, throughout the ablation process, as shown in Figure 4b. As the frequency increases, the rate of temperature rises also accelerates. However, this change is relatively gradual and small, with almost no thermal energy generated throughout the process. This phenomenon also indicates that it is the electric field, rather than heat, that plays a direct and primary role in the ablation of A549 cells during the process.

3.3. *Pulse Number Effect*

In electroporation ablation technology, the number of pulses plays a crucial role, with its influence rooted in the complex interaction mechanism between electrical pulses and cells. When cells are exposed to a pulsed electric field, subtle changes occur in the phospholipid bilayer structure of the cell membrane, resulting in the formation of nanoscale pores. This process is known as electroporation. When the electric field intensity is maintained within a specific range, these pores can self-recover, manifesting as reversible electroporation (RE). However, if the electric field intensity exceeds a certain threshold, the pores in the cell membrane will undergo irreversible changes, leading to cell death. This is referred to as irreversible electroporation (IRE). As the number of pulses increases, the effect of IRE becomes more pronounced. This enhanced effect stems from the accumulation of pulse numbers, meaning that more energy is delivered to the cell membrane, resulting in an increased number of pores that are harder to recover from, thereby increasing the probability of cell death. It is worth noting; however, that this enhancement is not unlimited and there is a saturation limit. When the number of pulses increases to a certain value, the enhancement trend of the IRE effect tends to plateau, with no significant further improvement. This implies that, upon reaching the saturation point, whether increasing the number of pulses, adjusting the voltage, or changing the frequency, the cumulative effect will stabilize. Currently, there is relatively little research on electroporation ablation in the low-frequency range, with most studies focusing on the field of square wave medium- to high-frequency pulses. Therefore, combining our newly developed pulse generator to study the impact of pulse number on cancer cell ablation effects is expected to provide more comprehensive and in-depth scientific support for the optimization and practical application of electroporation ablation technology.

To specifically investigate the influence of pulse number on the effect of electroporation ablation, we designed a series of carefully controlled experiments where the frequency and electric field intensity were consistently maintained at 10 Hz and 15 kV/cm, respectively. These parameters were based on the optimal ablation conditions determined in previous studies. To comprehensively assess the role of pulse number, we selected four different pulse number levels for testing: 0 (as a control group), 300, 600, and 900. This experimental design aimed to precisely determine the optimal pulse parameters by comparing the ablation effects at different pulse numbers. The experimental results are shown in Figure 5. Under the conditions of a frequency of 10 Hz and an electric field intensity of 15 kV/cm, when the pulse number was 300, the cell viability reflected by the CCK8 value after 12 h of culture post-ablation reached 9.7%. This result is highly consistent with previous research data of 11.3%, validating the stability and reliability of the experiment. Furthermore, when the pulse number increased to 600 and 900, the cell viability significantly decreased to 1.3% and 0.7%, respectively, indicating that the ablation effect gradually strengthened with the increase in pulse number, as shown in Figure 5b.

It is particularly noteworthy that when the pulse number increased from 300 to 600, there was a significant reduction in cell viability from 9.7% to 1.3%, indicating a strong effect. In contrast to similar PEF ablation studies, the results of one of the studies indicated that following the completion of pulsed electric field (PEF) treatment, an electrochemical assessment of the ablation of colorectal liver metastases revealed that 23 out of 27 metastases achieved complete remission; on the other hand, Dr. Jing Wang found that the proportion of cell apoptosis could reach around 80% in a study on the ablation of human oral squamous cell carcinoma using nanosecond pulsed electric fields. Combined with the results of other similar studies, the ablation efficiency is about 80–95% [39,43–45]. In contrast, our study showed that only 0.7% of the cells in the treatment group survived within 12 h after the ablation. This figure indicates an ablation efficiency exceeding 90%, a result attributed to the high-voltage pulse generator we employed. However, when the pulse number continued to increase from 600 to 900, although the ablation effect still increased, the magnitude of the increase was relatively small, with cell viability only decreasing from 1.3% to 0.7%. This phenomenon suggests that after the pulse number increases to a certain level, the enhancement trend of the ablation effect begins to plateau, possibly reaching a saturation point. On the other hand, in studies examining the effects of electrical pulses on cells, there have been previous reports using almost the same pulse parameters as ours, yet the ablation effect on the cells was significantly weaker compared to our results [46]. By comparing each of the pulse parameters used in both studies, it was found that the pulse width provided by our pulse source was only 10 ns, while the other study used a pulse width of 300 ns. This significant difference in pulse width is a crucial factor contributing to the disparity in ablation outcomes. Existing research has shown that narrower pulse widths result in better penetration of the pulse during cell ablation [40,47,48]. Therefore, our pulses are more likely to penetrate deeply into the cells and act on the intracellular organelles, leading to more effective cell destruction and consequently a more pronounced ablation effect. In summary, this experiment systematically studied and revealed the significant impact of pulse number on the effect of electroporation ablation and initially explored its potential saturation effect. Future research should further investigate the changes in the ablation effect over a wider range of pulse numbers and how to optimize pulse parameters to achieve more efficient and safer electroporation ablation therapy.

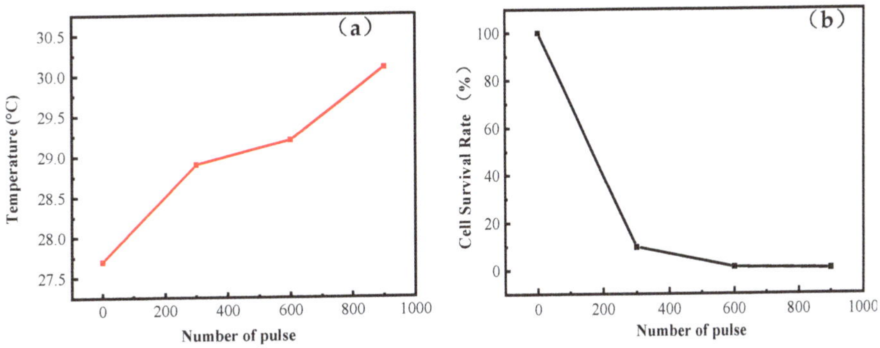

Figure 5. Temperature variation over time during the experiment and cell viability under ablation with different pulse numbers. (**a**) Temperature changes in the samples after ablation with different pulse numbers; (**b**) cell viability after 12 h of subculture following ablation with different pulse numbers.

In conclusion, during the process of the electroporation ablation of liver cancer cells using our newly developed pulse generator, the ablation effect is positively correlated with the cumulative effect of the applied number of electrical pulses. At the same time, when the number of electrical pulses reaches 900, further increasing the pulse number no longer has a significant impact on the final ablation effect. We refer to this as the "cumulative saturation effect of pulse number". This saturation differs from the saturation caused by increasing the electric field intensity. When treatment saturation is reached due to the electric field intensity, increasing the electric field will expand the ablation area, but the ablation effect within the original area between the electrodes does not significantly improve. On the other hand, the saturation caused by increasing the pulse number results in the number and size of pores on the cells reaching saturation and stabilization, without significantly affecting the ablation range.

4. Conclusions

This paper reports on a novel pulse generator, which is based on a silicon carbide (SiC) photoconductive switch. Through cell ablation experiments, we investigated the impact of various pulse parameters on the final ablation effect when using this device for liver cancer cell ablation. The pulse generator can reach a maximum voltage of 15 kV and a maximum frequency of 10 Hz, with the number of pulses flexibly adjustable according to demand. In our cell ablation studies, we found that within the low-frequency range (<10 Hz), voltage has a greater influence on the ablation of liver cancer cells than frequency. Furthermore, optimal ablation effects were achieved when the electric field intensity was set to 15 kV/cm, the frequency to 10 Hz, and the number of pulses to 900. Under these parameters, the cell survival rate after 12 h of subculture following ablation was only 0.7% of that of the control group under the same conditions, indicating a high level of ablation efficiency compared to similar studies. On the other hand, during the ablation process, the temperature change near the ablation probe was less than three centigrade, demonstrating that this device generates minimal heat during operation and thus exhibiting high medical safety. This research is significant for broadening the application fields of SiC devices and actively promotes the exploration of SiC devices in interdisciplinary applications.

Author Contributions: Conceptualization, Y.W. and H.G.; Methodology, Y.W. and Y.Z.; Software, Z.Z.; Validation, H.Y., Y.L. and Y.W.; Formal analysis, Z.Z.; Investigation, H.Y., Z.Z. and Y.L.; Resources, Y.W., H.G. and Y.Z.; Writing—original draft, H.Y.; Writing—review & editing, H.Y. and Z.Z.; Project administration, H.Y. and H.G.; Funding acquisition, Y.Z. All authors have read and agreed to the published version of the manuscript.

Funding: This research received no external funding.

Data Availability Statement: Data is contained within the article.

Conflicts of Interest: The authors declare no conflict of interest.

References

1. Tsuji, W.; Takuwa, H. Surgical treatment for stage IV breast cancer: A single-center, retrospective study. *Asian J. Surg.* **2020**, *43*, 713–715. [CrossRef] [PubMed]
2. Shirai, K.; Aoki, S.; Endo, M.; Takahashi, Y.; Fukuda, Y.; Akahane, K.; Musha, A.; Sato, H.; Wakatsuki, M.; Ishikawa, H.; et al. Recent developments in the field of radiotherapy for the management of lung cancer. *Jpn. J. Radiol.* **2024**. [CrossRef] [PubMed]
3. Domagała-Haduch, M.; Gorzelak-Magiera, A.; Michalecki, Ł.; Gisterek-Grocholska, I. Radiochemotherapy in Pancreatic Cancer. *Curr. Oncol.* **2024**, *31*, 3291–3300. [CrossRef]
4. Ferroni, C.; Varchi, G. Editorial: Small molecules and smart drug delivery systems for combination cancer immunotherapy. *Front. Chem.* **2024**, *12*, 1412624. [CrossRef] [PubMed]
5. Huang, J.H.; Guo, W.; Liu, Z. Discussion on gemcitabine combined with targeted drugs in the treatment of pancreatic cancer. *World J. Gastroenterol.* **2023**, *29*, 579–581. [CrossRef] [PubMed]
6. Huimin, Z.; Xueting, W.; Qi, Q.; Lingxin, F.; Xue, Y.; Zhuang, Y.; Jing, W. Multiple Primary Lung Cancers with ALK Rearrangement: A Case Report and Literature Review. *Front. Oncol.* **2022**, *12*, 897451. [CrossRef]
7. Wyld, L.; Audisio, R.A.; Poston, G.J. The evolution of cancer surgery and future perspectives. *Nat. Rev. Clin. Oncol.* **2015**, *12*, 115–124. [CrossRef]
8. Riis, M. Modern surgical treatment of breast cancer. *Ann. Med. Surg.* **2020**, *56*, 95–107. [CrossRef]
9. Raoof, S.; Clarke, C.A.; Hubbell, E.; Chang, E.T.; Cusack, J. Surgical resection as a predictor of cancer-specific survival by stage at diagnosis and cancer type, United States, 2006–2015. *Cancer Epidemiol.* **2023**, *84*, 102357. [CrossRef] [PubMed]
10. Mollaei, M.; Hassan, Z.M.; Khorshidi, F.; Langroudi, L. Chemotherapeutic drugs: Cell death- and resistance-related signaling pathways. Are they really as smart as the tumor cells? *Transl. Oncol.* **2021**, *14*, 101056. [CrossRef]
11. Brianna; Lee, S.H. Chemotherapy: How to reduce its adverse effects while maintaining the potency? *Med. Oncol.* **2023**, *40*, 88. [CrossRef]
12. Tilsed, C.M.; Fisher, S.A.; Nowak, A.K.; Lake, R.A.; Lesterhuis, W.J. Cancer chemotherapy: Insights into cellular and tumor microenvironmental mechanisms of action. *Front. Oncol.* **2022**, *12*, 960317. [CrossRef]
13. Chandra, R.A.; Keane, F.K.; Voncken, F.E.M.; Thomas, C.R., Jr. Contemporary radiotherapy: Present and future. *Lancet* **2021**, *398*, 171–184. [CrossRef]
14. Abdel-Wahab, M.; Gondhowiardjo, S.S.; Rosa, A.A.; Lievens, Y.; El-Haj, N.; Rubio, J.A.P.; Ben Prajogi, G.; Helgadottir, H.; Zubizarreta, E.; Meghzifene, A.; et al. Global Radiotherapy: Current Status and Future Directions—White Paper. *JCO Glob. Oncol.* **2021**, *7*, 827–842. [CrossRef]

15. Strand, N.H.; Hagedorn, J.M.; Dunn, T.; Johnson, B.; Abd-Elsayed, A.; Covington, S.; Freeman, J.; Dawodu, A.; Maloney, J. Advances in radiofrequency ablation: Mechanism of action and technology. *Ann. Palliat. Med.* **2024**, *13*, 1028–1034. [CrossRef]
16. Gómez-Barea, M.; García-Sánchez, T.; Ivorra, A. A computational comparison of radiofrequency and pulsed field ablation in terms of lesion morphology in the cardiac chamber. *Sci. Rep.* **2022**, *12*, 16144. [CrossRef]
17. Liu, Z.-L.; Chen, H.-H.; Zheng, L.-L.; Sun, L.-P.; Shi, L. Angiogenic signaling pathways and anti-angiogenic therapy for cancer. *Signal Transduct. Target. Ther.* **2023**, *8*, 198. [CrossRef] [PubMed]
18. Wu, C.; Li, X.; Lv, Z.; Chen, Q.; Lou, Y.; Mao, W.; Zhou, X. Second-generation cryoballoon versus contact force radiofrequency ablation for atrial fibrillation: An updated meta-analysis of evidence from randomized controlled trials. *Sci. Rep.* **2021**, *11*, 17907. [CrossRef]
19. Thai, J.N.; Sevrukov, A.B.; Ward, R.C.; Monticciolo, D.L. Cryoablation Therapy for Early-Stage Breast Cancer: Evidence and Rationale. *J. Breast. Imaging* **2023**, *5*, 646–657. [CrossRef] [PubMed]
20. De Vita, E.; De Tommasi, F.; Massaroni, C.; Iadicicco, A.; Faiella, E.; Carassiti, M.; Grasso, R.F.; Schena, E.; Campopiano, S. Investigation of the Heat Sink Effect During Microwave Ablation in Hepatic Tissue: Experimental and Numerical Analysis. *IEEE Sens. J.* **2021**, *21*, 22743–22751. [CrossRef]
21. Fan, Y.; Xu, L.; Liu, S.; Li, J.; Xia, J.; Qin, X.; Li, Y.; Gao, T.; Tang, X. The State-of-the-Art and Perspectives of Laser Ablation for Tumor Treatment. *Cyborg Bionic Syst.* **2024**, *5*, 0062. [CrossRef] [PubMed]
22. Zhang, L.; Zhou, W.; Zhou, J.Q.; Shi, Q.; Rago, T.; Gambelunghe, G.; Zou, D.Z.; Gu, J.; Lu, M.; Chen, F.; et al. Expert consensus on the use of laser ablation for papillary thyroid microcarcinoma. *Int. J. Hyperth.* **2022**, *39*, 1254–1263. [CrossRef] [PubMed]
23. Kok, E.N.D.; Eppenga, R.; Kuhlmann, K.F.D.; Groen, H.C.; van Veen, R.; van Dieren, J.M.; de Wijkerslooth, T.R.; van Leerdam, M.; Lambregts, D.M.J.; Heerink, W.J.; et al. Accurate surgical navigation with real-time tumor tracking in cancer surgery. *NPJ Precis. Onc.* **2020**, *4*, 8. [CrossRef] [PubMed]
24. Geboers, B.; Scheffer, H.J.; Graybill, P.M.; Ruarus, A.H.; Nieuwenhuizen, S.; Puijk, R.S.; van den Tol, P.M.; Davalos, R.V.; Rubinsky, B.; de Gruijl, T.D.; et al. High-Voltage Electrical Pulses in Oncology: Irreversible Electroporation, Electrochemotherapy, Gene Electrotransfer, Electrofusion, and Electroimmunotherapy. *Radiology* **2020**, *295*, 254–272. [CrossRef]
25. Maan, A.; Koruth, J. Pulsed Field Ablation: A New Paradigm for Catheter Ablation of Arrhythmias. *Curr. Cardiol. Rep.* **2022**, *24*, 103–108. [CrossRef] [PubMed]
26. Shaheen, N.; Shaheen, A.; Ramadan, A.; Nashwan, A.J. Efficacy and safety of novel pulsed field ablation (PFA) technique for atrial fibrillation: A systematic review and meta-analysis. *Health Sci. Rep.* **2023**, *6*, 1079. [CrossRef]
27. Reddy, V.Y.; Gerstenfeld, E.P.; Natale, A.; Whang, W.; Cuoco, F.A.; Patel, C.; Mountantonakis, S.E.; Gibson, D.N.; Harding, J.D.; Ellis, C.R.; et al. ADVENT Investigators. Pulsed Field or Conventional Thermal Ablation for Paroxysmal Atrial Fibrillation. *N. Engl. J. Med.* **2023**, *389*, 1660–1671. [CrossRef] [PubMed]
28. Gudvangen, E.; Kim, V.; Novickij, V.; Battista, F.; Pakhomov, A.G. Electroporation and cell killing by milli- to nanosecond pulses and avoiding neuromuscular stimulation in cancer ablation. *Sci. Rep.* **2022**, *12*, 1763. [CrossRef]
29. Ezzeddine, F.M.; Asirvatham, S.J.; Nguyen, D.T. Pulsed Field Ablation: A Comprehensive Update. *J. Clin. Med.* **2024**, *13*, 5191. [CrossRef] [PubMed]
30. Zang, L.; Gu, K.; Zhou, T.; Si, P.; Ji, X.; Zhang, H.; Yan, S.; Wu, X. Investigate the relationship between pulsed field ablation parameters and ablation outcomes. *J. Interv. Card. Electrophysiol.* **2024**. [CrossRef]
31. Palmeri, N.O.; Alyesh, D.; Keith, M.; Greenhaw, E.; Erickson, C.; Choe, W.; Sundaram, S. Pulsed-field ablation for atrial fibrillation without the use of fluoroscopy. *J. Interv. Card. Electrophysiol.* **2024**. [CrossRef]
32. Verma, A.; Haines, D.E.; Boersma, L.V.; Sood, N.; Natale, A.; Marchlinski, F.E.; Calkins, H.; Sanders, P.; Packer, D.L.; Kuck, K.H.; et al. PULSED AF Investigators. Pulsed Field Ablation for the Treatment of Atrial Fibrillation: PULSED AF Pivotal Trial. *Circulation* **2023**, *147*, 1422–1432. [CrossRef] [PubMed]
33. Alshahat, M.A.; Elgenedy, M.A.; Aboushady, A.A.; Williams, M.T.S. Cancer Treatment: An Overview of Pulsed Electric Field Utilization and Generation. *Appl. Sci.* **2023**, *13*, 10029. [CrossRef]
34. Lutz, J.; Scheuermann, U.; Schlangenotto, H.; De Doncker, R. Power Semiconductor Devices—Key Components for Efficient Electrical Energy Conversion Systems. In *Semiconductor Power Devices*; Springer: Cham, Switzerland, 2018. [CrossRef]
35. Zhang, Y.; Udrea, F.; Wang, H. Multidimensional device architectures for efficient power electronics. *Nat. Electron.* **2022**, *5*, 723–734. [CrossRef]
36. Pakhomova, O.N.; Gregory, B.W.; Semenov, I.; Pakhomov, A.G. Two Modes of Cell Death Caused by Exposure to Nanosecond Pulsed Electric Field. *PLoS ONE* **2013**, *8*, e70278. [CrossRef]
37. Kiełbik, A.; Szlasa, W.; Novickij, V.; Szewczyk, A.; Maciejewska, M.; Saczko, J.; Kulbacka, J. Effects of high-frequency nanosecond pulses on prostate cancer cells. *Sci. Rep.* **2024**, *11*, 15835. [CrossRef] [PubMed]
38. Gianulis, E.C.; Lee, J.; Jiang, C.; Xiao, S.; Ibey, B.L.; Pakhomov, A.G. Electroporation of mammalian cells by nanosecond electric field oscillations and its inhibition by the electric field reversal. *Sci. Rep.* **2015**, *5*, 13818. [CrossRef]
39. Wang, J.; Guo, J.; Wu, S.; Feng, H.; Sun, S.; Pan, J.; Zhang, J.; Beebe, S.J. Synergistic Effects of Nanosecond Pulsed Electric Fields Combined with Low Concentration of Gemcitabine on Human Oral Squamous Cell Carcinoma In Vitro. *PLoS ONE* **2012**, *7*, e43213. [CrossRef]

40. Wang, Y.; Yin, S.; Zhou, Y.; Zhou, W.; Chen, T.; Wu, Q.; Zhou, L.; Zheng, S. Dual-function of Baicalin in nsPEFs-treated Hepatocytes and Hepatocellular Carcinoma cells for Different Death Pathway and Mitochondrial Response. *Int. J. Med. Sci.* **2019**, *16*, 1271–1282. [CrossRef]
41. Liu, Z.G.; Chen, X.H.; Yu, Z.J.; Lv, J.; Ren, Z.G. Recent progress in pulsed electric field ablation for liver cancer. *World J. Gastroenterol.* **2020**, *26*, 3421–3431. [CrossRef] [PubMed]
42. He, L.; Xiao, D.; Feng, J.; Yao, C.; Tang, L. Induction of apoptosis of liver cancer cells by nanosecond pulsed electric fields (nsPEFs). *Med. Oncol.* **2017**, *34*, 24. [CrossRef]
43. Sano, M.B.; Arena, C.B.; Dewitt, M.R.; Saur, D.; Davalos, R.V. In-vitro bipolar nano- and microsecond electro-pulse bursts for irreversible electroporation therapies. *Bioelectrochemistry* **2014**, *100*, 69–79. [CrossRef]
44. Munavalli, G.S.; Zelickson, B.D.; Selim, M.M.; Kilmer, S.L.; Rohrer, T.E.; Newman, J.; Jauregui, L.; Knape, W.A.; Ebbers, E.; Uecker, D.; et al. Safety and Efficacy of Nanosecond Pulsed Electric Field Treatment of Sebaceous Gland Hyperplasia. *Dermatol. Surg.* **2019**, *46*, 803–809. [CrossRef]
45. Edhemovic, I.; Brecelj, E.; Gasljevic, G.; Music, M.M.; Gorjup, V.; Mali, B.; Jarm, T.; Kos, B.; Pavliha, D.; Kuzmanov, B.G.; et al. Intraoperative electrochemotherapy of colorectal liver metastases. *Surg. Oncol.* **2014**, *110*, 320–327. [CrossRef] [PubMed]
46. Buescher, E.S.; Schoenbach, K.H. Effects of submicrosecond, high intensity pulsed electric fields on living cells-intracellular electromanipulation. *IEEE Trans. Dielectr. Electr. Insul.* **2003**, *10*, 788–794. [CrossRef]
47. Miller, L.; Leor, J.; Rubinsky, B. Cancer Cells Ablation with Irreversible Electroporation. *Technol. Cancer Res. Treat.* **2016**, *4*, 699–705. [CrossRef] [PubMed]
48. Reberšek, M.; Miklavčič, D. Concepts of Electroporation Pulse Generation and Overview of Electric Pulse Generators for Cell and Tissue Electroporation. In *Advanced Electroporation Techniques in Biology and Medicine*; CRC Press: Boca Raton, FL, USA, 2010.

electronics

Article

A Voltage Equalization Strategy for Series-Connected SiC MOSFET Applications

Peng Li [1], Jialin Liu [2], Shikai Sun [2,*], Wenhao Yang [1], Yuyin Sun [1] and Yuming Zhang [1]

1 School of Microelectronics, Xidian University, Xi'an 710071, China; 1711310387@stu.xidian.edu.cn (P.L.); yang_wenhao@stu.xidian.edu.cn (W.Y.); yuyinsun@stu.xidian.edu.cn (Y.S.); zhangym@xidian.edu.cn (Y.Z.)
2 Beijing Microelectronics Technology Institute, Beijing 100076, China; liujialin@mxtronics.com
* Correspondence: sunshikai@mxtronics.com

Abstract: A novel clamped voltage equalization strategy is presented for series-connected Silicon Carbide (SiC) Metal–Oxide semiconductor Field-Effect transistors (MOSFETs) in this paper. Differences in device parameters and circuit asymmetry result in the uneven voltage distribution of series-connected SiC MOSFETs, which threatens the safe operation of the circuit. Dynamic voltage equalization is difficult to achieve due to the fast switching speed of SiC MOSFETs. This paper analyzes the switching characteristics and dynamic voltage equalization characteristics of SiC MOSFETs. Based on the analysis, an energy recovery strategy based on the clamping auxiliary circuit is proposed. A 2.8 kW (50 KHz) prototype is fabricated and tested to verify the strategy. Measurement results show that the maximum voltage stress is suppressed from 600 V to less than 320 V in the experimental condition.

Keywords: SiC MOSFETs; series-connected; voltage equalization; overvoltage

Citation: Li, P.; Liu, J.; Sun, S.; Yang, W.; Sun, Y.; Zhang, Y. A Voltage Equalization Strategy for Series-Connected SiC MOSFET Applications. *Electronics* **2024**, *13*, 3766. https://doi.org/10.3390/electronics13183766

Academic Editor: Djuradj Budimir

Received: 19 August 2024
Revised: 16 September 2024
Accepted: 20 September 2024
Published: 22 September 2024

1. Introduction

Silicon Carbide (SiC) material can push the power density and efficiency of semiconductor devices and power systems to higher limits due to its wide band gap, high critical field, and high thermal conductivity [1–3]. With the development of SiC technology, the application of SiC MOSFETs (Metal–Oxide semiconductor Field-Effect transistors) is increasingly popular [4,5]. In high-voltage applications, there are two solutions, individual high-voltage and series-connected low-voltage MOSFETs. The series-connected low-voltage SiC MOSFET method has low on-state impedance [6], high current density, high switching speed and high radiation immunity [7], and has great application prospects in power electronics. Although the advantages of the series SiC MOSFET structure are numerous, the device characteristics and the gate drive circuit differences will result in uneven voltage distribution. Balancing the drain–source voltages between the series MOSFETs is a pressing issue for the applications.

Dynamic voltage equalization is the main problem in series SiC MOSFET voltage equalization. Dynamic voltage equalization ensures that the drain source voltage V_{DS} of the series-connected MOSFETs is maintained within the rated voltage during the switching process, which is the main limitation in series-connected applications.

The relative waveforms of the series-connected SiC MOSFETs analysis are shown in Figure 1. The switching sequence of the three SiC MOSFETs is directly related to the gate drive resistance value, which is uncertain. When M3 is turned off first and M1 is turned off last, M1 has to withstand the bus voltage during the switching process, which is the worst working condition. The analysis is based on the worst condition. M3 turns off before the other devices, which results in a rapid V_{DS} rise and suffers a larger V_{DS} overshoot. Similarly, during the turn-on interval, M3 turns on first and causes V_{DS3} drops. V_{DS1} and V_{DS2} rise to withstand the bus voltage. The overvoltage breakdown risks increase due to the switching delay. Overvoltage is even more dangerous for SiC MOSFETs due to their breakneck switching speeds. Therefore, dynamic voltage equalization is the most critical issue in series-connected SiC MOSFET applications.

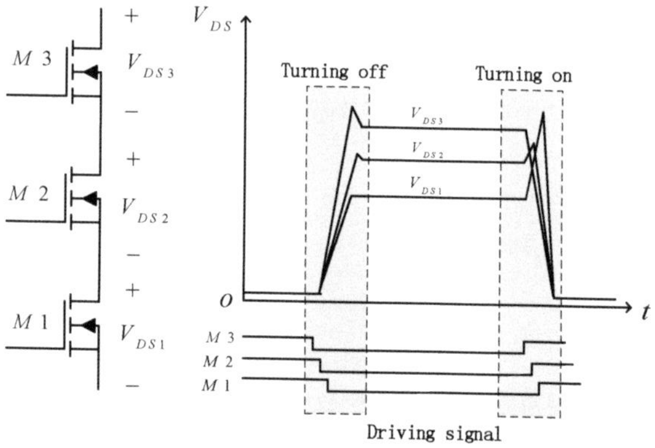

Figure 1. The schematic of three series-connected MOSFETs and their drain–source voltage waveforms.

Ref. [8] introduced an RCD (resistor–capacitor diode) snubber or RC (resistor capacitor) snubber. The principle is connecting, in parallel, a snubber capacitance with the drain and source of the MOSFET, and the capacity of the snubber capacitance is larger than the output capacitance of the MOSFET. This causes the rising rate of V_{DS} to decrease when the MOSFET is switched on and off, and the voltage inequality decreases accordingly. Refs. [9,10] realized the voltage equation based on an RC snubber circuit and analyzed the influence of the snubber circuit parameters on the switching loss and voltage equalization effect. The energy in the snubber capacitance is completely consumed in the resistor, which results in large losses. Refs. [11–13] introduced a diode clamping and voltage source energy recovery strategy, but the complexity of the auxiliary circuit and power loss limited their applications.

Ref. [14] introduced a voltage equalization strategy based on closed-loop feedback gate delay signal compensation. Ref. [15] proposed a high-frequency transformer-coupled gate realization method to achieve a consistent gate current and ensure the consistency of the gate signal. However, this method cannot improve the uneven voltage affected by the difference in device characteristics. The presence of high-frequency transformer coupling capacitance and leakage inductance also affects the operation of this circuit. The gate reference voltage control strategy is proposed in ref. [16]. The gate signal of each device is controlled by the drain–source voltage feedback and high-speed op-amp circuit to ensure that the gate signal of each device is consistent with the gate reference signal to achieve voltage equalization. A master–slave control-based voltage equalization strategy is proposed in ref. [17]. The drain–source voltage of one device in the series connection is used as a reference, and the voltage equalization is achieved by adjusting the gate voltage during the switching process of other devices. However, these methods require high-speed, high-precision integrated circuits with low reliability. An active clamp circuit based on a gate–drain Zener diode clamping is proposed in refs. [18–20]. When the device is subjected to excessive voltage, the Zener diode and the auxiliary circuit reverse breakdown, thus limiting the further rise of the drain–source voltage of the device. Refs. [21–23] proposed a single-driver series MOSFET structure based on capacitive coupling. All of the above methods need to be based on the accurate calculation of the device parameters, which makes it difficult to be applied in massive applications. Refs. [24,25] proposed a single-driver series-parallel SiC MOSFET module, which uses an additional power supply to assist in driving the top-end MOSFETs. However, the voltage equalization of the series MOSFETs in this method relies on the clamping buffer circuit, and the accumulated energy in the clamping circuit is dissipated in the resistor, resulting in a large loss.

The load-side buffer circuit has a simple structure, but the disadvantages of its large size and high loss are difficult to overcome. The gate active control strategy relies on high-speed integrated circuits with the disadvantages of many components, large cost, and low reliability. The gate delay control method requires large-scale digital integrated circuits such as FPGAs (Field Programmable Gate Arrays) or DSPs (Digital Signal Processings), and the circuit and feedback control algorithms are complex. The active clamp circuit extends the switching time of the MOSFET, which is prone to causing relatively large switching losses.

By analyzing the current research status of series power devices, this paper proposes a new series SiC MOSFET topology. The strategy is analyzed in detail, and the effectiveness is verified by an experiment. The experimental results show that the structure can suppress the SiC MOSFET drain–source voltage from 600 V to 320 V. The strategy is easy to apply to multiple MOSFETs in series and has a simple structure. The loss is relatively small due to the energy recovery circuit. At the same time, the circuit control strategy is relatively simple and can improve reliability.

2. The Analysis of the Dynamic Characteristics

2.1. The Switching Characteristics of the SiC MOSFET

Dynamic voltage equalization is affected by the switching characteristics of SiC MOSFETs; the switching process is quantitatively analyzed based on the inductive clamp circuit [26]. The topology is shown in Figure 2. Since the turn-on and turn-off processes of SiC MOSFETs are similar, only the turn-on process is analyzed in this paper.

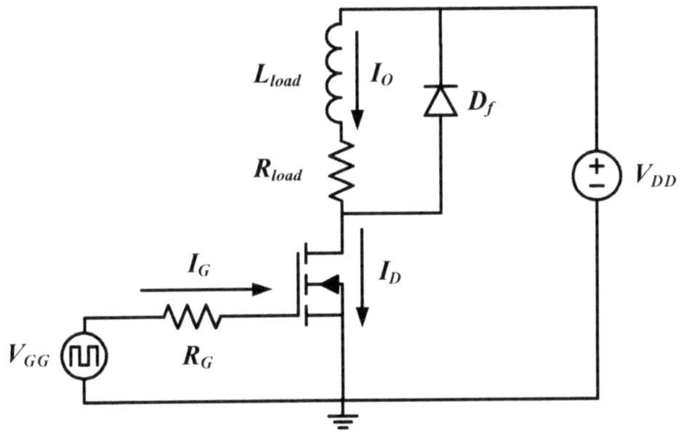

Figure 2. Inductive clamp circuit for analysing MOSFET switching characteristics.

The relative schematic waveforms and the equivalent circuit diagram of the turn-on process are shown in Figures 3 and 4, respectively. When the MOSFET is off, its gate–source voltage v_{GS} and drain current i_D are zero, and its drain–source voltage v_{DS} is the bus voltage V_{DD}. The clamp diode D_f turns on and flows the load current I_O. At t_0, the gate–source voltage v_{GS} rises.

$[t_0–t_1]$: v_{GS} rises from zero, the gate current i_G charges the input capacitor. Since gate–source capacitance C_{GS} is much larger than gate–drain capacitance C_{GD}, most of the gate current i_G charges C_{GS} and a small amount flows to C_{GD}. Since v_{GS} is below the MOSFET threshold voltage V_{TH}, i_D remains at zero, and v_{DS} remains constant. The time required for v_{GS} to rise from zero to the threshold voltage V_{TH} is called the turn-on delay time $t_{d(on)}$, which is derived as follows:

$$i_G(t) = \frac{V_{GG} - v_{GS}(t)}{R_G} = C_{GS}\frac{dv_{GS}(t)}{dt} - C_{GD}\frac{d(v_{GS}(t) - v_{DS}(t))}{dt} \quad (1)$$

v_{DS} is constant and equal to the bus voltage V_{DD} in this interval; the expression for i_G is shown in (2):

$$i_G(t) = (C_{GS} + C_{GD})\frac{dv_{GS}(t)}{dt} \tag{2}$$

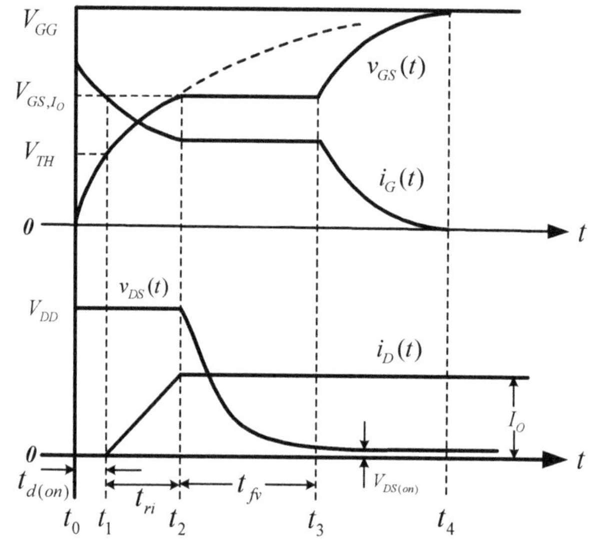

Figure 3. The schematic of the turning-on process.

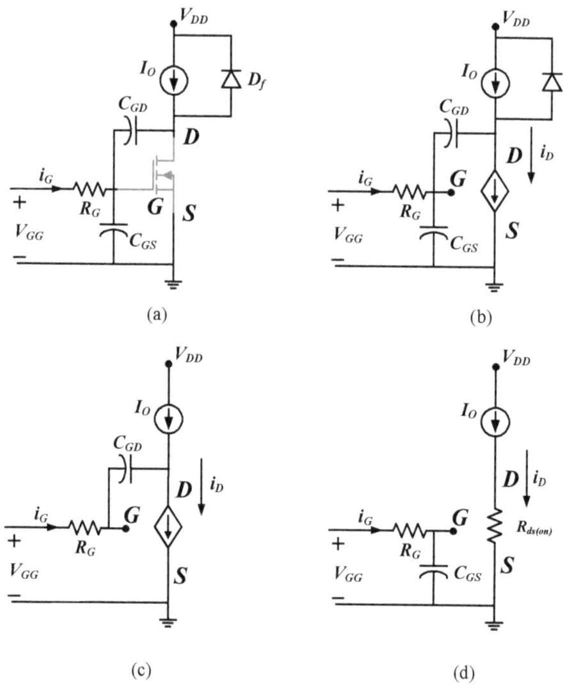

Figure 4. The equivalent circuit diagram of the MOSFET turn-on process: (**a**) $[t_0–t_1]$; (**b**) $[t_1–t_2]$; (**c**) $[t_2–t_3]$; (**d**) $[t_3–t_4]$.

Based on (1) and (2), and the initial condition $v_{GS}(t_0) = 0$, v_{GS} can be expressed by (3) and (4) in this interval:

$$v_{GS}(t) = V_{GG}(1 - e^{-(t-t_0)/\tau}) \tag{3}$$

$$\tau = R_G(C_{GS} + C_{GD}) \tag{4}$$

where R_G is the gate resistance of the MOSFE; i_G can be expressed as (5)

$$i_G = \frac{V_{GG}}{R_G} e^{-(t-t_0)/\tau} \tag{5}$$

[t_1–t_2]: This interval is also known as the current rise interval. After t_1, $v_{GS} > V_{TH}$, i_G continues to charge the input capacitance C_{iss}, and v_{GS} continues rising. i_D rises as a function of v_{GS} from the t_1 moment and the MOSFET enters the saturation operating region.

The transconductance g_m and i_D are given by (6) and (7), respectively:

$$g_m = \frac{\partial i_D}{\partial v_{gs}} = \frac{2\sqrt{I_{DSS} i_D}}{V_{TH}} \tag{6}$$

$$i_D(t) = g_m(v_{GS} - V_{TH}) \tag{7}$$

where I_{DSS} is the MOSFET drain current at zero gate bias.

Since $i_D < I_O$ in this interval, D_f is still conducting the load current and $v_{DS} = V_{DD}$. The expression for v_{GS} agrees with Expression (3). Expression (3) is associated with (7) to obtain Expression (8) for i_G, which can be seen to decrease exponentially in i_G:

$$i_G(t) = g_m(V_{GG} - V_{TH}) - g_m V_{GG} e^{-(t-t_1)/\tau} \tag{8}$$

At t_2, i_D reaches the load current I_O and D_f turns off. The current rise phase time interval t_{ri} satisfies (9):

$$t_{ri} = \tau \ln \frac{g_m V_{GG}}{g_m(V_{GG} - V_{TH}) - I_O} \tag{9}$$

[t_2–t_3]: This interval is also known as the Miller plateau interval. At t_2, i_D reaches the load current and remains constant. The MOSFET operates in the saturation region and v_{GS} remains constant. The Miller platform voltage V_{GS,I_O} is expressed as (10)

$$V_{GS,I_O} = \frac{I_O}{g_m} + V_{TH} \tag{10}$$

i_G (t_2) can be expressed as (11):

$$i_G(t_2) = \frac{V_{GG} - v_{GS(t_2)}}{R_G} = \frac{V_{GG} - (I_O/g_m) - V_{TH}}{R_G} \tag{11}$$

In this interval, v_{GS} is constant and i_G begins to charge C_{GD}. The Formula (12) for i_G can be obtained as follows:

$$i_G(t) = C_{GD}\frac{d(v_G - v_D)}{dt} \tag{12}$$

Since V_{GS} remains constant, Expression (13) can be obtained:

$$i_G(t) = -C_{GD}\frac{dv_{DS}}{dt} = \frac{V_{GG} - V_{GG,I_O}}{R_G} = \frac{V_{GG} - V_{TH} - I_O/g_m}{R_G} \tag{13}$$

The initial condition is $v_{DS}(t_2) = V_{DD}$ at t_2, and bringing in (13) yields an expression for v_{DS} over time within the interval:

$$v_{DS}(t) = -\frac{V_{GG} - V_{GS,I_O}}{R_G C_{GD}}(t - t_2) + V_{DD} \tag{14}$$

As the gate drive current charges C_{GD}, v_{DS} decreases linearly. At this point, the MOSFET gradually enters the linear operating region. The larger V_{DD} is, the longer the Miller platform phase lasts. At t_3, v_{DS} drops to the on-state drain–source voltage drop $V_{DS(on)}$, where $V_{DS(on)} = I_D R_{ds(on)}$. The Miller platform duration t_{fv} is given by (15):

$$t_{fv} = t_3 - t_2 = R_G C_{GD} \frac{(V_{DD} - I_D R_{ds(on)})}{V_{GG} - V_{TH} - I_O/g_m} \tag{15}$$

$[t_3–t_4]$: After t_3, v_{GS} continues to rise under the continued charging of i_G. Since v_{DS} remains constant at this point, the rate of rise of v_{GS} can be referred to the phase $[t_0–t_1]$.

$$v_{GS}(t) = V_{GG}(1 - e^{-(t-t_3)/\tau}) \tag{16}$$

During this interval, v_{GS} grows exponentially until the driving voltage V_{GG} is reached. At t_4, i_D becomes zero and the MOSFET turn-on process ends. The total turn-on delay t_{ON} is the summation of the three part-time intervals and is given by (17):

$$t_{ON} = t_{d(on)} + t_{ri} + t_{fv} \tag{17}$$

2.2. The Dynamic Voltage Distribution Analysis of the Series-Connected SiC MOSFETs

Ideally, series-connected SiC MOSFETs would achieve good dynamic and static voltage equalization. However, in actual circuits, due to the impossibility of achieving perfectly uniform devices and highly symmetrical driving circuits, the voltage equalization effect often differs from the ideal situation. The device and circuit factors affecting the dynamic voltage equalization characteristics of series-connected SiC MOSFETs are theoretically investigated in what follows.

Figure 5 shows the equivalent switching model for SiC MOSFETs, where C_{GD} and C_{DS} are related to the drain–source voltage V_{DS}, while C_{GS} can be considered to be constant. The input capacitance C_{iss} of the MOSFET consists of capacitance C_{GD} and C_{GS}. The output capacitance C_{oss} includes capacitance C_{DS} and C_{GD}. Furthermore, the feedback capacitance C_{rss} is the capacitance C_{GD}. The model includes gate parasitic inductance L_G and source-level parasitic inductance L_S. Parasitic inductance is introduced by the interconnections and affects the switching process of the devices.

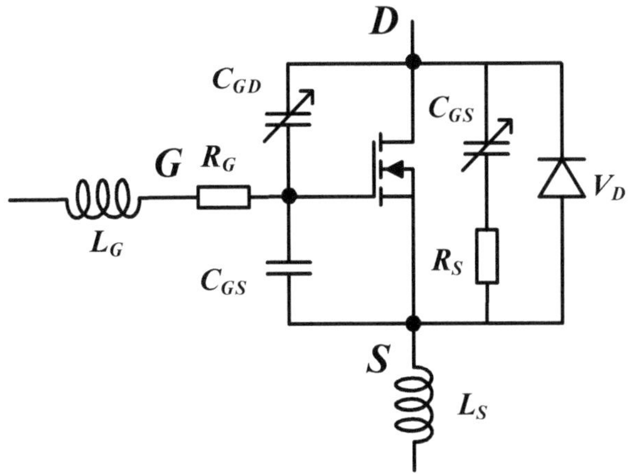

Figure 5. The switching model of the SiC MOSFET.

The dynamic voltage inhomogeneity of series SiC MOSFETs is mainly caused by two aspects, gate drive circuit delay differences and device parameter differences. Furthermore, differences in parasitic parameters on the main circuit may cause voltage unevenness. Among them, series SiC MOSFET static voltage equalization is mainly affected by the device turn-off leakage current. The device dynamic voltage equalization is affected by more factors, including device dynamic characteristic differences, gate drive delay mismatch, and different parasitic parameters on the main circuit. Device dynamic characteristics such as threshold voltage V_{TH}, inter-pole parasitic capacitance, transconductance g_m, parasitic inductance, etc., all affect the dynamic voltage equalization of series-connected SiC MOSFETs. The causes of voltage imbalance during turn-on and turn-off will be analyzed separately in the following section.

Based on the quantitative description of the SiC MOSFET switching process, it can be assumed that the following factors may be responsible for the voltage imbalance of series-connected SiC MOSFETs during the turn-on phase:

(1) The difference at the moment when the MOSFET enters the current rise phase (t_{ri}). There are two sources of this difference, one is the gate drive delay difference, which is due to the gate drive circuit difference; the other is the difference in the turn-on delay time ($t_{d(on)}$) of individual MOSFETs, which is mainly due to the threshold voltage V_{TH} and input capacitance C_{iss} differences. During the current rise interval, the MOSFET can be equated to a gate voltage-controlled current source. At this point, the MOSFET can be viewed as an ever-decreasing resistor from the drain–source point of view. If the MOSFET enters this stage later, its equivalent drain–source resistance is relatively large, and, thus, its drain–source voltage is large. The difference in the stage of current rise entered by each MOSFET can lead to uneven voltage.

(2) MOSFETs have different rates of current change during the current rise interval (t_{ri}). According to (8), the difference in transconductance of each MOSFET will result in a different rate of current rise. This is due to unavoidable differences in device fabrication. From the drain–source point of view, the equivalent resistance of each MOSFET will be different. Uneven voltage may occur at this stage.

(3) MOSFETs have different rates of voltage change during the voltage drop interval (t_{fv}). In this interval, MOSFET can be equivalent to a variable capacitor, and its capacitance rises as the drain–source voltage falls. Thus, the difference in output equivalent capacitance can lead to uneven voltage. However, since the total voltage to which the series MOSFETs are subjected at this stage has already begun to fall, voltage unevenness does not normally lead to serious problems.

In general, the main reasons for the uneven voltage of series-connected SiC MOSFETs in the turning-off process are as follows:

(1) The difference at the moment when the MOSFET enters the voltage rise phase (t_{rv}). There are two sources of this difference, one is the gate drive delay difference, which is due to the gate drive circuit difference; the second is the difference in the turn-off delay time ($t_{d(off)}$) of each MOSFET, which is mainly due to the difference in the threshold voltage V_{TH} and the input capacitance C_{iss}. In this interval, the MOSFET can be equated to a variable capacitor and its capacitance decreases as the drain–source voltage rises. The voltage rise interval can be considered as a charging process of the ever-smaller MOSFET output capacitance. MOSFETs that turn off early are necessarily subjected to higher drain–source voltages. If the difference in MOSFET turn-off delay is too large, it is easy to cause overvoltage damage to the MOSFET. According to theoretical analyses and experimental results, the turn-off delay difference [27] is considered to be the main factor affecting the turn-off voltage equalization of series-connected SiC MOSFETs. Thus, it is quite important to ensure that the series SiC MOSFET turn-off delay is synchronized.

(2) The difference in the rate of voltage change of MOSFETs during the voltage rise phase (t_{rv}). Even though the drain–source voltage of each MOSFET in the series starts to rise at the same time, the difference in the rate of voltage change of the MOSFETs at this stage is difficult to avoid. The MOSFET at this stage can be equated to the charging process of a variable capacitor. The rate of voltage change during the Miller plateau period dV/dt is mainly determined by the gate drive current and the MOSFET feedback capacitance C_{rss}. Due to device fabrication differences, the feedback capacitance is different, causing the MOSFET voltage imbalance at this stage.

(3) Circuit stray capacitance. Relevant studies have shown that [28] even if the driver circuit and the MOSFET are kept the same, the difference in stray parasitic capacitance will still cause uneven turn-off voltage. There are two main sources of stray capacitance. The first is the unavoidable stray capacitance that exists between the gate driver and ground, mainly the transformer coupling capacitance. The second is the parasitic stray capacitance between each port of the MOSFET on the main circuit and ground, which is related to device placement and PCB layout.

(4) MOSFETs have different rates of current change during the current drop phase (t_{fi}). At this point in this interval, the MOSFET can be regarded as an ever-increasing resistance from the drain–source point of view. The unavoidable difference in transconductance between the MOSFETs will result in a difference in the equivalent resistance of each MOSFET, which leads to an imbalance in the MOSFET drain–source voltage.

2.3. The Simulation of the Dynamic Voltage Distribution

Since the voltage imbalance in the turn-off transient is more critical than the voltage imbalance in the turn-on transient, in this section, the factors affecting the voltage equalization during turn-off was verified by PSpice simulation.

The simulation circuit is shown in Figure 6a. The bus voltage V_{DC} was set to 2000 V, and the MOSFET gate resistors were all set to 20 Ω. The MOSFETs were driven by a pulse signal source V_{pulse}, and their turn-on and turn-off voltages were +20 V and −2 V, respectively. SCT20280KE is a typical SiC MOSFET co-packaged with SiC-SBD. The co-packaged SiC SBD can protect the MOSFET from being broken down, which is suitable for serially connected applications. The SiC MOSFET applied in the simulation was Rohm's SCT2080KE. Table 1 lists the related parameters of the simulation model.

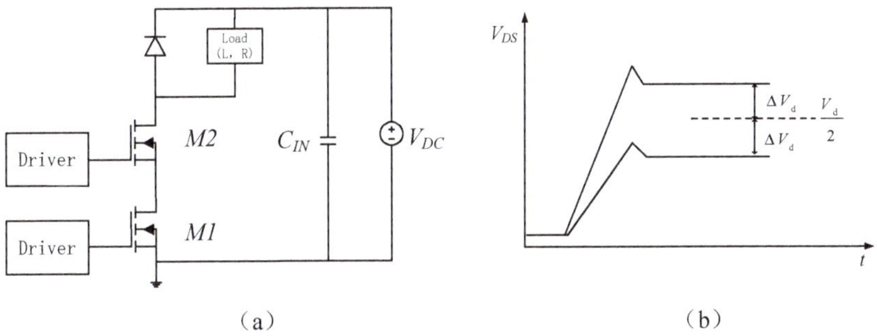

Figure 6. The simulation schematic: (**a**) the simulation circuit, and (**b**) the related waveforms in the turning-off interval.

Table 1. The main parameters of the simulation model.

Parameters		**Parameters**	
Power MOSFET	SCH2080KE	Freewheeling diode	SDT12S60
Bus voltage	2800 V	PWM frequency	50 kHz
Load resistance	200 Ω	Load inductance	200 μH

The V_{DS} voltage unbalanced waveform during turn-off in the theoretical analysis is shown in Figure 6b. To characterize the degree of series SiC MOSFET V_{DS} imbalance at turn-off, the turn-off voltage imbalance factor α_{off} is introduced in this paper. For a two-tube series SiC MOSFET, the voltage imbalance coefficient α_{off} at turn-off is defined as shown in (18):

$$\alpha_{off} = \frac{\Delta V_d}{V_d/2} \times 100\% \tag{18}$$

The Driver Delay Variance Affection

To verify the effect of the drive delay difference t_{off} on the turn-off voltage distribution of series-connected SiC MOSFETs, the delay of the gate drive signal V_{pulse} of a single MOSFET is adjusted to obtain the voltage distribution difference while keeping other parameters constant. Figure 7 shows the voltage imbalance coefficient α_{off} versus the delay difference t_{off}.

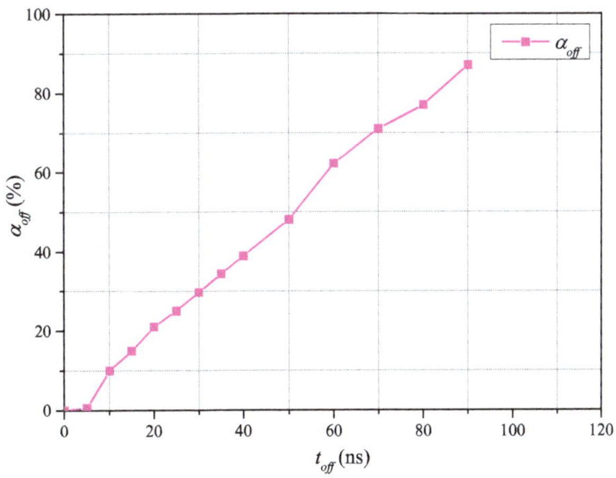

Figure 7. Voltage imbalance coefficient α_{off} versus t_{off}.

As shown in Figure 7, α_{off} has an approximately linear relationship with t_{off}. According to the previous analysis, the difference in gate signal delay will lead to the difference when the MOSFET enters the voltage rise phase (t_{rv}). According to the MOSFET switching characteristics, it can be assumed that the gate signal delay difference does not affect the MOSFET drain–source voltage rise rate. The voltage distribution and α_{off} can be obtained if the two MOSFET voltage rise rates are the same.

$$\Delta V_d = \frac{t_{off}}{2}\frac{dV_{DS}}{dt} \tag{19}$$

$$\alpha_{off} = \frac{t_{off}}{V_d}\frac{dV_{DS}}{dt} \tag{20}$$

The MOSFET voltage rise rate was calculated to be 19.3 kV/µs. Due to the fast switching speed of SiC MOSFET, the voltage rise rate was high. Therefore, gate delay drive differences can lead to severe turn-off voltage inhomogeneity.

3. The Novel Series SiC MOSFET Topology Design

The new series SiC MOSFET topology consisted of a series SiC MOSFET, a clamp circuit and its clamp auxiliary circuit, and an energy recovery circuit, which can be used as a power switch in power electronics. The topology is shown in Figure 8. In this section, the inductive clamp circuit was used as the main power circuit to illustrate the working principle and parameter selection principle of the series SiC MOSFET topology, where V_{DC} is the input bus voltage, and C_{BUS} is the bus capacitance.

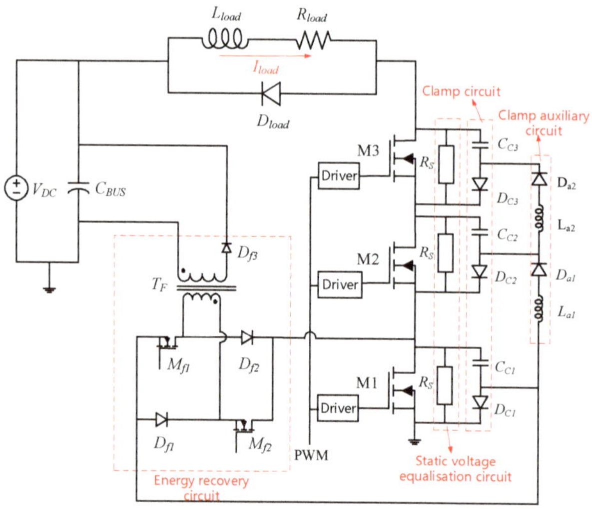

Figure 8. The topology of the proposed strategy.

The topology consists of series SiC MOSFETs M1, M2, and M3. PWM is the main switching tube control signal. The circuit load is an inductor L_{load} and a resistor R_{load}, and D_{load} is a continuity diode. D_{load} is the continuity diode, and R_S is the static voltage equalizing resistor. A clamp circuit is connected in parallel with the drain–source of each MOSFET, consisting of clamp diode D_C and clamp capacitor C_C. The current-limiting inductor L_a and the auxiliary diode D_a form the clamping auxiliary circuit. The two-tube flyback DC/DC circuit serves as an energy recovery circuit, and its input and output are clamp capacitor C_{C1} and bus capacitor C_{BUS}, respectively. M_{f1} and M_{f2} are the switching tubes of the two-tube flyback circuit. Diodes D_{f1} and D_{f2} protect the switching tubes. The T_F is a flyback transformer for voltage isolation and energy transfer.

As the designer was familiar with the clamp circuit and clamp auxiliary loop, this paper only focuses on the energy recovery circuit. The energy recovery circuit has two effects; the first one is to limit the clamp capacitance voltage, and the second one is to transfer the energy accumulated in the clamped capacitor back to the bus capacitor for reuse to reduce losses. The feedback control circuit will determine whether to turn on the energy recovery circuit switching tube based on the sampled capacitor C_{C1} voltage. The energy recovery circuit only works when the capacitor C_{C1}'s voltage exceeds the rated value.

3.1. The Energy Recovery Circuit Design

The energy recovery circuit is based on a double-tube flyback DC/DC circuit, as shown in Figure 9. Compared with the single-tube flyback circuit, the double-tube flyback circuit reduces the MOSFET withstand voltage requirement and effectively avoids the voltage spike caused by the transformer leakage inductance during turn-off. The flyback transformer T_F plays the role of electrical isolation and energy transfer, and its turns ratio is given by $N = N_P/N_S$. The switching of the switching tubes M_{f1} and M_{f2} is controlled by the output signal $V_{control}$ from the feedback control circuit. Since the energy recovery circuit operates in PFM mode with variable switching frequency, the two-tube flyback circuit designed in this paper operates in DCM mode. Compared with the CCM mode, the flyback circuit in the DCM mode is simple to calculate, easy to realize, and less prone to magnetic saturation.

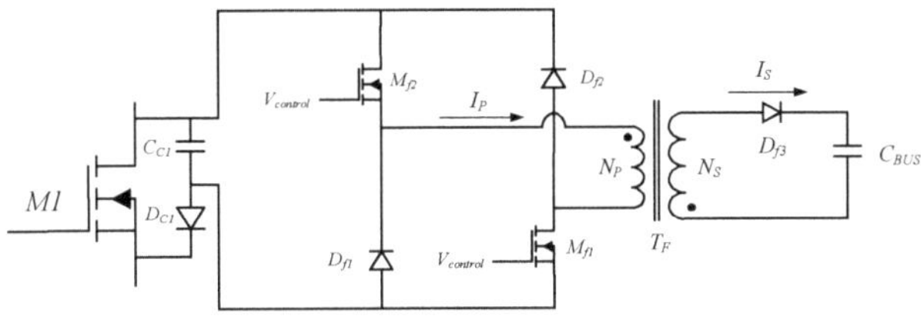

Figure 9. The energy recovery circuit based on a two-tube flyback circuit.

The working waveform of the energy recovery circuit is shown in Figure 10, and its working process circuit diagram is shown in Figure 11. Its working process is divided into two stages, the primary coil energy storage stage and the secondary coil discharge stage:

(1) Primary coil energy storage stage: this stage of the circuit is shown in Figure 11a. C_{C1} voltage is too high before t_0 and the feedback control circuit output $V_{control}$ becomes high at t_0. At this time, $V_{C1} = V_0$, and the primary coil current $i_P = 0$. The switching MOSFET M_{f1} and M_{f2} are turned on, and the primary coil N_P stores energy, and i_P continues to rise. C_{C1} releases energy and V_{C1} decreases from V_0. The pulse signal $V_{control}$ lasts for t_{pulse}. During the phase $[t_0–t_1]$, the inductance L_P of the primary coil of the flyback transformer resonates with the capacitance C_{C1} and i_P can be assumed to rise linearly during this period:

$$i_P(t) = \frac{V_0}{L_P}(t - t_0) \tag{21}$$

At this stage, the secondary coil current $i_S = 0$. D_{f3} is off and it is subjected to a reverse voltage V_d:

$$V_d = \frac{V_{C1}}{N} + V_{DC} \tag{22}$$

(2) Secondary coil discharge stage: this stage of the circuit is shown in Figure 11b. At t_1, $V_{control}$ becomes low and i_P reaches its maximum value I_{Pmax}. M_{f1} and M_{f2} are switched off and the primary coil N_P releases energy to the secondary coil N_S.

$$I_{P\max} = t_{pulse}\frac{V_0}{L_P} \tag{23}$$

During the phase $[t_0–t_1]$, V_{C1} continues to fall, and, at the moment t_1, V_{C1} falls to V_1.

$$V_1 = V_0\sqrt{(1 - \frac{t_{pulse}^2}{2L_P C_{C1}})} \tag{24}$$

The secondary coil N_S back-excites the current at t_1, when i_S reaches its maximum value I_{Smax} = N · I_{Pmax}, as shown in Figure 11b. In the $[t_1–t_2]$ stage, D_{f3} conducts, the secondary coil N_S releases energy to the bus capacitance C_{BUS}, and the i_S decreases linearly:

$$is_S(t) = t_{pulse}\frac{V_0 \cdot N}{L_P} - \frac{V_{DC}}{L_P}(t - t_1) \tag{25}$$

where L_S is the transformer secondary coil inductance. The $[t_1–t_2]$ phase duration t_S is

$$t_S = \frac{I_{S\max} \cdot L_S}{V_{DC}} = t_{pulse}\frac{V_0}{V_{DC} \cdot N} \tag{26}$$

During the $[t_1–t_2]$ stage, there is a reflected voltage V_{NP} on the primary coil of the transformer, which satisfies (27). Due to the presence of transformer leakage inductance, relatively large voltage spikes will be generated when M_{f1} and M_{f2} are switched off at t_1. The leakage inductance energy of the primary coil N_P can be returned to the capacitor C_{C1} through D_{f1} and D_{f2} to avoid voltage spikes to protect the switching tube.

$$V_{NP} = N(V_{DC} + V_D) \tag{27}$$

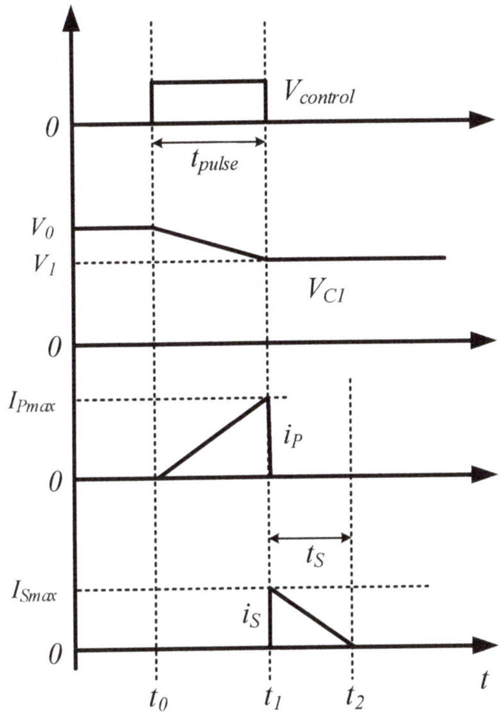

Figure 10. The working waveforms of the energy recovery circuit.

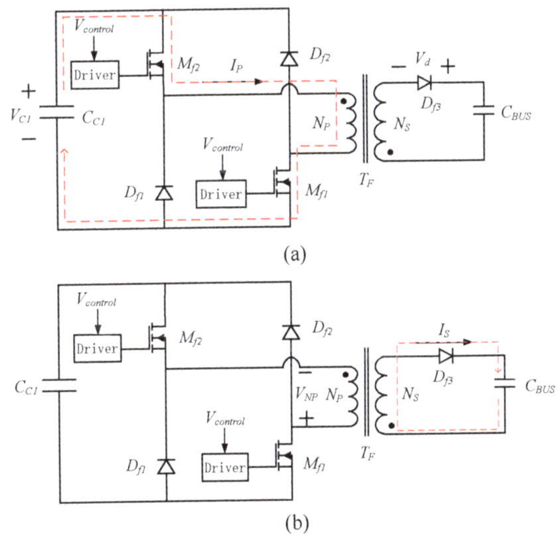

Figure 11. The equivalent circuit in the working process.

I_S drops to zero at t_2, D_{f3} will turn off, and the circuit operates in DCM mode. To ensure that the flyback circuit operates in intermittent mode, (28) needs to be satisfied:

$$t_S + t_{pulse} = t_{pulse}(1 + \frac{V_0}{V_{DC} \cdot N}) < \frac{1}{f_{\max}} \tag{28}$$

Without considering losses, if the flyback circuit operates at a frequency of f_{max}, the power delivered is P_S, which is also the energy released by C_{C1}. The increase in C_{C1} energy due to the action of the clamp circuit and the clamp auxiliary circuit will be passed back to the bus capacitor through the flyback circuit, ensuring that V_{C1} does not become too high.

$$P_S = \frac{1}{2}t_{pulse}^2 \cdot V_0^2 \cdot \frac{f_{max}}{L_P} \tag{29}$$

3.2. The Feedback Control Circuit Analysis

The feedback control circuit of the energy recovery circuit designed in this paper is shown in Figure 12, where V_{sample} is the voltage V_{C1} of the clamp capacitor C_{C1} sampled by the Hall voltage sensor. To suppress the noise interference of the power circuit on the sampled output signal, V_{sample} is followed by an active low-pass filter to filter out the high-frequency noise. The monostable trigger circuit will be triggered at the rising edge of the input signal and output a control signal $V_{control}$ with a certain pulse width. The $V_{control}$ is the input signal of the energy recovery circuit driver circuit, which is used to control the energy recovery circuit switching tube. The maximum voltage allowed by the clamping capacitor is V_{C1max}; when V_{C1} exceeds V_{C1max}, the sampled signal V_{sample} will be larger than the reference signal V_{ref}. At this time, COMP1's output is high, indicating that the capacitor C_{C1}'s voltage is too high.

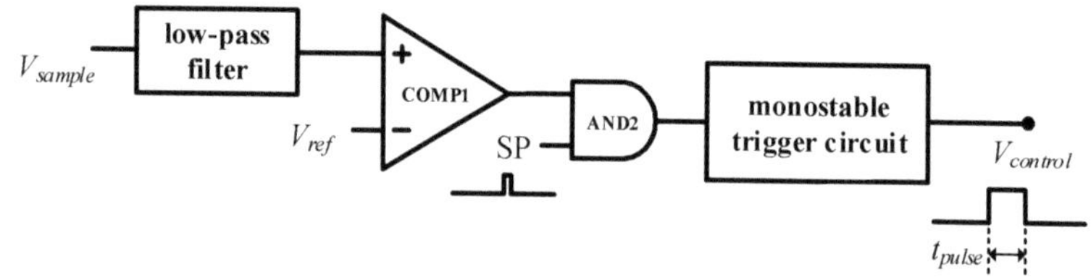

Figure 12. The diagram of the feedback control circuit.

The SP signal is a high-level pulse width small period square wave that determines whether the comparator COMP1's output is valid or not. When the short pulse signal SP goes high, there are two cases for the comparator COMP1's output. If COMP1's output is low, it indicates that V_{C1} is lower than the allowed maximum voltage V_{C1max}, and the output of AND2 is low. The monostable trigger circuit output is low. If COMP1's output is high, it indicates that V_{C1} is higher than the allowed maximum voltage V_{C1max}. The output of AND2 changes from low to high and triggers the monostable trigger circuit. The monostable trigger circuit will output a high-level pulse to control the energy recovery circuit. As the comparator COMP1's output is only effective when the short pulse signal SP is high, this can effectively reduce the false trigger caused by sampling error or comparator output noise. The frequency of the SP signal is the maximum operating frequency of the energy recovery circuit. The relative waveforms are shown in Figure 13.

(1) Before t_0, since V_{C1} is high enough, the V_{sample} will be higher than V_{ref}, and then, COMP1's output is high. However, SP is low and AND2's output is still low.

(2) At t_0, SP becomes high and lasts for a short time. Since the COMP1 output is high, the AND2 output changes from low to high. This rising edge will trigger a monostable trigger to generate a $V_{control}$ signal with a pulse width of t_{pulse} for controlling the energy recovery circuit. The operation of the energy recovery circuit is shown in Figure 10. During phase [t_0–t_2], the switching tube of the energy recovery circuit is switched on. During this phase, the V_{C1} continues to drop and energy is transferred to the primary coil of the flyback transformer.

(3) At t_2, $V_{control}$ becomes low. The switching tube of the energy recovery circuit turns off, and the transformer's secondary coil back-excites the current. Although V_{C1} has dropped to the permissible value, the comparator COMP1's output is still high. This is due to the delay of the Hall voltage sensor, which does not reduce the COMP1

output until the moment t_3 after the t_{SD} time. During the [t_2–t_4] phase, V_{C1} may rise due to the activation of the clamp capacitor and the clamp auxiliary circuit.

(4) At t_4, the SP signal becomes high and the period of the SP signal is t_{SP}. If V_{C1}'s voltage rises more in the phase [t_2–t_4], causing the COMP1 output to be high at t_4, the monostable flip-flop will be activated next, repeating the process (1)–(3). If V_{C1} is small and does not exceed the predetermined voltage in the phase [t_2–t_4], the COMP1 output is low at t_4, the AND2 output is low and the monostable flip-flop will not operate. The monostable circuit operation will be decided Until the next rising edge of the SP signal according to the COMP1 output.

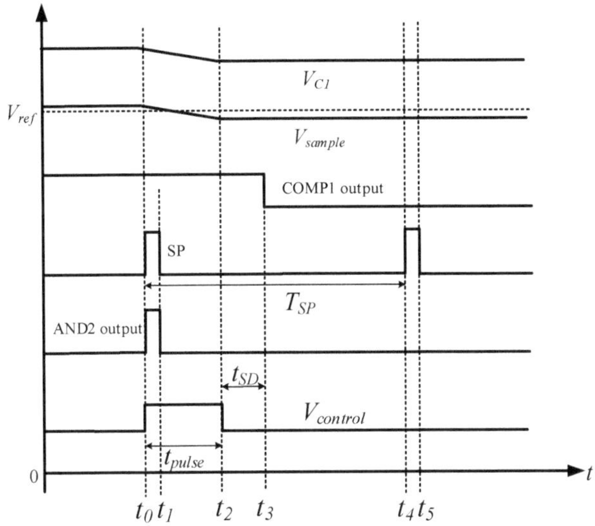

Figure 13. The relative waveforms of the feedback control circuit.

If the switching process of the series-connected MOSFETs and the driver circuit varies greatly, it will result in a large ripple voltage of each clamp capacitor during the switching process. In this case, V_{C1} will rise rapidly. To limit the rise of V_{C1}, the energy recovery circuit switching frequency must be larger; in contrast, if the series-connected MOSFETs and drive circuits are highly consistent, the switching process of each clamp capacitor ripple voltage is smaller. In this situation, V_{C1} rises slowly, and the energy recovery circuit switching frequency is small. Ideally, if each MOSFET switching process is identical, V_{C1} will not rise. At this time, the energy recovery circuit operating frequency is zero. In summary, the operating frequency of the energy recovery circuit depends on the degree of difference in the switching process of each MOSFET in the series-connected MOSFET.

Based on the above analysis, it can be seen that, if COMP1's output signal is high every time the SP signal is high, the monostable flip-flop will continuously output pulses, which occurs when the difference in the switching process of each MOSFET is too large. Therefore, the maximum frequency of the flyback circuit operation is equal to the frequency of the SP signal $f_{max} = 1/t_{SP}$. According to (29), the maximum power delivered by the flyback circuit is P_{Smax}.

$$P_{S\max} = \frac{1}{2} t_{pulse}^2 \cdot V_0^2 \cdot \frac{f_{\max}}{L_P} \quad (30)$$

The maximum power P_{Smax} in the worst case needs to be taken into account when the flyback circuit operates at the maximum frequency V_{C1}. Consideration needs to be given to the value of the energy that will be accumulated by the clamp capacitor due to the clamping action when the difference in drive delay between the MOSFETs is at its maximum, and P_{Smax} needs to be ensured to be greater than the amount of energy that will be accumulated due to the clamping action. For the three-tube tandem SiC MOSFET topology in this paper, when only the turn-off process is considered, it is assumed that the M3 drive signal is delayed t_{D1} to M1, while the M2 drive signal is delayed t_{D2} to M1. Then, the maximum power P_{Smax} of the flyback circuit needs to satisfy (31):

$$(t_{D1} + t_{D2}) \cdot I_{load} \cdot \frac{V_{DC}}{3} \cdot f_{PWM} < P_{S\max} \quad (31)$$

If t_{D1} and t_{D2} are assumed to be 100 ns and 200 ns, respectively, it is a serious case for series-connected SiC MOSFETs. When the bus voltage is 2800 V and the load current is 10 A, the maximum power of the flyback circuit needs to be higher than 140 W. With the turn-on phase considered, the maximum power of the flyback circuit should be greater than the calculated value in (31), which can be taken to be about 1.5–2 times the calculated value. For series n-MOSFETs, the maximum capacitance of V_{C1max} needs to be set to be slightly larger than the equilibrium voltage V_{DC}/n, and, at the same time, V_{C1max} needs to be smaller than the breakdown voltage of the MOSFET. In this paper, V_{C1max} is larger than V_{DC}/n by 30–50 V. If $V_{control}$ pulse width is set to 10 μs, and the inductance of the primary coil of the flyback transformer is set to 1.5 mH, the maximum current of the primary coil is 6.5 A. The maximum power of the designed flyback transformer is 210 W, and it can meet the design requirements. Flyback circuit design needs to consider the switching device voltage stress. This paper takes the margin of switching device voltage stress as 20%. According to (22), the diode D_{f3} is subjected to the maximum reverse voltage at t_0. Its rated reverse voltage V_F needs to satisfy the relation (32):

$$0.8V_F > \frac{V_0}{N} + V_{DC} \quad (32)$$

According to (27), M_{f1} and M_{f2} are subjected to the maximum reverse voltage at t_1. The rated voltage V_{MOS} of M_{f1} and M_{f2} needs to satisfy the relation (33):

$$2 \times 0.8V_{MOS} > N(V_{DC} + V_D) + V_1 \quad (33)$$

In this paper, we took N = 0.316, and set V_0 = 930 V and V_1 = 950 V. According to (22), the diode reverse voltage V_F is >7200 V, which was realized by connecting six 1200 V SiC SBDs in series. According to (27), V_{MOS} > 1146 V, which was realised by using 1200 V SiC MOSFETs.

4. The Experimental Results

A prototype was also fabricated and tested to verify its effectiveness. The overall circuit of the test is shown in Figure 14, where M1 and M2 are two SiC MOSFETs connected in series. The prototype is shown in Figure 15. The main parameters of the converter are in Table 2, and the parameters of the SiC MOSFET are listed in Table 3.

In this paper, the input voltage of the main Buck circuit was set to 600 V, and the duty cycle of the Buck circuit was set to 50%, which was used to complete the 300–600 V Buck conversion. The average current of the main inductor L_{load} was 10 A; then, the power of the main circuit design was 2.8 kW.

To verify the effectiveness of the strategy, the gate resistors of M1 and M2 were set differently; the M1 gate resistance was set to be 33 ω, while the M2 was 0 ω. M1 and M2 were directly connected in series without applying other circuits, and the drain–source voltages of the two MOSFETs, V_{DS1} and V_{DS2}, are shown in Figure 16.

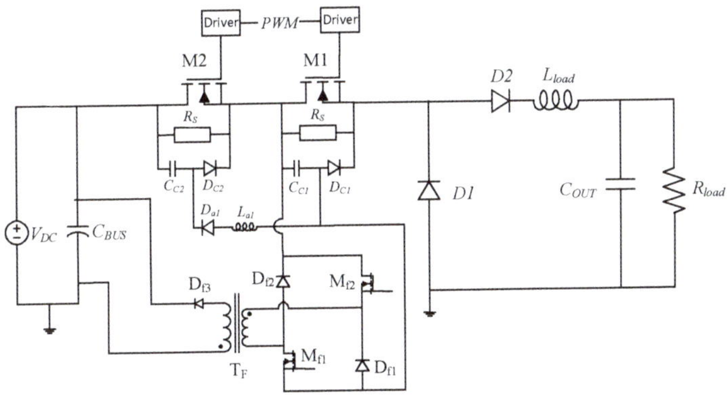

Figure 14. Dual tube series SiC MOSFET's main Buck power circuit.

Figure 15. Dual tube series SiC MOSFET's main Buck power circuit.

Table 2. The relative parameters of the prototype.

Circuit Parameters		Circuit Parameters	
The main MOSFET	SCH2080KE	Diode D_1 and D_2	SCS240KE2
Bus voltage	600 V	Working frequency	50 kHz
Load inductance	400 μH	Output capacitance	340 μF
Clamp capacitance	2 μF	Clamp diode D_C	ES1J
Auxiliary diode	ES1J	Current limiting Inductance L_a	15 μH
SP working frequency	10 kHz	Static Resistor R_S	100 kΩ

Table 3. The parameters of the SiC MOSFET.

Break voltage	1200 V	Leakage current	400 μA
Threshold voltage	4.0 V	on-state resistance@v_{GS} = 18 V	80 mΩ
Input capacitance	1850 pF	Output capacitance	175 pF

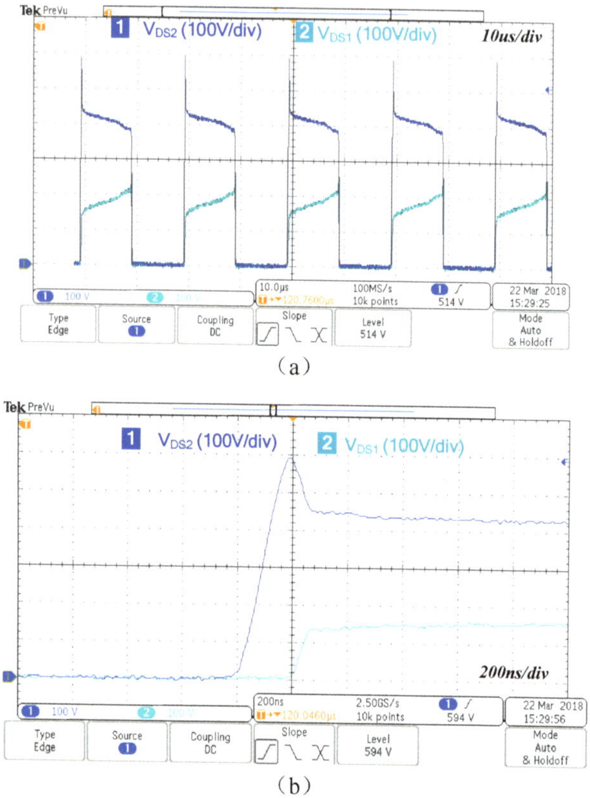

Figure 16. The test results without the proposed strategy: (**a**) the whole waveforms, and (**b**) the working details.

Figure 16b shows the voltage distribution during turn-off. Due to the small M2 gate resistance, M2 turns off first and V_{DS2} rises first during the turn-off interval. The delay difference between the two MOSFETs during turn-off is 200 ns. In this case, V_{DS2} rises directly to the bus voltage of 600 V. M1 is then turned off, resulting in severe voltage unevenness during turn-off. The difference between the two V_{DS} in the static state is slightly reduced due to the static voltage equalizing resistor. This experiment verifies that different driving delays will cause voltage unevenness, which seriously affects the safe operation of series-connected SiC MOSFETs.

The tested results in the same case with the proposed strategy are shown in Figure 17. The voltage distribution during turn-off is slightly different from cycle to cycle because the clamp capacitor voltage changes from cycle to cycle. During both turn-on and turn-off intervals, the maximum MOSFET voltage does not exceed 320 V. Comparing with Figure 16, it can be seen that this topology can effectively limit the drain–source voltage rise, and ensure that the dynamic and static voltages of series-connected SiC MOSFETs are balanced.

The variation curves of the two MOSFET V_{DS} during switching are shown in Figure 18. Figure 18b shows the turn-off detail. It can be seen that M2 turns off first, causing V_{DS2} to start rising. When V_{DS2} rises to the clamp capacitor voltage, the clamp circuit turns on, and V_{DS2} stops rising. Then, M1 turns on, and V_{DS1} starts to rise. It can be seen that this topology can limit the MOSFET drain–source voltage from being too high during the shutdown.

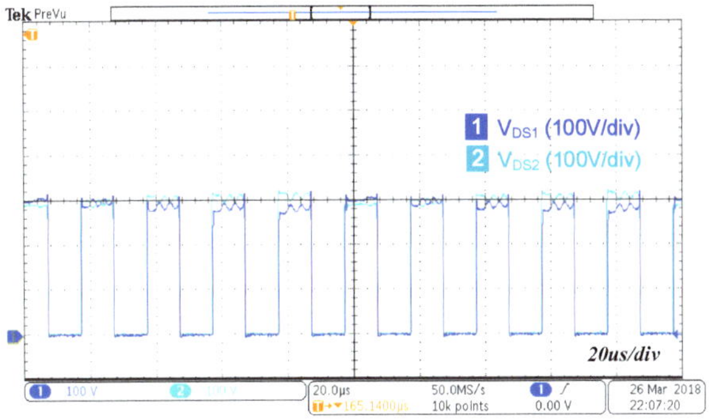

Figure 17. The test results with the proposed strategy.

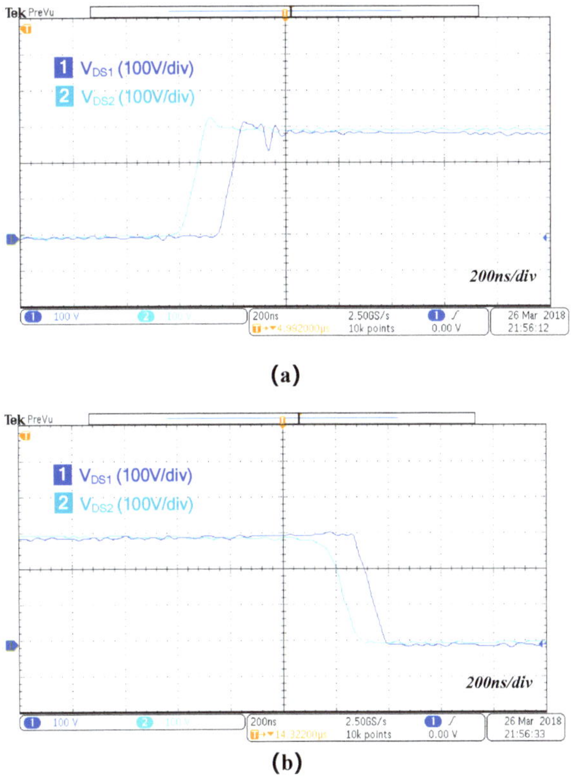

Figure 18. The working details with the proposed strategy: (**a**) the turn-on details; (**b**) the turn-off details.

Figure 18b shows the curve of V_{DS}'s change during the turn-on process. M2 turns off first, leading V_{DS2} to start falling. There is no voltage spike in the turn-on process.

The experimental results are the same as the analysis above. The voltage spike in the turn-off interval is suppressed from 600 V to 320 V.

5. Conclusions

A comparison was made between state-of-the-art approaches and the approach proposed in this work; the results are shown in Table 4. Compared to other approaches, this work achieved energy recovery at the cost of six power devices. Due to the relative simplicity of the control method in this work, the number of analog devices used was low.

Table 4. The comparison among state-of-the-art approaches.

The Approaches	Extra Devices	Complexity	Energy Recovery
This work	10 power devices and 2 analog devices	Moderate	Yes
Ref. [29]	11 analog devices	High	No
Ref. [30]	7 analog devices	High	No
Ref. [31]	3 power devices and 6 analog devices	High	No
Ref. [32]	3 power devices and 6 analog devices	High	No

Based on the analysis above, when increasing the number of series-connected SiC MOSFETs, this method only requires adding simple auxiliary circuits to the device without changing the energy recovery circuitry, which has a small impact on the circuit complexity. The energy recovery circuit uses a transformer and is the largest part of the auxiliary circuit. Since the energy recovery circuit operates at a lower frequency and processes less energy per cycle, it has the possibility of magnetic integration with the magnetic devices of the main circuit under certain conditions.

Based on the analysis above, this strategy has some advantages. Firstly, the approach is effective; the voltage spike is suppressed from bus voltage 600 V to 320 V. Secondly, the strategy realizes energy recovery, which is beneficial for efficiency. Thirdly, the drive strategy is simple; fewer analog devices are used.

Author Contributions: Conceptualization, S.S. and J.L.; methodology, P.L.; validation, P.L., S.S. and J.L.; formal analysis, S.S.; investigation, J.L.; resources, Y.Z.; data curation, W.Y.; writing—original draft preparation, P.L.; writing—review and editing, Y.S.; visualization, S.S.; supervision, Y.Z.; project administration, Y.Z.; funding acquisition, Y.Z. All authors have read and agreed to the published version of the manuscript.

Funding: This research was funded by Key R&D Program of Shaanxi Province grant number 2023-YBGY-005.

Data Availability Statement: Data are contained within the article.

Conflicts of Interest: The authors declare no conflicts of interest.

References

1. Millán, J.; Godignon, P.; Perpiñà, X.; Pérez-Tomás, A.; Rebollo, J. A Survey of Wide Bandgap Power Semiconductor Devices. *IEEE Trans. Power Electron.* **2014**, *29*, 2155–2163. [CrossRef]
2. Palmour, J.W. Silicon carbide power device development for industrial markets. In Proceedings of the 2014 IEEE International Electron Devices Meeting, San Francisco, CA, USA, 15–17 December 2014; pp. 1.1.1–1.1.8. [CrossRef]
3. Song, Q.; Yang, S.; Tang, G.; Han, C.; Zhang, Y.; Tang, X.; Zhang, Y.; Zhang, Y. 4H-SiC Trench MOSFET With L-Shaped Gate. *IEEE Electron Device Lett.* **2016**, *37*, 463–466. [CrossRef]
4. She, X.; Huang, A.Q.; Lucía, O.; Ozpineci, B. Review of Silicon Carbide Power Devices and Their Applications. *IEEE Trans. Ind. Electron.* **2017**, *64*, 8193–8205. [CrossRef]
5. Song, Q.; Yuan, H.; Sun, Q.; Han, C.; Tang, X.; Zhang, Y.; Yuan, L.; Yang, S.; Zhang, Y. Reverse-Bias Stress-Induced Electrical Parameters Instability in 4H-SiC JBS Diodes Terminated Nonequidistance FLRs. *IEEE Trans. Electron Devices* **2019**, *66*, 3935–3939. [CrossRef]
6. Yu, L.; Sheng, K. Modeling and Optimal Device Design for 4H-SiC Super-Junction Devices. *IEEE Trans. Electron Devices* **2008**, *55*, 1961–1969. [CrossRef]

7. Bolotnikov, A.; Losee, P.; Permuy, A.; Dunne, G.; Kennerly, S.; Rowden, B.; Nasadoski, J.; Harfman-Todorovic, M.; Raju, R.; Tao, F.; et al. Overview of 1.2–2.2 kV SiC MOSFETs targeted for industrial power conversion applications. In Proceedings of the 2015 IEEE Applied Power Electronics Conference and Exposition (APEC), Charlotte, NC, USA, 15–19 March 2015; pp. 2445–2452. [CrossRef]
8. Ferraro, A. An overview of low-loss snubber technology for transistor converters. In Proceedings of the 1982 IEEE Power Electronics Specialists conference, Cambridge, MA, USA, 14–17 June 1982; pp. 466–477. [CrossRef]
9. Vechalapu, K.; Negi, A.; Bhattacharya, S. Comparative performance evaluation of series connected 15 kV SiC IGBT devices and 15 kV SiC MOSFET devices for MV power conversion systems. In Proceedings of the Energy Conversion Congress and Exposition (ECCE), Milwaukee, WI, USA, 18–22 September 2016; pp. 1–8.
10. Busatto, G.; Cascone, B.; Fratelli, L.; Luciano, A. Series connection of IGBTs in hard-switching applications. In Proceedings of the Industry Applications Conference, 1998. Thirty-Third IAS Annual Meeting, St. Louis, MO, USA, 12–15 October 1998; Volume 2, pp. 825–830.
11. Robinson, F.V.; Hamidi, V. Series connecting devices for high-voltage power conversion. In Proceedings of the 2007 42nd International Universities Power Engineering Conference, Brighton, UK, 4–6 September 2007; pp. 1134–1139. [CrossRef]
12. Lim, T.C.; Williams, B.W.; Finney, S.J. Active Snubber Energy Recovery Circuit for Series-Connected IGBTs. *IEEE Trans. Power Electron.* **2011**, *26*, 1879–1889. [CrossRef]
13. Zheng, S.; Wang, Y.; Wu, X.; Qian, Z.; Peng, F.Z. The voltage sharing of commercial IGBTS in series with passive components. In Proceedings of the 2011 IEEE Energy Conversion Congress and Exposition, Phoenix, AZ, USA, 17–22 September 2011; pp. 3008–3012. [CrossRef]
14. Gerster, C. Fast high-power/high-voltage switch using series-connected IGBTs with active gate-controlled voltage-balancing. In Proceedings of the Applied Power Electronics Conference and Exposition—APEC'94, Orlando, FL, USA, 13–17 February 1994; Volume 1, pp. 469–472. [CrossRef]
15. Sasagawa, K.; Abe, Y.; Matsuse, K. Voltage-balancing method for IGBTs connected in series. *IEEE Trans. Ind. Appl.* **2004**, *40*, 1025–1030. [CrossRef]
16. Palmer, P.R.; Zhang, J.; Zhang, X. SiC MOSFETs connected in series with active voltage control. In Proceedings of the 2015 IEEE 3rd Workshop on Wide Bandgap Power Devices and Applications (WiPDA), Blacksburg, VA, USA, 2–4 November 2015; pp. 60–65. [CrossRef]
17. Raciti, A.; Belverde, G.; Galluzzo, A.; Greco, G.; Melito, M.; Musumeci, S. Control of the switching transients of IGBT series strings by high-performance drive units. *IEEE Trans. Ind. Electron.* **2001**, *48*, 482–490. [CrossRef]
18. Bruckmann, M.; Sommer, R.; Fasching, M.; Sigg, J. Series connection of high voltage IGBT modules. In Proceedings of the Conference Record of 1998 IEEE Industry Applications Conference. Thirty-Third IAS Annual Meeting (Cat. No.98CH36242), St. Louis, MO, USA, 12–15 October 1998; Volume 2, pp. 1067–1072. [CrossRef]
19. Saiz, J.; Mermet, M.; Frey, D.; Jeannin, P.O.; Schanen, J.L.; Muszicki, P. Optimisation and integration of an active clamping circuit for IGBT series association. In Proceedings of the Conference Record of the 2001 IEEE Industry Applications Conference. 36th IAS Annual Meeting (Cat. No.01CH37248), Chicago, IL, USA, 30 September–4 October 2001; Volume 2, pp. 1046–1051. [CrossRef]
20. Nakatake, H.; Iwata, A. Series connection of IGBTs used multilevel clamp circuit and turn off timing adjustment circuit. In Proceedings of the Power Electronics Specialist Conference, 2003. PESC'03, Acapulco, Mexico, 15–19 June 2003; Volume 4, pp. 1910–1915. [CrossRef]
21. Hess, H.L.; Baker, R.J. Transformerless capacitive coupling of gate signals for series operation of power MOS devices. *IEEE Trans. Power Electron.* **2000**, *15*, 923–930. [CrossRef]
22. Kim, J.H.; Min, B.D.; Shenderey, S.V.; Rim, G.H. High voltage pulsed power supply using IGBT stacks. *IEEE Trans. Dielectr. Electr. Insul.* **2007**, *14*, 921–926. [CrossRef]
23. Teerakawanich, N.; Johnson, C.M. Design optimization of quasi-active gate control for series-connected power devices. *IEEE Trans. Power Electron.* **2014**, *29*, 2705–2714. [CrossRef]
24. Wu, X.; Cheng, S.; Xiao, Q.; Sheng, K. A 3600 V/80 A Series–Parallel-Connected Silicon Carbide MOSFETs Module with a Single External Gate Driver. *IEEE Trans. Power Electron.* **2014**, *29*, 2296–2306. [CrossRef]
25. Xiao, Q.; Yan, Y.; Wu, X.; Ren, N.; Sheng, K. A 10 kV/200 A SiC MOSFET module with series-parallel hybrid connection of 1200 V/50 A dies. In Proceedings of the Power Semiconductor Devices & IC's (ISPSD), Hong Kong, China, 10–14 May 2015; pp. 349–352.
26. Maniktala, S. Chapter 8—Conduction and Switching Losses. In *Switching Power Supplies A–Z*, 2nd ed.; Maniktala, S., Ed.; Newnes: Oxford, UK, 2012; pp. 311–341. [CrossRef]
27. Van Nguyen, T.; Jeannin, P.O.; Vagnon, E.; Frey, D.; Crebier, J.C. Series connection of IGBT. In Proceedings of the Applied Power Electronics Conference and Exposition (APEC), Palm Springs, CA, USA, 21–25 February 2010; pp. 2238–2244.
28. Han, Y.; Chen, W.; Chen, X.; Ma, X.; Sha, Y.; Yang, X.; Li, X. A 4000 V input auxiliary power supply with series connected SiC MOSFETs for MMC-based HVDC system. In Proceedings of the Power Electronics and Motion Control Conference (IPEMC-ECCE Asia), Hefei, China, 22–26 May 2016; pp. 279–284.
29. Zhao, M.; Lin, H.; Wang, T. An Adaptive Driving Signals Delay Control for Voltage Balancing of Multiple Series-Connected SiC MOSFETs. In Proceedings of the 2022 IEEE Applied Power Electronics Conference and Exposition (APEC), Houston, TX, USA, 20–24 March 2022; pp. 1619–1624. [CrossRef]

30. Zhao, M.; Lin, H.; Wang, T. The Application of Series-Connected SiC MOSFETs in Buck Converter Based on Active Driving Signal Delay Control for Voltage Balancing. In Proceedings of the 2022 IEEE Applied Power Electronics Conference and Exposition (APEC), Houston, TX, USA, 20–24 March 2022; pp. 1614–1618. [CrossRef]
31. Gao, Z.; Pei, X.; Cui, W.; Wu, H.; Shao, S.; Zhang, J. Voltage Balancing Method for Series-Connected SiC MOSFETs Under No Load Condition. In Proceedings of the 2023 20th China International Forum on Solid State Lighting & 2023 9th International Forum on Wide Bandgap Semiconductors (SSLCHINA: IFWS), Xiamen, China, 27–30 November 2023; pp. 157–160. [CrossRef]
32. Gao, Z.; Shao, S.; Cui, W.; Wu, Y.; Zhang, J.; Sheng, K. A Control Strategy for Reducing Voltage Ripples in Series-Connected SiC MOSFETs Using Active Clamping Modules in High Power VSCs. *IEEE Trans. Power Electron.* **2024**, *39*, 12149–12155. [CrossRef]

electronics

Article

A Single-Stage Electronic Lighting Driver Circuit Utilizing SiC Schottky Diodes for Supplying a Deep Ultraviolet LED Disinfection and Sterilization Lamp

Chun-An Cheng, Chien-Hsuan Chang, Hung-Liang Cheng, En-Chih Chang *, Long-Fu Lan, Sheng-Hong Hou and Cheng-Kuan Lin

Department of Electrical Engineering, I-Shou University, Kaohsiung City 84001, Taiwan; cacheng@isu.edu.tw (C.-A.C.); chchang@isu.edu.tw (C.-H.C.); hlcheng@isu.edu.tw (H.-L.C.); isu11001004m@cloud.isu.edu.tw (L.-F.L.); isu11001001m@cloud.isu.edu.tw (S.-H.H.); isu11001014m@cloud.isu.edu.tw (C.-K.L.)

* Correspondence: enchihchang@isu.edu.tw; Tel.: +886-7-6577711 (ext. 6642)

Abstract: Recently, a new type of lighting source, deep ultraviolet light-emitting diode (LED), has appeared in the markets of space purification and surface sterilization. In this paper, a new type of electronic lighting driver for supplying a deep-ultraviolet LED sterilization lamp is proposed and developed. The main circuit combines a buck converter and a flyback converter into a single-stage single-switch buck-flyback AC-DC power converter with power factor correction. In addition, the proposed electronic lighting driver leverages a wide bandgap SiC Schottky diode as the output diode to lower the power diode losses and recycles the energy stored in the leakage inductance of the transformer in order to improve the circuit efficiency. The magnetizing inductor inside the presented AC-DC power converter is designed to operate in discontinuous conduction mode (DCM), which naturally enables power factor correction (PFC). A single-stage prototype driver with a power rating of 3.6 W (90 V/40 mA) was developed and implemented for providing a deep ultraviolet LED disinfection and sterilization lamp. Experimental results show that the measured power factor (PF) is greater than 0.9 and the measured total harmonic distortion (THD) of the input current is less than 18% at an input utility voltage of 110 V. Furthermore, the measured output voltage ripple factor is less than 1% and the output current ripple factor is less than 4%. In addition, the proposed single-stage electronic lighting driver for supplying a deep ultraviolet LED disinfection and sterilization lamp achieves high circuit efficiency (greater than 90%), low circuit component count, and low circuit cost.

Keywords: electronic lighting driver circuit; deep ultraviolet LED; disinfection and sterilization lamp; power factor correction; total harmonic distortion

Citation: Cheng, C.-A.; Chang, C.-H.; Cheng, H.-L.; Chang, E.-C.; Lan, L.-F.; Hou, S.-H.; Lin, C.-K. A Single-Stage Electronic Lighting Driver Circuit Utilizing SiC Schottky Diodes for Supplying a Deep Ultraviolet LED Disinfection and Sterilization Lamp. *Electronics* **2024**, *13*, 3048. https://doi.org/10.3390/electronics13153048

Academic Editors: Yimeng Zhang, Lejia Sun and Yuming Zhang

Received: 24 June 2024
Revised: 27 July 2024
Accepted: 29 July 2024
Published: 1 August 2024

1. Introduction

Microorganisms such as bacteria, viruses, cysts and molds use germination or spore production to reproduce. When microorganisms are irradiated or exposed to disinfecting and sterilizing lamps, the light penetrates through the cell walls and destroys their DNA, changing their DNA so they become harmless and incapable of reproducing, or even die, thus achieving the effect of disinfecting and sterilizing. Sterilization and disinfection lamps are used in a wide range of applications in a variety of sectors, including healthcare facilities, the food and beverage industry, public spaces and transportation, commercial and office buildings, residential buildings, hospitality, educational institutions, retail and grocery stores, water treatment and agriculture. These diverse applications highlight the versatility and effectiveness of disinfecting and sterilizing lamps in promoting health and safety in different environments [1–4]. The traditional light source for sterilization and disinfection is the ultraviolet (UV) mercury lamp, which has the advantages of low price and high luminous power and is commonly used in large places such as hospitals, factories, restaurants and schools. However, the disadvantages of UV mercury lamps are long warm-up time, short lamp life, and that the mercury content in the lamps does not meet environmental requirements. UV light is invisible to the human eye and has a wavelength of 100 to 400 nanometers. Depending on the wavelength, UV can be subdivided into three types: long-wave UV-A (wavelengths of 320–400 nm), medium-wave UV-B (wavelengths of 280–320 nm) and short-wave UV-C (wavelengths of 100–280 nm). Short-wave UV-C, which can also be referred to as deep ultraviolet, has disinfecting and germicidal properties. The use of short-wave ultraviolet UV-C or deep ultraviolet light destroys the DNA or RNA molecular structure of microbial cells, thereby preventing cell growth or regeneration, and ultimately achieving the effect of disinfection and sterilization. Deep UV LEDs are characterized by environmental protection and energy saving, long life, no warm-up time, no mercury, and miniaturization of light source. Traditional UV mercury lamps have been gradually replaced by deep UV LEDs, which have become a new light source for space purification and surface sterilization markets such as homes, schools, hospitals, office buildings, supermarkets or department stores. In addition, deep ultraviolet disinfection and sterilization lamps can easily kill bacteria, but due to the low penetrating power of deep ultraviolet disinfection and sterilization lamps, they can be affected by dust particles; if the cells damaged by deep ultraviolet disinfection and sterilization lamps are not completely destroyed, the cells are still able to be repaired and revitalized. Therefore, it is important to provide the appropriate distance and sufficient irradiation time and intensity of the deep ultraviolet disinfection and sterilization lamp to the disinfected objects in order to achieve a good sterilization effect [5–8].

Figure 1 shows the two-stage circuit topology of an existing electronic lighting driver for supplying deep ultraviolet LED sterilization and disinfection lamps. The first stage is an AC-DC power converter with power factor correction (PFC), and the second stage is a DC-DC converter and delivers the required DC power to the deep-UV LED sterilization lamp [8]. Two-stage electronic lighting driver circuits that can be applied to deep ultraviolet light-emitting diode (LED) sterilization lamps require separate controllers for both the front and rear stages, making the circuits more complex and requiring a larger number of circuit components. In order to reduce the number of power switches and circuit components and to improve the overall efficiency of the conventional two-stage version of the circuit, the literature [9–20] has developed a number of single-stage, single-switch LED driver circuits that integrate an AC-DC power converter and a DC-DC power converter suitable for powering deep ultraviolet LED sterilization and disinfection lamps.

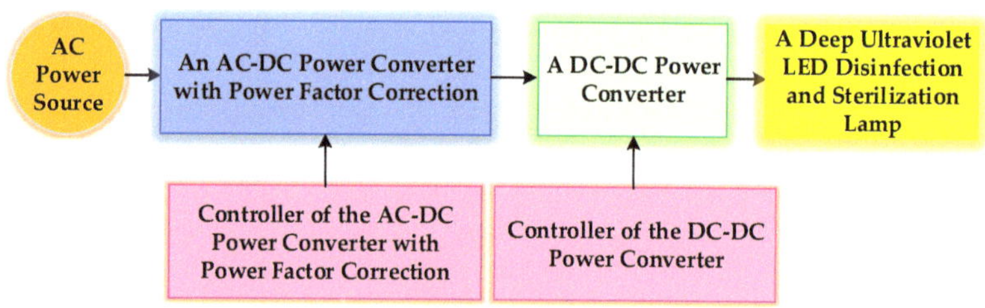

Figure 1. Block diagram of the existing two-stage electronic lighting driver for supplying a deep ultraviolet LED disinfection and sterilization lamp [8].

As the environmental issues derived from global warming and carbon emissions have become more and more serious in recent years, human beings have taken energy saving, carbon reduction, and love for the earth as the common primary development direction. Thus, fossil fuels must be gradually reduced and rapidly introduced into green energy and power saving applications, and, therefore, high energy efficiency and low energy consumption are also gradually adopted as the goal in daily life products. In the Paris Agreement of the United Nations Conference on Climate Change, it is declared that the global warming rate must be kept within 2 °C. Based on the current economic development trend, even if the warming is kept within 2 °C in 2050, CO_2 emissions will still increase by 21%, and up to 50% more electricity will have to be generated to cope with and provide for various human activities. Therefore, it is increasingly important to significantly upgrade and improve existing energy sources. The largest raw material for semiconductors is the production of first-generation silicon (Si) wafers. However, the physical properties of this material have reached their limits in existing Si-based products and it is no longer possible to increase power, reduce heat loss and increase speed. As a result, there is a need to evolve to other materials that can better utilize the efficiency of electron transfer and low energy consumption, and it is in this context that the third-generation wide-bandgap semiconductors that are currently hotly debated in the marketplace, namely, silicon carbide (SiC) and gallium nitride (GaN), which are characterized by high energy efficiency and low energy consumption, have come into being. In addition, power devices made of wide-gap semiconductor materials can operate at higher voltages, generate more power, and have lower energy losses, while being significantly smaller than power devices made of silicon semiconductor materials. Silicon carbide (SiC) is a compound semiconductor material consisting of silicon (Si) and carbon (C). SiC has a lower resistance of the drift layer than Si, does not require conduction modulation, has an energy gap about 3-times wider than that of Si, and has an insulation breaking field strength about 10-times higher than that of Si. Compared to the high-voltage Si-PN diodes and high-speed Si-Fast Recovery diodes, SiC Schottky diodes made of wide-gap semiconductor materials have excellent high-speed and high-voltage characteristics. Since the reverse recovery time of SiC Schottky diodes is extremely short, the energy lost during reverse recovery can be greatly reduced, which is suitable for applications with high switching frequency, and is conducive to the miniaturization of the overall circuit as well as the improvement of power density [21,22]. Supplementary notes on the long-term reliability and potential degradation of silicon carbide Schottky diodes under thermal and electrical stresses are given below. According to the literature [21], silicon carbide Schottky diodes do not have minority carrier injection and their recovery time does not increase with temperature compared to conventional silicon diodes. In addition, SiC Schottky diodes are majority carrier devices and therefore the reverse current is smaller and temperature independent. In addition, compared to silicon diodes, SiC Schottky diodes have the advantage of faster recovery, which reduces switching losses and noise, and lower forward voltage, which reduces conduction losses. Furthermore, the literature [22] mentions the long-term reliability of SiC Schottky diodes with respect to dV/dt failure and dI/dt failure. Furthermore, in silicon diodes, if dI/dt is large, a pattern of increased recovery current (Irr) occurs, which can damage the device through current concentration. In contrast, since the recovery current in SiC Schottky diodes is very small, it can be assumed that such a pattern is unlikely to occur.

To address the above issues, a novel electronic lighting driver for providing a deep ultraviolet LED disinfection and sterilization lamp is proposed and developed in this paper, which is an extension and further improvement of the literature [20]. The main circuit combines a buck converter and a flyback converter into a single-stage single-switch buck-flyback AC-DC power converter with input-current shaping. In addition, the proposed driver recycles the energy stored in the leakage inductance of the transformer and exploits a wide bandgap silicon carbide Schottky diode as the output diode to enhance the circuit efficiency.

This paper is organized and introduced as follows. Section 2 describes and analyzes the mode of operation of a single-stage electronic lighting driver circuit utilizing a SiC Schottky diode to supply a deep ultraviolet LED disinfection and sterilization lamp, and provides design guidelines for some circuit parameters. Section 3 shows the experimental results of a prototype electronic lighting driver circuit utilizing a SiC Schottky diode to supply a deep ultraviolet LED sterilization and disinfection lamp. Finally, some conclusions are given in Section 4.

2. Descriptions and Analysis of Operational Modes in the Proposed Single-Stage Electronic Lighting Driver Circuit Utilizing SiC Schottky Diodes for Supplying a Deep Ultraviolet LED Disinfection and Sterilization Lamp

The electronic lighting driver with PFC for powering a deep ultraviolet LED disinfection and sterilization lamp developed in this paper is shown in Figure 2, which integrates a buck converter and a flyback converter into a single-stage AC-DC power conversion and includes a filter inductor L_F, a filter capacitor C_F, a full-bridge rectifier (including D_1, D_2, D_3 and D_4), a power switch S_B, two diodes D_B and D_F, a transformer T_R with a magnetizing inductor L_M and a leakage inductor L_{lk}, two output capacitors C_{O1} and C_{O2}, and the deep ultraviolet LED disinfection and sterilization lamp. By designing the magnetizing inductor L_M in the proposed electronic lighting driver circuit to operate in discontinuous conduction mode (DCM), PFC can be realized naturally. In addition, the stored energy in the leakage inductance L_{lk} of the transformer T_R in the proposed electronic lighting driver circuit can be recovered and the circuit efficiency can be enhanced by exploiting a wide-bandgap SiC Schottky diode as an output diode.

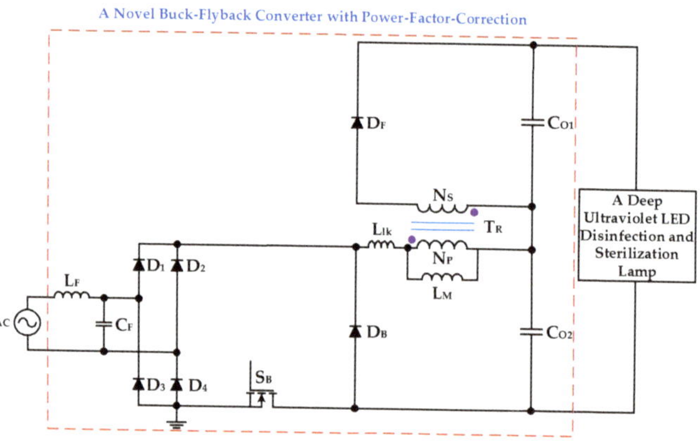

Figure 2. The proposed single-stage electronic lighting driver utilizing SiC Schottky didoes for supplying a deep ultraviolet LED disinfection and sterilization lamp.

Figure 3 is the equivalent diagram of the single-stage electronic lighting driver circuit utilizing SiC Schottky diodes for supplying a deep ultraviolet LED disinfection and sterilization lamp developed in this paper with the power factor correction function. Theoretical waveforms of the proposed single-stage electronic lighting driver utilizing SiC Schottky didoes for supplying a deep ultraviolet LED disinfection and sterilization lamp are shown in Figure 4. When analyzing the proposed single-stage electronic lighting driver circuit for the deep ultraviolet LED disinfection and sterilization lamp, assumptions are made during descriptions and explanations of the operation mode in the presented driver circuit and are shown as follows:

(a) The equivalent voltage source of the input utility-line voltage after passing through the full-wave rectifier circuit is denoted by V_{REC}. Since the switching frequency f_s is much larger than the line frequency f_{AC}, the rectified voltage V_{REC} can be regarded as a constant value during one switching cycle in the circuit mode analysis.
(b) The leakage inductance L_{lk} of the transformer T_R is considered.
(c) The magnetizing inductor L_M of the transformer T_R is designed to operate in discontinuous conduction mode (DCM).
(d) Neglecting the conduction voltage drops and their equivalent resistances of all diodes.
(e) The remaining circuit components are considered as ideal components.

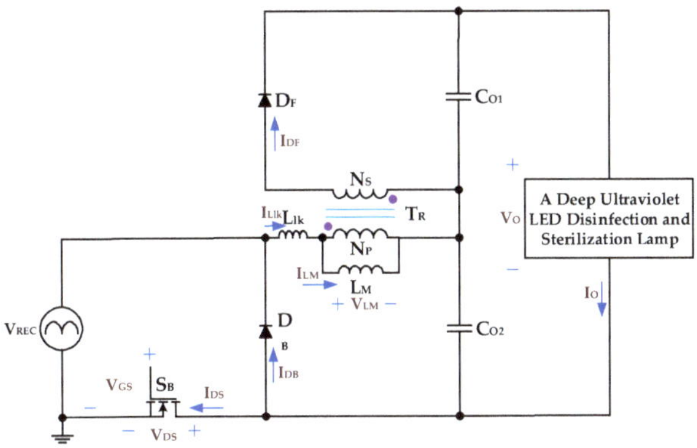

Figure 3. Equivalent circuit of the proposed electronic lighting driver circuit utilizing SiC Schottky didoes for supplying a deep ultraviolet LED disinfection and sterilization lamp while analyzing the operational modes.

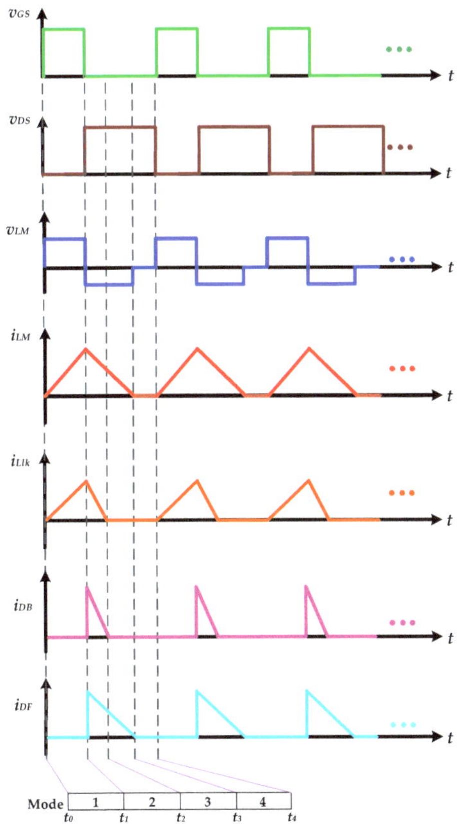

Figure 4. Theoretical waveforms of the proposed single-stage electronic lighting driver utilizing SiC Schottky didoes for supplying a deep ultraviolet LED disinfection and sterilization lamp.

Operation Mode 1 ($t_0 \le t < t_1$): Figure 5 shows the equivalent circuit of the proposed single-stage electronic lighting driver circuit for supplying a deep ultraviolet LED disinfection and sterilization lamp during Mode 1. The power switch S_B is driven on, and the equivalent voltage source V_{REC} is connected to both ends of the diode D_B, making the diode D_B a reverse biased non-conducting state. The equivalent voltage source V_{REC} provides energy to the output capacitor C_{O2} and the magnetizing inductor L_M of the transformer T_R as well as the leakage inductor through the switch S_B, which results in a linear increase in the current of the magnetizing inductor L_M, and the voltage polarity on both sides of the magnetization inductor L_M and the leakage inductor L_{lk} is positive left and negative right. The output capacitors C_{O1} and C_{O2} provide energy to the deep ultraviolet LED disinfection and sterilization lamp. When the switch S_B turns off, the current of the magnetizing inductor L_M reaches its maximum value and the mode ends.

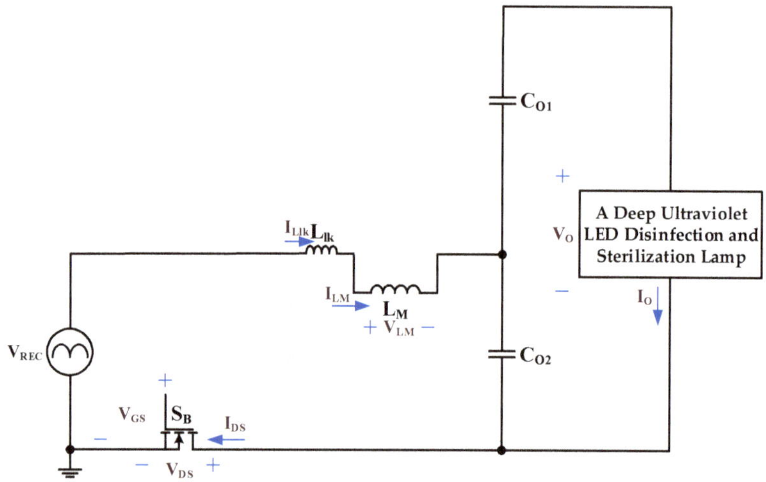

Figure 5. Equivalent circuit of the proposed single-stage electronic lighting driver circuit utilizing SiC Schottky didoes for supplying a deep ultraviolet LED disinfection and sterilization lamp during Mode 1.

Operational Mode 2 ($t_1 \leq t < t_2$): Figure 6 shows the equivalent circuit of the proposed single-stage electronic lighting driver circuit for supplying a deep ultraviolet LED disinfection and sterilization lamp during Mode 2. When the power switch S_B turns off, the voltage polarity of the magnetizing inductor L_M and the leakage inductor L_{lk} is changed to right positive and left negative according to the Lenz's law, which turns the diode D_B to be the on state of forward bias. The current of the magnetizing inductor L_M flows into the position where the primary-side winding N_P of the transformer T_R does not have a dot end, and the voltage polarity of the two ends of the winding N_P is right-positive and left-negative. Therefore, the polarity of the voltage reflected to both ends of the secondary-side winding N_S is positive left and negative right, which makes the diode D_F work in the forward-biased conduction state. At this time, the magnetizing inductor L_M and the leakage inductor L_{lk} are in the state of releasing energy, causing the current of the magnetizing inductor L_M to decrease linearly. Since the diode D_B is in the forward-biased conduction state, the energy of the magnetizing inductor L_M and the leakage inductor L_{lk} is supplied to the output capacitor C_{O2} through the diode D_B, and the current of the magnetizing inductor L_M shows a linear decrease. Due to the forward-biased conduction of the diode D_F, the energy of the magnetizing inductor L_M is supplied through the transformer T_R and the diode D_F to the output capacitance C_{O1} and C_{O2} as well as to the deep ultraviolet LED disinfection and sterilization lamp. In addition, the output capacitors C_{O1} and C_{O2} continue to provide energy to the deep ultraviolet LED disinfection and sterilization lamp. The mode ends when the leakage inductor L_{lk} finishes releasing energy and the current I_{Llk} of the leakage inductor L_{lk} drops linearly to zero.

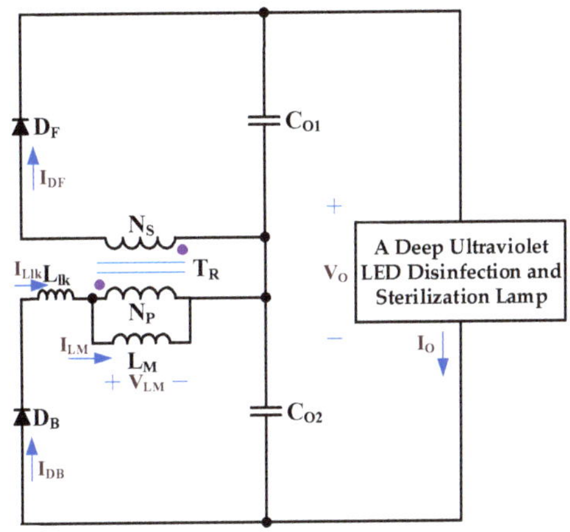

Figure 6. Equivalent circuit of the proposed single-stage electronic lighting driver circuit utilizing SiC Schottky didoes for supplying a deep ultraviolet LED disinfection and sterilization lamp during Mode 2.

Operational Mode 3 ($t_2 \leq t < t_3$): Figure 7 shows the equivalent circuit of the proposed single-stage electronic lighting driver circuit for supplying a deep ultraviolet LED disinfection and sterilization lamp during Mode 3. The energy of the magnetizing inductor L_M is continuously supplied through the transformer T_R and the diode D_F to the output capacitors C_{O1} and C_{O2} and to the deep ultraviolet LED disinfection and sterilization lamp. In addition, the output capacitors C_{O1} and C_{O2} continue to provide energy to the deep ultraviolet LED disinfection and sterilization lamp. This mode ends when the magnetizing inductor L_M has finished releasing energy and the magnetizing inductor current I_{LM} drops linearly to zero.

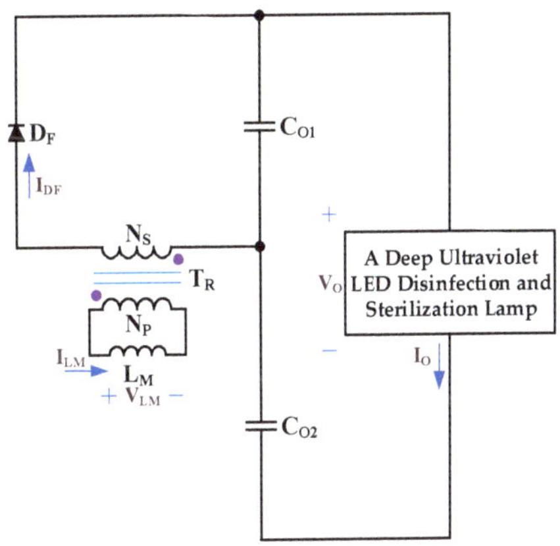

Figure 7. Equivalent circuit of the proposed single-stage electronic lighting driver circuit utilizing SiC Schottky didoes for supplying a deep ultraviolet LED disinfection and sterilization lamp during Mode 3.

Operational Mode 4 ($t_3 \leq t < t_4$): Figure 8 shows the equivalent circuit of the proposed single-stage electronic lighting driver circuit for supplying a deep ultraviolet LED disinfection and sterilization lamp during Mode 4. The output capacitors C_{O1} and C_{O2}

continuously provide energy to the deep ultraviolet LED disinfection and sterilization lamp. When the power switch S_B is turned on again, this mode ends and the circuit reverts to the first mode of operation.

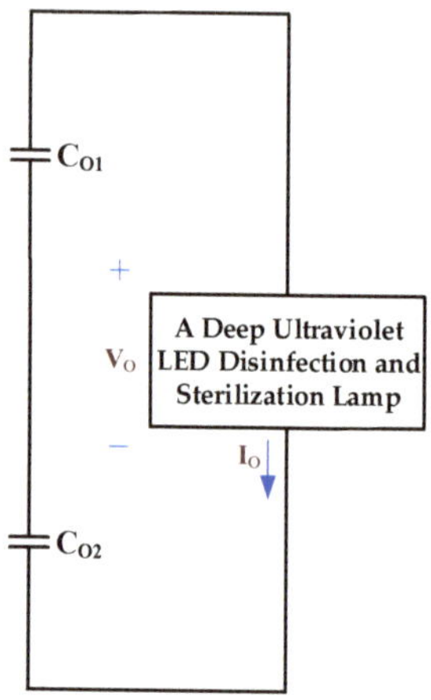

Figure 8. Equivalent circuit of the proposed single-stage electronic lighting driver circuit utilizing SiC Schottky didoes for supplying a deep ultraviolet LED disinfection and sterilization lamp during Mode 4.

2.1. *Design Guideline of the Magnetizing Inductor L_M in the Transformer T_R*

In order to show the design guideline of the magnetizing inductor L_M in the transformer T_R, Figure 9 illustrates theoretical waveforms of the magnetizing inductor current $i_{LM(t)}$, the peak level of $i_{LM\text{-}pk(t)}$ and the input utility-line current $i_{AC(t)}$ in the positive half-cycle of the utility-line voltage $v_{AC(t)}$.

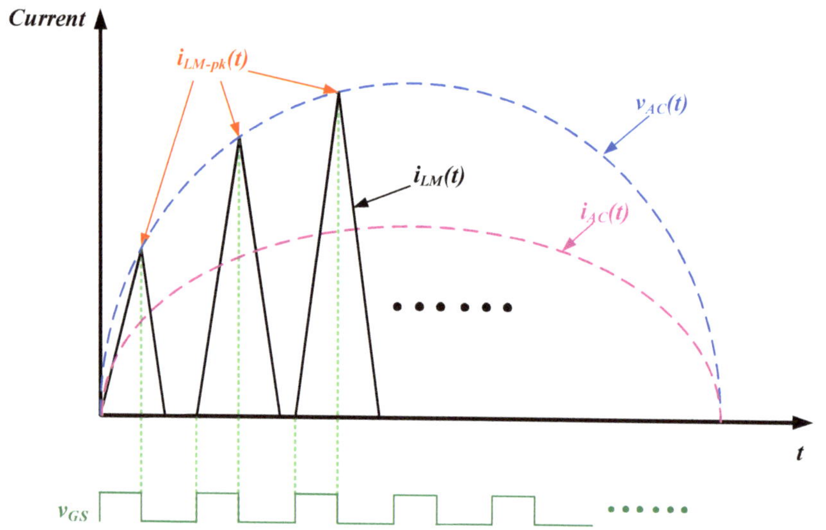

Figure 9. Illustrative waveforms of the magnetizing inductor current $i_{LM(t)}$, the peak level of $i_{LM\text{-}pk(t)}$ and the input utility-line current $i_{AC(t)}$ in the positive half-cycle of the utility-line voltage $v_{AC(t)}$.

Referring to Figure 9, the positive half-cycle of the instantaneous utility-line voltage $v_{AC(t)}$ can be expressed by

$$|v_{AC}(t)| = \sqrt{2}v_{AC-rms}|\sin(2\pi f_{AC}t)| \quad (1)$$

The peak level of the magnetizing inductor current i_{LM}, denoted as $i_{LM,peak}$, can be represented by

$$i_{LM,peak}(t) = \frac{\sqrt{2}v_{AC-rms}|\sin(2\pi f_{AC}t)|Duty T_S}{2L_M} \quad (2)$$

By filtering the high-frequency components of the peak level of the magnetizing inductor current $i_{LM,peak(t)}$, the input utility-line current i_{AC} is equal to the average level of $i_{LM,peak(t)}$ during one switching period and can be expressed as

$$i_{AC}(t) = \frac{1}{T_{AC}}\int_0^{T_{AC}} i_{LM,peak}(t)dt = \frac{\sqrt{2}v_{AC-rms}Duty^2 T_S}{4L_M}\sin(2\pi f_{AC}t) \quad (3)$$

where T_{AC} is the utility-line period.

The average value of the input utility line power P_{IN} is obtained by multiplying the instantaneous value of the utility line voltage $v_{AC(t)}$ by the instantaneous value of the utility line current $i_{AC(t)}$, and then averaging it over a period of one cycle, which is calculated as follows

$$P_{IN} = \frac{1}{T_{AC}}\int_0^{T_{AC}} v_{AC}(t)i_{AC}(t)dt = \frac{{v_{AC-rms}}^2 Duty^2 T_S}{4L_M} \quad (4)$$

The rated output power P_O of the deep ultraviolet LED disinfection and sterilization lamp is related to the estimated efficiency of the driver circuit η multiplied by the input power P_{IN}, which is expressed as follows

$$P_O = \eta P_{IN} = \frac{\eta {v_{AC-rms}}^2 Duty^2 T_S}{4L_M} \quad (5)$$

Rearranging (5), the design formula for the inductance of the magnetizing inductor L_M is given by

$$L_M = \frac{\eta {v_{AC-rms}}^2 Duty^2 T_S}{4P_O} \quad (6)$$

2.2. Design Guideline of the Output Capacitors C_{O1} and C_{O2}

The design consideration of the output capacitors, C_{O1} and C_{O2}, depends on the DC output voltage, the allowable overvoltage, the output power and the desired voltage ripple. The voltage ripple, ΔV_O, is half the peak-to-peak value of the output voltage at twice the mains frequency, and is a function of the capacitance C_O and the peak capacitor current (which is equal to the output current I_O) and can be expressed in the following equation [23].

$$\Delta V_O = I_O\sqrt{\left(\frac{1}{2\pi \times 2f_{AC} \times C_O}\right)^2 + {R_{ESR}}^2} \quad (7)$$

Neglecting the equivalent series resistance, which is denoted as R_{ESR}, of the output capacitor, the capacitance C_O of the output capacitor can be expressed as

$$C_O \geq \frac{I_O}{4\pi f_{AC} \times \Delta V_O} = \frac{P_O}{4\pi f_{AC} \times V_O \times \Delta V_O} \quad (8)$$

In addition, the voltage ripple ΔV_O is typically selected from 1% to 5% of the output voltage.

3. Experimental Results of the Proposed Electronic Lighting Driver Circuit Utilizing SiC Schottky Didoes for Supplying a Deep Ultraviolet LED Disinfection and Sterilization Lamp

Figure 10 shows a photograph of the deep ultraviolet LED disinfection and sterilization module used in this paper with specifications that include a wavelength of 275 nm and a point angle of 120 degrees. Moreover, the manufacturer has connected a constant current control IC in series and a Schottky diode in parallel for reverse polarity protection on the aluminum substrate in each deep ultraviolet LED disinfection and sterilization module. The specification of the deep ultraviolet LED disinfection and sterilization lamp used in this paper consists of ten deep ultraviolet LED disinfection and sterilization modules connected in series with a rated voltage of 90 volts, a rated current of 40 milliamps, and a rated power of 3.6 watts. Table 1 shows a comparison between the deep ultraviolet LED disinfection and sterilization module in [24] and the module used in this paper. As can be seen from Table 1, the deep ultraviolet LED disinfection and sterilization module used in this paper has a larger peak wavelength, a slightly narrower viewing angle, and a slightly larger optical output power and spectral half-width than the module in [24]. Table 2 shows the electrical parameters of the single-stage electronic lighting prototype driver circuit for the deep ultraviolet LED disinfection and sterilization lamp developed in this paper. The RMS AC input voltage was 110 volts, and the specifications of the deep ultra LED disinfection and sterilization lamp were used as the output parameters, including: output power of 3.6 watts, output voltage of 90 volts, and output current of 40 milliamps. Moreover, safety concerns and ensuring compliance with regulatory standards for UV exposure and electrical safety include: Do not look directly at the UV light source when using a deep ultraviolet disinfection and sterilization lamp as deep UV radiation may cause skin erythema, conjunctival irritation and fatigue. In addition, do not overuse deep UV disinfection and sterilization lamps as prolonged use or contact with UV lamps or direct exposure to UV radiation may cause skin and eye damage.

Figure 10. Photograph of the deep ultraviolet LED disinfection and sterilization module used in this paper with specifications that include a wavelength of 275 nm and a point angle of 120 degrees.

3.1. Calculating the Magnetizing Inductor L_M in the Transformer T_R

Referring to (6) with a η of 0.8, a $Duty$ of 0.25, a P_O of 3.6 W, a switching frequency f_S of 50 kHz (which is the reciprocal of the switching period T_S), and a $v_{AC\text{-}rms}$ of 110 V, the inductance of the magnetizing inductor L_M is calculated by

$$L_M = \frac{\eta v_{AC-rms}{}^2 Duty^2 T_S}{4P_O} = \frac{0.8 \times 110^2 \times 0.25^2 \times \left(\frac{1}{50{,}000}\right)}{4 \times 3.6} = 834\ \mu H \tag{9}$$

Table 1. Comparisons between the deep ultraviolet LED disinfection and sterilization module in [24] and the one utilized in this paper.

Item	The Deep Ultraviolet LED Disinfection and Sterilization Module in [24]	The Deep Ultraviolet LED Disinfection and Sterilization Module Utilized in this Paper
Typical value of the peak wavelength	255 nm	275 nm
Typical value of the optical output power	3.5 mW	4 mW
viewing angle	125 degree	120 degree
Spectrum half width	11 nm	12 nm

Table 2. Specifications of the proposed electronic lighting driver circuit utilizing SiC Schottky didoes for supplying a deep ultraviolet LED disinfection and sterilization lamp.

Parameter	Value
Input AC voltage v_{AC}	110 V
Rated output power P_O	3.6 W
Rated output voltage V_O	90 V
Rated output current I_O	40 mA

In order to operate the magnetizing inductor current in a discontinuous conduction mode so as to allow the driver circuit to have an input current shaping function and to improve the power factor, the magnetizing inductor L_M is selected as 800 μH for the prototype driver circuit.

3.2. Calculating of the Output Capacitors C_{O1} and C_{O2}

Substituting the circuit parameters into Equation (8), with a P_O of 3.6 W, an f_{AC} of 60 Hz, a V_O of 90 V, a ΔV_O of 2.7 V (which is 3% of the output voltage V_O), the capacitance C_O can be obtained by:

$$C_O \geq \frac{P_O}{4\pi f_{AC} \times V_O \times \Delta V_O} = \frac{3.6}{4\pi \times 60 \times 90 \times 2.7} = 196.49\ \mu F \quad (10)$$

In implementing the prototype circuit, the capacitance values of the output capacitors C_{O1} and C_{O2} were selected to be 220 μF and the withstand voltage to be 250 V.

Table 3 shows the component specifications of the single-stage electronic lighting prototype driver circuit for the deep ultraviolet LED disinfection and sterilization lamp developed in this paper. In addition, majority carrier diodes with Schottky technology on silicon carbide wide bandgap material have higher performance and are used as output diodes D_B and D_F. By varying the number of series-connected deep ultraviolet disinfection and sterilization modules used, a scalable design of the proposed electronic lighting driver circuits can be achieved to accommodate higher power applications and varying load conditions. As a result, the output power P_O and output voltage V_O can be varied while the output current I_O remains constant and can be used in a variety of practical application scenarios. Key circuit parameters of the proposed drive circuit, such as the magnetizing inductor and output capacitor, can then be redesigned.

Figure 11 shows the measured waveform of the magnetizing inductor current i_{LM}. Figure 12 shows the measured unfolded waveform of the magnetizing inductor current i_{LM}. From Figure 12, it can be seen that the magnetizing inductor current i_{LM} operates in the discontinuous conduction mode. Figure 13 shows the measured waveforms of output voltage V_O and output current I_O, and the average values of output voltage and current were 91.35 volts and 38.57 milliamps, respectively. Figure 14 shows the ripple waveforms of output voltage V_O and output current I_O.

Table 3. Key components used in the proposed electronic lighting driver circuit utilizing SiC Schottky didoes for supplying a deep ultraviolet LED disinfection and sterilization lamp.

Component	Value
Filter inductor L_f	1 mH
Filter capacitor C_f	220 nF
Diodes D_1, D_2, D_3, D_4	MUR460
Power switch S_B	IRF730
Magnetizing inductor L_M	800 μH
Leakage inductor L_{lk}	12 μH
Diodes D_B, D_F	VS-3C04ET07T-M3
Output capacitor C_{O1}, C_{O2}	220 μF/200 V

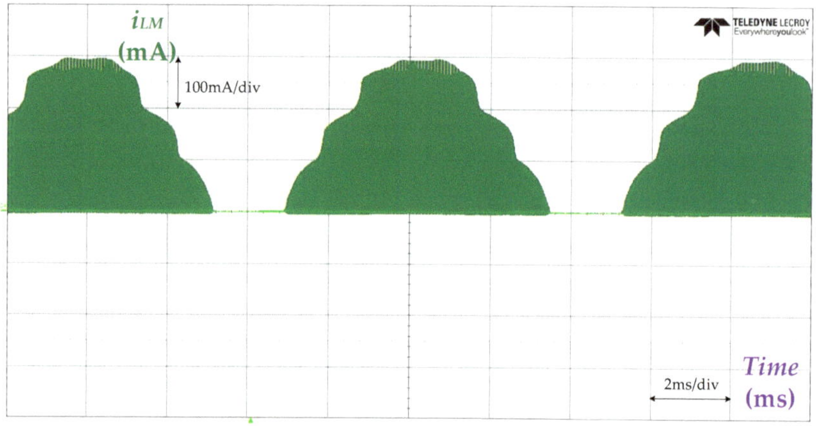

Figure 11. Measured magnetizing inductor current i_{LM} (100 mA/div); time scale: 2 ms/div.

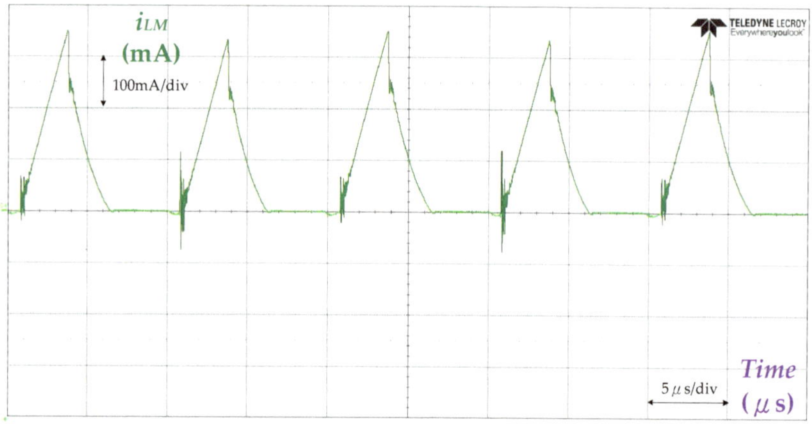

Figure 12. Measured unfolded waveform of the magnetizing inductor current i_{LM} (100 mA/div); time scale: 5 μs/div.

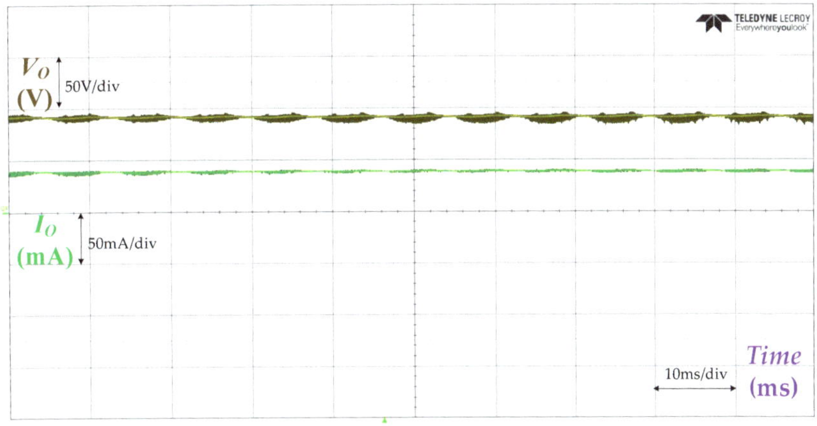

Figure 13. Measured waveforms of the output voltage V_o (50 V/div) and the output current I_o (50 mA/div); time scale: 10 ms/div.

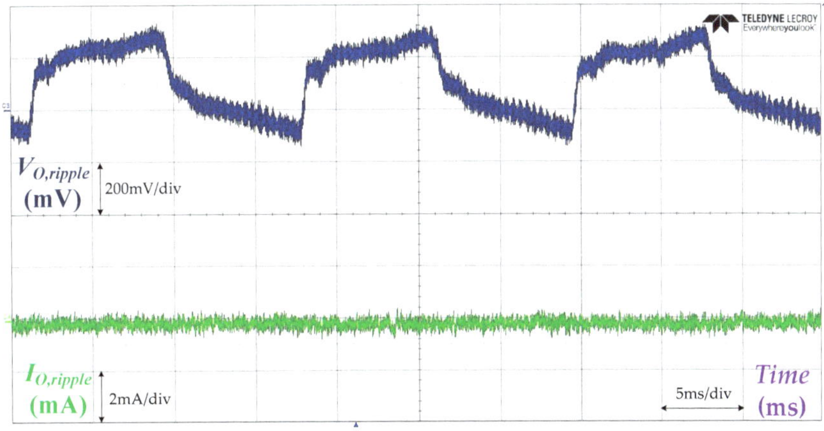

Figure 14. Measured waveforms of the output voltage ripple $V_{o,ripple}$ (200 mV/div) and the output current ripple $I_{o,ripple}$ (2 mA/div); time scale: 5 ms/div.

Table 4 shows the output voltage ripple factor and output current ripple factor of the prototype electronic lighting driver circuit for a deep ultraviolet LED disinfection and sterilization lamp at 110 volts RMS AC input voltage. As can be seen in Table 4, dividing the peak-to-peak output voltage of 143.8 millivolts by the average value of 91.35 volts yields a voltage ripple factor of 0.16%; dividing the peak-to-peak output current of 1.1658 milliamps by the average value of 38.57 milliamps yields a current ripple factor of 3.02%.

Figure 15 shows the measured waveforms of AC input voltage v_{AC} and input current i_{AC}. From the figure, it can be indicated that the AC input current follows the input voltage and the phases of the two waveforms are the same; therefore, the prototype of the proposed single-stage driver circuit for supplying the deep ultraviolet LED disinfecting and sterilizing lamp has the effect of power factor correction. The power factor and the total harmonic distortion factor of the input current were measured to be 0.9236 and 17.401%, respectively, using a power analyzer. In addition, the measured input power of the prototype drive circuit is 3.901 W and the output power is 3.583 W, and the efficiency of the prototype circuit is 91.85%.

Table 4. Measurement of output voltage ripple factor and output current ripple factor of electronic lighting prototype driver circuit utilizing SiC Schottky didoes for a deep ultraviolet LED disinfection and sterilization lamp.

Parameters	Values
Peak-to-peak value of the output voltage	143.8 mV
Mean value of the output voltage	91.35 V
Ripple factor of the output voltage	0.16%
Peak-to-peak value of the output current	1.1658 mA
Mean value of the output current	38.57 mA
Ripple factor of the output current	3.02%

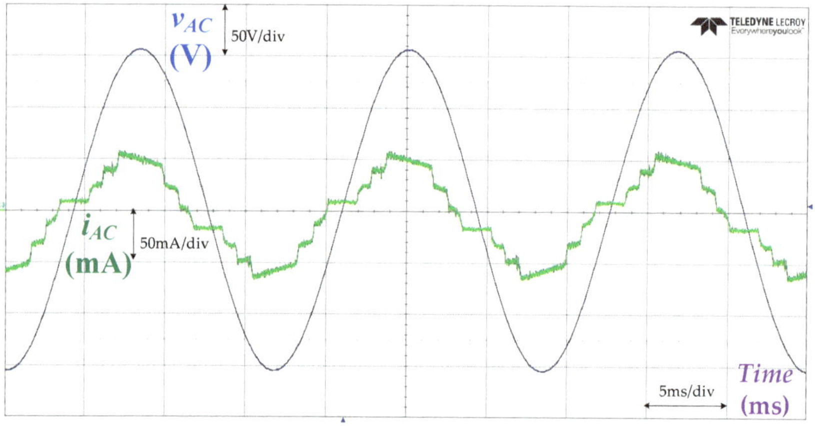

Figure 15. Measured input utility-line voltage v_{AC} (50 V/div) and current i_{AC} (50 mA/div); time scale: 5 ms/div.

Table 5 shows a comparison between the existing AC-DC LED driver in [19] which integrates an inverse buck-boost converter with a lossless snubber and supplies an 18 W-rated (60 V/0.3 A) power, and the proposed one which integrates a buck converter and a flyback converter and supplies a 3.6 W-rated (90 V/0.04 A) power. As can be seen from Table 3, both LED drivers are supplied by an input AC voltage of 110 V and use a single power switch. The proposed AC-DC LED driver circuit saves a capacitor, a magnetic element and a diode compared to the ones in [19]. In addition, the current THD and circuit efficiency of the proposed AC-DC LED driver is better than that of the existing driver.

Table 5. Comparisons between the existing AC-DC LED driver in [19] and the proposed one.

Item	Existing AC-DC LED Driver in Reference [19]	ProposedAC-DC LED Driver
Circuit topology	Integration of an inverse buck-boost converter with a lossless snubber	Integration of a buck converter and a flyback converter
Input AC voltage	110 V	110 V
Output power	18 W (60 V/0.3 A)	3.6 W (90 V/0.04 A)
Number of required switch	1	1
Number of required capacitors	4	3
Number of required magnetic element	3	2
Number of required diodes	7	6
Measured power factor	0.9737	0.9236
Measured current THD	18.422%	17.401%
Measured circuit efficiency	85.01%	91.85%

4. Conclusions

This paper proposes a single-stage electronic lighting prototype driver circuit utilizing SiC Schottky didoes for a deep ultraviolet LED disinfection and sterilization lamp, which integrates a buck converter and a flyback converter to form a single-stage, single-power-switch AC-DC power converter circuit architecture with a factor correction function. By designing the inductor in the AC-DC power converter to operate in discontinuous conduction mode, the effect of power factor correction can be achieved naturally. In addition, the proposed single-stage electronic lighting driver circuit employs a majority-carrier Schottky diode made of silicon carbide wide-bandgap semiconductor material and recovers the energy from the leakage inductance of the transformer, thereby improving the conversion efficiency of the driver circuit. In this paper, a 3.6 watt deep ultraviolet LED electronic lighting prototype driver circuit for a disinfection and sterilization lamp has been developed and tested. At an AC input voltage of 110 volts RMS and at rated output power, the results were: output voltage ripple factor less than 1%, output current ripple factor less than 4%, power factor greater than 0.9, input current total harmonic distortion factor less than 18%, and circuit efficiency greater than 90%. The single-stage deep ultraviolet LED electronic lighting prototype driver circuit proposed in this paper can simplify the number of power switches required in the driver circuit and has high circuit efficiency.

Author Contributions: C.-A.C. developed the circuit topology of the electronic lighting driver and designed the methodology; C.-H.C., H.-L.C. and E.-C.C. arranged the software resources and set up simulation along with performed circuit simulations; L.-F.L., S.-H.H. and C.-K.L. implemented the prototype electronic lighting driver circuit, and carried out the measurements of the circuit as well as analyzed experimental results with the guidance from C.-A.C.; C.-A.C. prepared the first draft of the article and revised the manuscript for submission. All authors have read and agreed to the published version of the manuscript.

Funding: This research was funded by the National Science and Technology Council (NSTC) of Taiwan for its grant numbers MOST 110-2221-E-214-014, NSTC 111-2221-E-214-011 and NSTC 112-2221-E-214-005.

Data Availability Statement: The data presented in this study are available in the article.

Acknowledgments: The authors would like to express their sincerest gratitude to the National Science and Technology Council (NSTC) of Taiwan for its grant numbers MOST 110-2221-E-214-014, NSTC 111-2221-E-214-011 and NSTC 112-2221-E-214-005.

Conflicts of Interest: The authors declare no conflicts of interest.

References

1. Kebbi, Y.; Muhammad, A.I.; Sant'Ana, A.S.; Prado-Silva, L.D.; Liu, D.; Ding, T. Recent advances on the application of UV-LED technology for microbial inactivation: Progress and mechanism. *Compr. Rev. Food Sci. Food Saf.* **2020**, *19*, 3501–3527. [CrossRef] [PubMed]
2. Shatalov, M.; Jain, R.; Saxena, T.; Dobrinsky, A.; Shur, M. Development of deep UV LEDs and current problems in material and device technology. In *Semiconductors and Semimetals*; Elsevier: Amsterdam, The Netherlands, 2017; Volume 96, pp. 45–83.
3. Koutchma, T.; Popović, V.; Green, A. Overview of ultraviolet (UV) LEDs technology for applications in food production. In *Ultraviolet LED Technology for Food Applications*; Academic Press: Cambridge, MA, USA, 2019; pp. 1–23.
4. Akgün, M.P.; Ünlütürk, S. Effects of ultraviolet light emitting diodes (LEDs) on microbial and enzyme inactivation of apple juice. *Int. J. Food Microbiol.* **2017**, *260*, 65–74. [CrossRef] [PubMed]
5. Kaya, Z.; Yıldız, S.; Ünlütürk, S. Effect of UV-C irradiation and heat treatment on the shelf life stability of a lemon–melon juice blend: Multivariate statistical approach. *Innov. Food Sci. Emerg. Technol.* **2015**, *29*, 230–239. [CrossRef]
6. Guan, Z.; Liu, P.; Zhou, T.; Zhou, L.; Zhang, D.; Xie, Q.; Yu, Q.; He, Y.; Wang, S.; Wang, X.; et al. Study on the Light Field Regulation of UVC-LED Disinfection for Cold Chain Transportation. *Appl. Sci.* **2022**, *12*, 1285. [CrossRef]
7. Liang, J.J.; Liao, C.C.; Chang, C.S.; Lee, C.Y.; Chen, S.Y.; Huang, S.B.; Yeh, Y.F.; Singh, K.J.; Kuo, H.C.; Lin, Y.L.; et al. The effectiveness of far-ultraviolet (UVC) light prototype devices with different wavelengths on disinfecting SARS-CoV-2. *Appl. Sci.* **2021**, *11*, 10661. [CrossRef]
8. Juarez-Leon, F.A.; Soriano-Sanchez, A.G.; Rodriguez-Licea, M.A.; Perez-Pinal, F.J. Design and Implementation of a Germicidal UVC-LED Lamp. *IEEE Access* **2020**, *8*, 196951–196962. [CrossRef] [PubMed]

9. Alonso, J.M.; Costa, M.A.D.; Ordiz, C. Integrated Buck-Flyback Converter as a High-Power-Factor Off-Line Power Supply. *IEEE Trans. Ind. Electron.* **2008**, *55*, 1090–1100. [CrossRef]
10. Gacio, D.; Alonso, J.M.; Calleja, A.J.; Garcia, J.; Rico-Secades, M. A universal-input single-stag2e high-power-factor power supply for HB-LEDs based on integrated buck–flyback converter. *IEEE Trans. Ind. Electron.* **2010**, *58*, 589–599. [CrossRef]
11. Abdelmessih, G.Z.; Alonso, J.M.; Costa, M.A.D. Loss Analysis for Efficiency Improvement of the Integrated Buck–Flyback LED Driver. *IEEE Trans. Ind. Appl.* **2018**, *54*, 6543–6553. [CrossRef]
12. Abdelmessih, G.Z.; Alonso, J.M.; Tsai, W.T. Analysis and experimentation on a new high power factor off-line LED driver based on interleaved integrated buck flyback converter. *IEEE Trans. Ind. Appl.* **2019**, *55*, 4359–4369. [CrossRef]
13. Luz, P.C.; Bolán, P.E.; Maboni, T.; Cosetin, M.R.; da Silva, M.F.; do Prado, R.N. An integrated insulated Buck-Boost-Flyback converter to feed LED's lamps to street lighting with reduced capacitances. In Proceedings of the 2014 11th IEEE/IAS International Conference on Industry Applications, Juiz de Fora, Brazil, 7–10 December 2014; pp. 1–6.
14. Divya, K.M.; Parackal, R. High power factor integrated buck-boost flyback converter driving multiple outputs. In Proceedings of the 2015 Online International Conference on Green Engineering and Technologies (IC-GET), Coimbatore, India, 27 November 2015; pp. 1–5.
15. Cheng, C.A.; Chung, T.Y. A Novel Integrated Driver with PFC for Supplying Light-Emitting Diodes. In Proceedings of the 41th Annual Conference of the IEEE Industrial Electronics Society (IECON 2015), Yokohama, Japan, 9–12 November 2015; pp. 3514–3518.
16. Poorali, B.; Adib, E. Analysis of the integrated SEPIC-flyback converter as a single-stage single-switch power-factor-correction LED driver. *IEEE Trans. Ind. Electron.* **2016**, *63*, 3562–3570. [CrossRef]
17. Cheng, C.-A.; Chang, C.-H.; Cheng, H.-L.; Chang, M.-T. A Novel LED Driver with Power Factor Correction Suitable for Streetlight Applications. In Proceedings of the International Future Energy Electronics Conference 2017 (IFEEC 2017)—ECCE Asia, Kaohsiung, Taiwan, 3–7 June 2017; pp. 1–4.
18. Cheng, C.A.; Chang, C.H.; Cheng, H.L.; Chang, E.C.; Lai, C.C. A novel LED tube lamp driver with power-factor correction for indoor lighting applications. In Proceedings of the 2018 IEEE 7th Global Conference on Consumer Electronics (GCCE), Nara, Japan, 9–12 October 2018; pp. 700–701.
19. Cheng, C.A.; Chen, Y.L.; Chen, Y.J.; Lai, S.W. A novel cost-effective AC-DC power supply for indoor LED lighting applications. In Proceedings of the 2019 IEEE PES Asia-Pacific Power and Energy Engineering Conference (APPEEC), Macao, China, 1–4 December 2019; pp. 1–3.
20. Cheng, C.A.; Lan, L.F.; Hou, S.H.; Lin, C.K. A Single-Stage Electronic Lighting Driver for a Deep Ultraviolet LED Disinfection and Sterilization Lamp. In Proceedings of the 2023 IEEE Workshop on Wide Bandgap Power Devices and Applications in Asia (WiPDA Asia), Hsinchu, Taiwan, 27–29 August 2023; pp. 1–3.
21. Effective Use of Power Devices—Silicon Carbide Power Devices Understanding & Application Examples Utilizing the Merits. ROHM Semiconductor, 2017. Available online: https://pages.rohm.com.tw/rs/247-PYD-578/images/TechWeb_E_DL_SiC.pdf (accessed on 31 July 2024).
22. SiC Power Devices and Modules Application Note, ROHM Semiconductor, 2020. Available online: https://fscdn.rohm.com/en/products/databook/applinote/discrete/sic/common/sic_appli-e.pdf (accessed on 31 July 2024).
23. AN966 APPLICATION NOTE, STMicroelectronics, "L6561, ENHANCED TRANSITION MODE POWER FACTOR CORRECTOR", pp. 1–23, March 2003. Available online: https://www.st.com/resource/en/application_note/cd00004002-l6561-enhanced-transition-mode-power-factor-corrector-stmicroelectronics.pdf (accessed on 31 July 2024).
24. SEOULVIOSYS. UV CA3535 Series (CUD5GF1B). Available online: https://media.digikey.com/pdf/Data%20Sheets/Seoul%20 Semiconductor/CUD5GF1B_R0.1_Prelliminary_2-3-20.pdf (accessed on 31 July 2024).

electronics

MDPI

Article

A Multiscale Simulation on Aluminum Ion Implantation-Induced Defects in 4H-SiC MOSFETs

Yawen Wang [1], Haipeng Lan [1], Qiwei Shangguan [1], Yawei Lv [1,*] and Changzhong Jiang [2,*]

[1] School of Physics and Electronics, Hunan University, Changsha 410082, China; wangyawen2021@hnu.edu.cn (Y.W.); lanhaipeng@hnu.edu.cn (H.L.); shangguanqiwei@hnu.edu.cn (Q.S.)
[2] College of Materials Science and Engineering, Hunan University, Changsha 410082, China
* Correspondence: lvyawei@hnu.edu.cn (Y.L.); czjiang@hnu.edu.cn (C.J.)

Abstract: Aluminum (Al) ion implantation is one of the most important technologies in SiC device manufacturing processes due to its ability to produce the p-type doping effect, which is essential to building p–n junctions and blocking high voltages. However, besides the doping effect, defects are also probably induced by the implantation. Here, the impacts of Al ion implantation-induced defects on 4H-SiC MOSFET channel transport behaviors are studied using a multiscale simulation flow, including the molecular dynamics (MD) simulation, density functional theory (DFT) calculation, and tight-binding (TB) model-based quantum transport simulation. The simulation results show that an Al ion can not only replace a Si lattice site to realize the p-doping effect, but it can also replace the C lattice site to induce mid-gap trap levels or become an interstitial to induce the n-doping effect. Moreover, the implantation tends to bring additional point defects to the 4H-SiC body region near the Al ions, which will lead to more complicated coupling effects between them, such as degrading the p-type doping effect by trapping free hole carriers and inducing new trap states at the 4H-SiC bandgap. The quantum transport simulations indicate that these coupling effects will impede local electron transports, compensating for the doping effect and increasing the leakage current of the 4H-SiC MOSFET. In this study, the complicated coupling effects between the implanted Al ions and the implantation-induced point defects are revealed, which provides new references for experiments to increase the accepter activation rate and restrain the defect effect in SiC devices.

Keywords: 4H-SiC; Al; implantation; defects; molecular dynamics (MD); density functional theory (DFT); tight-binding (TB) model; quantum transport

Citation: Wang, Y.; Lan, H.; Shangguan, Q.; Lv, Y.; Jiang, C. A Multiscale Simulation on Aluminum Ion Implantation-Induced Defects in 4H-SiC MOSFETs. *Electronics* **2024**, *13*, 2758. https://doi.org/10.3390/electronics13142758

Academic Editors: Yimeng Zhang, Yuming Zhang and Lejia Sun

Received: 16 June 2024
Revised: 4 July 2024
Accepted: 11 July 2024
Published: 13 July 2024

1. Introduction

Although silicon (Si) is still the most widely used material in power electronics due to its mature manufacturing technique, drawbacks stemming from its physical properties have already stimulated research on the replacement of Si material by wide bandgap materials. As a representative and compared with Si, silicon carbide (SiC) shows obvious advantages in bandgaps, breakdown field, thermal conductivity, and saturation drift velocity [1–5], making it suitable for a variety of novel device applications [6,7]. SiC is characterized by strong covalent bonds between Si and carbon (C) atoms, and the flexible directions of these bonds can lead to different spatial arrangements, resulting in different crystal structures, such as the hexagonal and cubic morphologies [8–10]. Among them, the hexagonal 4H-SiC emerges due to its comprehensive quality, which can realize a balance among excellent physical properties, low defect density, and moderate manufacturing cost [4,11].

Due to the high hardness and negligible impurity diffusion rate, ion implantation, followed by a mandatory post-annealing treatment, is almost the only way to tune the conductivity of SiC before its device applications. For example, the realization of p-type 4H-SiC relies on the aluminum (Al) ion implantation [12] since Al ions exhibit relatively low thermal ionization energy, high solubility limit, and good defect-suppression ability.

Compared with the n-type doping technology, Al ion implantation still suffers from electrical activation and defect problems today. Kawahara et al. had already observed two deep-level transient spectroscopy (DLTS) peaks, IN3 and IN9 (E_c—0.6 eV and E_c—1.6 eV), after low-dose Al ion implantation and post-annealing at 1700 °C. They assigned these trap levels to the well-known $Z_{1/2}$ and $EH_{6/7}$ levels, which could be induced by C vacancies (V_C). At high dose levels, the electrical activations of implanted Al become another problem, as it will reach a limit along the increasing annealing temperature and time [13–15]. There is a contradiction between the electrical activation of the Al ion and the control of the V_C concentration. Nipoti et al. suggested that maintaining a post-implantation temperature of around 1600–1650 °C was beneficial to low V_C concentrations while reaching a 69% electrical activation of 1×10^{20} cm^{-3} Al ion implantation needed an annealing temperature of up to 1950 °C [15]. To realize an excellent p-type doping effect, theoretical models and simulations are seriously needed to investigate the atomic environments around the implanted Al ions after complicated cascade collisions and thermal treatments.

At the atomic scale level, Hornos et al. carried out early density functional theory (DFT) calculations on 4H-SiC supercells and found some stable and metastable Al ion-related complex defects [16], but their originations and stabilities are questionable due to the lacking of molecular dynamics (MD) simulations. Wu et al. studied the defect evolution and doping efficiency of Al ion implantation in 3C-SiC using the MD method, and they found surface recrystallizations that were closely related to compressive stresses [17]. Their defect descriptions stayed at the atom cluster level, leading to indistinguishable defect details, let alone their couplings with Al ions.

Not only the Al ion implantation-related defect structures studied above but also the defect-induced trap levels and their impacts on transport behaviors of 4H-SiC devices are also desired. Megherbi et al. carried out a numerical simulation study on 4H-SiC p-i-n diodes with electrically activated Al acceptors and implantation-induced trap concentrations [18]. By adding the charges due to traps and defects into their drift-diffusion (DD) models, their detrimental effects on carrier transport behaviors at different bias regions were observed. Aside from the DD model, the quantum tight-binding (TB) model had already been established 20 years ago [19]. Although it is believed to be more accurate to describe the defect effect, no work has used the TB model to directly assess the impacts of defects on carrier transport in 4H-SiC devices. We infer that the reason may be the complex model revisions towards building trap levels in the bandgap.

Here, we combine the advanced theoretical models and simulation methods above to implement a multiscale simulation of the Al ion implantation and post-annealing processes. We focus on the implantation-induced point defects in 4H-SiC and their couplings with the Al ions. The attractions between Al ions and point defects and their coupling effects are revealed by DFT calculations. Moreover, TB models representing small parts of the 4H-SiC MOSFET channels are established, including the trap levels due to these defects. From the Poisson-nonequilibrium Green's function (NEGF) solver, the impacts of Al ion implantation on local carrier transports are finally obtained. From the simulation flow, the important Al ion-point defect coupling effects, which can seriously degrade local transport behaviors, are highlighted, offering new theoretical evidence for experiments to explain the complex doping results.

2. Simulation Methods

2.1. MD Simulation

We used the MD simulation software Large-scale Atomic/molecular Massively Parallel Simulator (LAMMPS) version (2 August 2023) to simulate the implantation and annealing process of Al ions [20]. The Open Visualization Tool (OVITO) was used to visualize the simulation model and analyze the data [21]. The size of the implantation and annealing model of 4H-SiC is 5.23 nm × 5.33 nm × 9.06 nm and contains 24,480 atoms. The whole system is divided into three distinct layers along the deep direction as follows: a 7.04 nm Newton layer, a 1.01 nm thermostat layer, and a 1.01 nm boundary layer. In the Newton layer, atomic evolution was carried out according to the classical laws of dynamics using a microcanonical ensemble (NVE). The atoms directly involved in the displacement cascade followed the actual energy exchange process during implantation. In the thermostat layer, the Berendsen thermostat was used to simulate the heat exchange process inside the material under nonequilibrium conditions, as well as the temperature dissipation during the ion bombardment of the target [22]. The boundary layer limits the degree of freedom of the simulated system by fixing the boundary atoms. At the beginning of the simulation, the canonical ensemble (NVT) was used to relax the entire system at 300 K.

Then, we used a 1 keV Al ion to implant the 4H-SiC (0001) plane with a 7° angle from the z-axis to prevent the channeling effect, corresponding to the implantation region of 2 nm × 2 nm, and it was enough to cover the range of Al ion motion. It is also noted that 1 keV energy is enough to inject the Al ion into the bulk 4H-SiC, leading to abundant defect types that deserve further studies. To prevent the serious overestimation of the Al ion range, different energies obtained through the Stopping and Range of Ions in Matter (SRIM) software were used to calculate the corresponding electronic stopping force through linear interpolation [23]. The periodic boundary condition was adopted for the x- and y-directions and the non-periodic boundary condition was adopted in the z (deep)-direction. This allows atoms in the x- and y-directions to interact at the boundaries, exiting from one edge of the simulation box and re-entering from the opposite edge. Instead of the non-periodic surface energy, the surface energy under periodic boundary conditions can accurately simulate the surface changes in the target material in actual processing. Timestep Δt can vary between 0.002 fs and 1 fs, based on the speed of the fastest particle in the system, which is specified to move no more than 0.02 Å in a single integration, ensuring the accuracy of the simulation under high-speed cascade collision.

Then, the annealing process was adopted in the Nose–Hoover thermostat NPT system to achieve temperature control. All boundaries were set as periodic. The time step was 0.5 fs from 300 K to 3300 K at a rate of 10 K/ps, and the temperature was held for 200 ps to achieve self-repair of the lattice damage and point defects [17]. The high annealing temperature of 3300 K was chosen due to the overestimation of the 4H-SiC melting point by the Tersoff potential. The whole implantation and annealing process simulations were repeated 10 times to find out the defect types as much as possible.

The accuracy of the potential function is the core of the MD to correctly describes the interaction between particles. For SiC systems, the Tersoff potential is often used to describe the covalent interaction between SiC atoms [24]. The Zigler–Biersack–Littmark (ZBL) repulsion potential is mainly used to describe the short-range interaction between C and Si atoms [25]. Previous studies have shown that the Tersoff force field cannot only accurately describe the amorphous phase, liquid phase, and quenching of the covalent bond system but can also obtain the solid-phase epitaxial growth simulation results of the amorphous covalent bond system with good consistency with the experimental regrowth rate. In addition, the SiC atom departure threshold energy, calculated by the Tersoff potential function, is basically consistent with the experimental results, which is the key to accurately calculating the cascade collision process and simulating the damage evolution. Secondly, Lampin et al. proved that there is a linear relationship between the recrystallization rate calculated using the Tersoff potential and the experimental value, and the actual corresponding temperature value can be obtained indirectly by calculating the temperature under the Tersoff potential. Therefore, the Tersoff/ZBL potential is selected in this paper to describe the interaction between SiC atoms during p-type doping ion implantation and post-annealing:

$$E = \frac{1}{2}\sum_{i}\sum_{i \neq j} V_{ij} \tag{1}$$

$$V_{ij} = \left(1 - f_F\left(r_{ij}\right)\right)V_{ij}^{ZBL} + f_F\left(r_{ij}\right)V_{ij}^{Tersoff} \tag{2}$$

$$f_F(r_{ij}) = \frac{1}{1 + e^{-A_F(r_{ij} - r_c)}} \tag{3}$$

Among them, V_{ij}^{ZBL} and $V_{ij}^{Tersoff}$ are the functional parts of the Tersoff and ZBL potentials, respectively; r_{ij} is the bond length between atom i and atom j; r_c is the cutoff range of the ZBL potential function; A_F is used to determine the smoothness between the ZBL and Tersoff potentials; f_F is the Fermi-like function of smoothing the ZBL repulsion potential; and E is the total energy of the system.

In addition, for the interaction of metal atoms, the embedded-atom method (EAM), developed by Winey et al., is used to describe the interaction between Al ions [26]:

$$E_{ij} = \sum_i F_i(\rho_i) + \frac{1}{2}\sum_{i \neq j} \phi_{ij}(R_{ij}) \tag{4}$$

where $F_i(\rho_i)$ is the atomic embedding energy, representing the energy required for atom i to embed electron density; ρ_i is the total electron density around atom i; and $\phi_{ij}(R_{ij})$ is the distance-dependent Al atomic pair potential. Dandekar and Shin developed the Morse potential between Al and 3C-SiC based on the potential energy curve obtained by Zhao et al. by using the ab initio method [27,28]. The form is as follows:

$$E_{ij} = D_0\left[e^{-2\alpha(r_{ij} - r_0)} - 2e^{-\alpha(r_{ij} - r_0)}\right] \tag{5}$$

where r represents the distance between particle pairs; r_0 is the equilibrium bond length; and D_0 and α represent the depth and width of the potential well, respectively. D_0, α, and r_0 of Al and C are 0.4691 eV, 1.738 Å^{-1}, and 2.246 Å, respectively. D_0, α, and r_0 of Al and Si are 0.4824 eV, 1.322 Å^{-1}, and 2.92 Å, respectively.

2.2. Defect Statistics

The Wigner–Seitz (WS) analysis method was considered to calculate the change in the number of point defects before and after ion implantation and annealing, which is believed to accurately extract various point defects, and we needed two configurations of the atomic model as inputs as follows: the reference configuration and the displacement configuration. For the reference configuration, we selected the model before implantation and defined the position of the atoms in the crystal without defects. The initial sites of the atoms were divided by three-dimensional Voronoi space, and each atom was precisely assigned to a site. Displacement configuration is the configuration that needs to be analyzed and often contains some point defects such as vacancies, interstitials, and antisites. Eventually, some sites may not be occupied by atoms at all, which are called vacancies. Other sites may be occupied by multiple atoms, which are called interstitials, and the sites occupied by the atoms that are not the original ones are antisites. However, the error of this method in identifying point defects increases with significant changes in the volume of the system, as many atoms are almost removed from the original site of the reference configuration due to the expansion or contraction of the volume.

Next, we combined the identify diamond structure (IDS) method to visualize the defect steps further, considering that the WS method could be affected by the volume of the system [29]. The atoms of cubic diamond and hexagonal diamond structures indicate that the first- and second-nearest neighbor lattices are perfect. Specifically, the first-neighbor lattices in both the cubic diamond (first neighbor) and hexagonal diamond (first neighbor) are perfect, but the second-neighbor lattices are damaged. Conversely, atoms of cubic diamond (second neighbor) and hexagonal diamond (second neighbor) structures suggest a perfect second-neighbor lattice with a damaged first-neighbor lattice. If an atom belongs to another structure (none of the structures above), it is highly likely to be a defective atom. We used the IDS method to identify and extract atoms belonging to other structures and their nearby atoms with distances smaller than 5 Å (including their first- and second-nearest neighbors). Then, the WS method was carried out to identify the point defect types in these atoms, thereby significantly reducing the identification error.

2.3. DFT Calculation

The DFT calculations using the Perdew–Burke–Ernzerhof (PBE) function are realized by the QUANTUM ESPRESSO software [30]. The ultrasoft pseudopotential (USPP) is selected for Si, and projector augmented wave (PAW) pseudopotentials are applied to C and Al. To facilitate the subsequent transport calculations, we used a $3 \times 5 \times 1$ supercell containing 240 atoms to calculate the defect effects, which is derived from a 16-atom rectangular unit cell. During the crystal force optimization process, the supercell variations are allowed until the force on each atom is smaller than 1×10^{-3} Ry/bohr [31]. The energy cutoff of 30 Ry is used in the band structure calculations along the $\Gamma - X - K - \Gamma - Y$ high symmetry direction [12,32].

2.4. CI–NEB Calculation

The main purpose of the climbing image–nudged elastic band (CI–NEB) method is to find the minimum energy path (MEP) on the potential energy surface [33]. The highest energy point on the MEP connecting the initial and final states in the chemical reaction and atomic diffusion of solids is called the transition state (TS). We utilized seven images to pinpoint the TS, including the initial and final states. The energy convergence threshold of the images is 1×10^{-6} Ry, and the forces convergence threshold is 1×10^{-4} Ry/bohr. Additionally, the freezing parameter was set to true in order to halt the optimization when the error in the intermediate states fell below 0.05 eV/Å, thereby enhancing the accuracy and efficiency of the calculations.

2.5. MLWF Transformation

Then, the Wannier90 software package was employed to generate a set of maximally localized Wannier functions (MLWFs) to transform the electron wavefunctions [34]. This process enables obtaining the tight-binding (TB) model and determining the hopping parameters between the defect-induced trap levels and the conduction/valence band (CB and VB) states of the defect structures. The trap levels can be induced into the 4H-SiC TB model by expanding and filling the matrix with new lines and rows, with the scattering rates being included in the TB models through the imaginary parts of the corresponding onset energies [35].

2.6. Quantum Transport Simulation

In order to investigate the impacts of the Al ion implantation-induced defects on 4H-SiC MOSFET channel transport behaviors, a local 10 nm channel region containing defects was considered (local channel), as shown in Figure 1. Since point defects like the C vacancies and C interstitials in 4H-SiC are highly localized, showing an extension of less than 1 nm [15], a 10 nm channel region containing these point defects is enough to study their local electrical behaviors. The oxide and gate thicknesses are 1 nm and 0.5 nm. To reduce the influence of the gate metal type, the work function of the gate was assumed to be equal to the perfect 4H-SiC crystal. The quantum transport problem was solved by the iterations between Schrodinger's equation in the NEGF form and Poisson's equation using the NanoTCAD ViDES code [36]. The gate voltages (V_g) ranging from 0–6 V were applied, and the drain-to-source voltage (V_d) was set to 0.3 V. The convergence criterion for the equation solution is that the magnitude of the difference in the grid potential vector between adjacent iterations is less than 0.01 V.

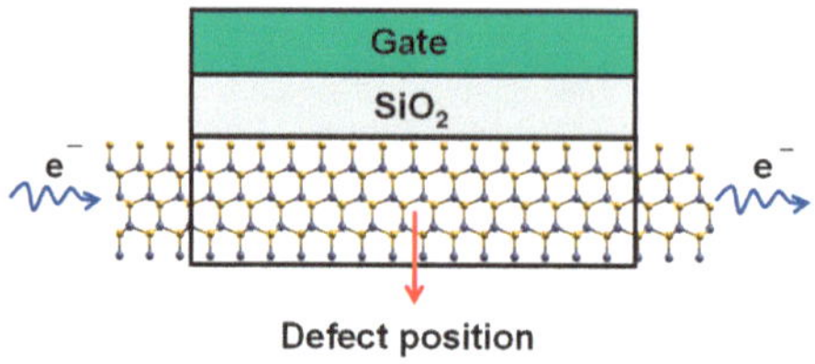

Figure 1. Schematic illustration of the quantum transport simulation model, in which a local channel region containing defects in 4H-SiC is considered.

3. Results and Discussions

3.1. Molecular Dynamics Simulations

The study starts from the MD simulations on the Al ion implantation and post-annealing processes. Before the implantation, the local structure of the implantation region only contained cubic diamond and hexagonal diamond atoms, as shown in Figure 2a. Then, the Al ion (red) was injected into the 4H-SiC (0001) plane at 300 K. It can be seen in Figure 2b that near the Al ion, there are many other structure atoms and defect clusters after implantation, indicating that the implantation has caused obvious lattice damage and point defects. After 3300 K high-temperature annealing, the number of other structure atoms decreases obviously, and defect clusters disappear entirely, as shown in Figure 2c. Figure 2d depicts the defect numbers of different types during the MD simulation using the WS and IDS method. The number of point defects increases significantly after the implantation. For example, the number of C vacancies (V_C) and C interstitials (C_i) are 9 and 7, respectively, while the number of other point defects, like Si replacing C (Si_C), C replacing Si (C_{Si}), Si interstitial (Si_i) and Si vacancy (V_{Si}), are all less than 4. After the annealing and the atoms cooled to 300 K, the number of V_C and C_i apparently decreased to 3 and 1, respectively. Other types of point defects were almost reduced to zero.

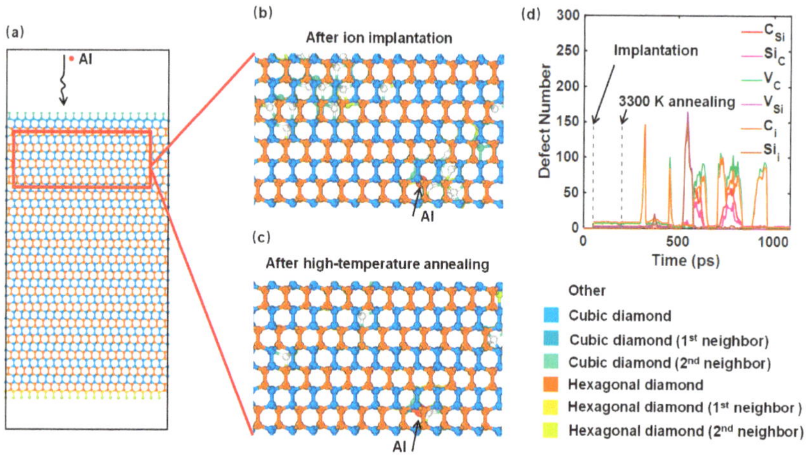

Figure 2. Schematic illustration of the MD simulations on the Al ion implantation and post-annealing processes. (**a**) 4H-SiC Newton layer before implantation. (**b**,**c**) 4H-SiC crystal structures after the implantation and annealing. (**d**) Statistic defect numbers during the two processes.

In order to explore the defect evolution, we extracted the damaged lattice atoms (pink) identified by the IDS method in Figure 3 at the moments just after implantation (0 ps), annealing (300 ps and 600 ps), and after annealing (900 ps). At 0 ps, it had the largest number of damaged lattice atoms and damaged area, in which the damaged atoms gather along the implantation trajectory of the Al ion. Then, they experienced a self-repairing stage during the annealing (300 ps and 600 ps images), resulting in the continuously reduced damaged area. Finally, the damaged area and the number of damaged lattice atoms became the smallest after 900 ps annealing, which proves that a 3300 K high temperature is appropriate to repair the implantation damage.

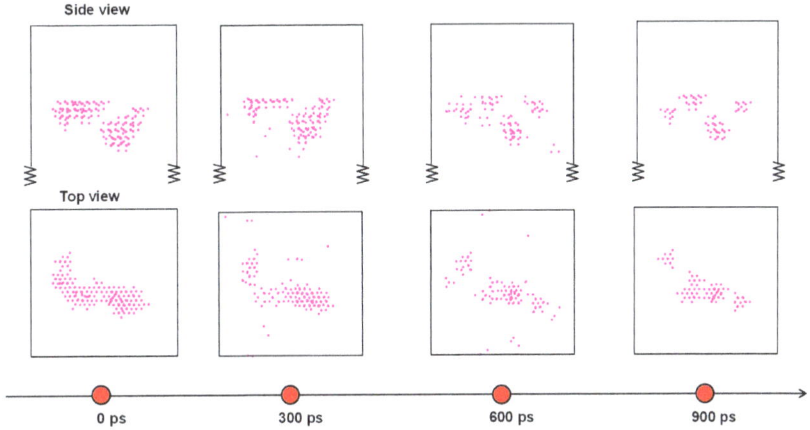

Figure 3. Defect evolution during 3300 K annealing process.

In both Figures 2 and 3, there are still many defects clustered in the implantation trajectory of Al ion that failed to self-repair after high-temperature annealing. For example, the Al atom fails to perfectly replace Si atoms in this case. To catch all the stubborn defects that are important to 4H-SiC device performance, we repeated the MD simulation 10 times under the same condition. In each case, the Al ion and its nearby point defects were extracted using a supercell, as illustrated in Figure 4a, and Figure 4b–e shows four typical supercell results. It was found that after Al ion implantation and post-annealing, three configurations existed as follows: (i) Al replacing Si (Al_{Si}), which is the perfect Al position to induce the p-type doping effect and the majority results in our MD simulations, (ii) Al replacing C (Al_C), as shown in Figure 4c, and (iii) Al being an interstitial (Al_i), as shown in Figure 4d. No matter whether the Al ion was perfectly doped, there were always many point defects around, including the frequent Si_C, V_C, C_{Si}, antisites, C_i, and unfrequent C-clusters. We also examined the cases under 350 eV and 2 keV implantation energies and 2900 K and 3700 K annealing temperatures. The results indicate that the defects all belong to the above types.

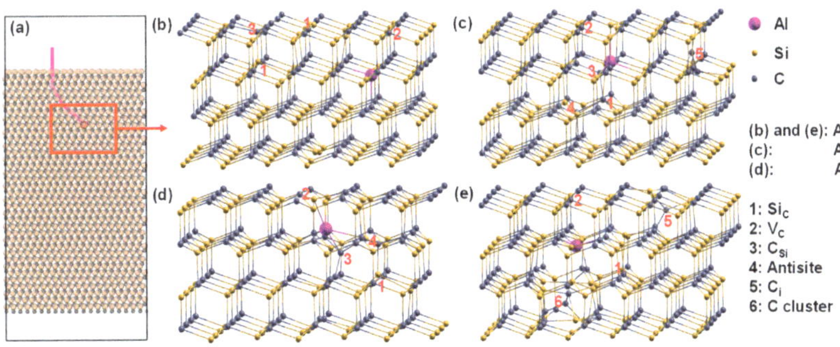

Figure 4. Schematic illustration of the implanted Al ion trajectory and the induced point defects in 4H-SiC. (**a**) The Al ion trajectory. (**b**–**e**) The final local crystal environments around Al.

3.2. DFT Calculation

The MD results reveal that the Al ion implantation and post-annealing processes will induce a variety of compound defects formed by the coupling of Al ion and point defects. In order to investigate the Al ion-point defect coupling effects, we constructed defect supercell models containing about 240 atoms and different defect types to calculate the electrical properties using the DFT method.

Before the consideration of Al ions, the intrinsic defects in 4H-SiC were studied first. As shown in Figure 5, among these defects, C_{Si} has a negligible impact on the band structure of the perfect 4H-SiC crystal. The band structures of Si_C and antisite cases are similar to each other, all showing shallow hole trap levels near the VB. Therefore, it is inferred that these levels are induced solely by the Si_C. Both V_C and C_i introduce a mid-gap trap and a shallow electron trap near the CB. They both can affect the transport behaviors by trapping electrons.

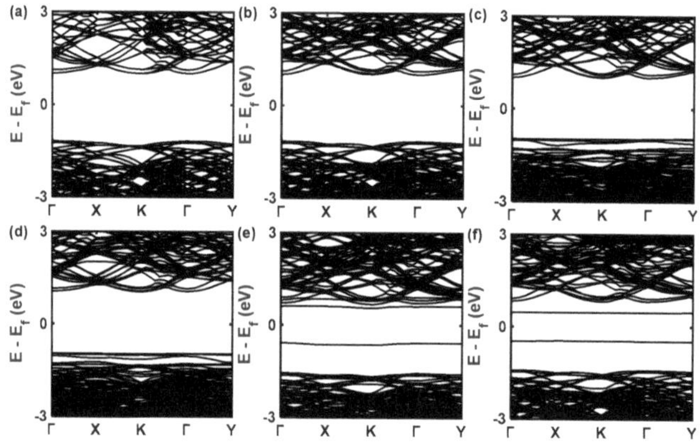

Figure 5. Band structures of 4H-SiC supercells containing 240 atoms and different point defects. (**a**) Perfect crystal supercell. (**b–f**) Supercells containing C_{Si}, Si_C, antisite, V_C, and C_i defects.

When a Si atom is replaced by an Al ion, the perfect p-type doping is achieved, and the Fermi level goes into the VB edge of 4H-SiC, as shown in Figure 6a. The presence of Csi around Al_{Si} has little impact on the p-type doping properties, as evidenced by the comparison between Figure 6a,c. However, Figure 6b,d illustrate that Si_C and antisite around Al_{Si} will induce additional trap levels near the VB. Although they maintain the p-type characteristic of the system, the free holes released by Al doping are trapped and will not contribute to the currents when bias is applied. It is also noted that these trap levels can be solely attributed to the Si_C. In contrast, as shown in Figure 6e,f, the V_C and C_i directly transform the system into an n-type property due to the half-filled trap state near the CB edge, which pins the Fermi level. Additionally, the C-cluster defect induces a near-neutral condition, with the half-filled trap levels located at the mid-gap in Figure 6g. The different defect couplings around Al_{Si} may cause opposite doping effects, which greatly affects the doping results during the device manufacturing process.

When the Al substitutes for the C lattice site, it does not produce a doping effect but introduces three closely spaced trap levels in the mid-gap, as shown in Figure 7a. Meanwhile, Figure 7c illustrates that the presence of C_{Si} leads to a downward shift of the two electron-occupied trap levels, indicating that their abilities to trap electrons during electron transport are weakened. When Si_C, antisite, and V_C appear around Al_C, as indicated by Figure 7b,d,e, an additional trap level is induced. Moreover, the highest half-filled trap level pins the Fermi level near the VB edge, showing the n-type doping effect.

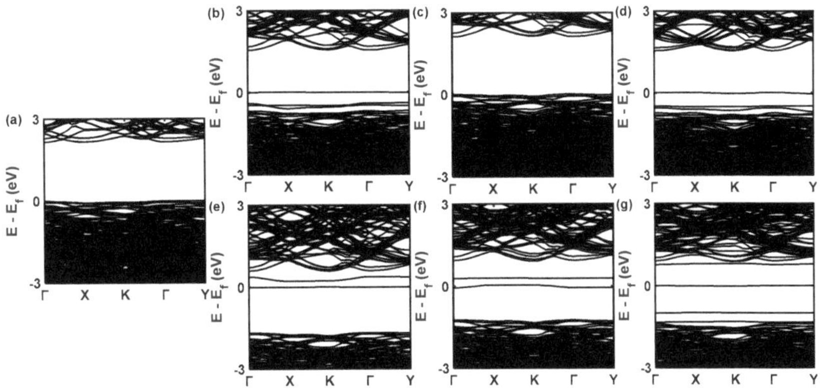

Figure 6. Band structures of 4H-SiC supercells containing 240 atoms and an Al ion. (**a**) Supercell only containing Al_{Si} doping. (**b**–**g**) Supercells containing Al_{Si}–Si_C, C_{Si}, antisite, V_C, C_i, and C-cluster coupling defects.

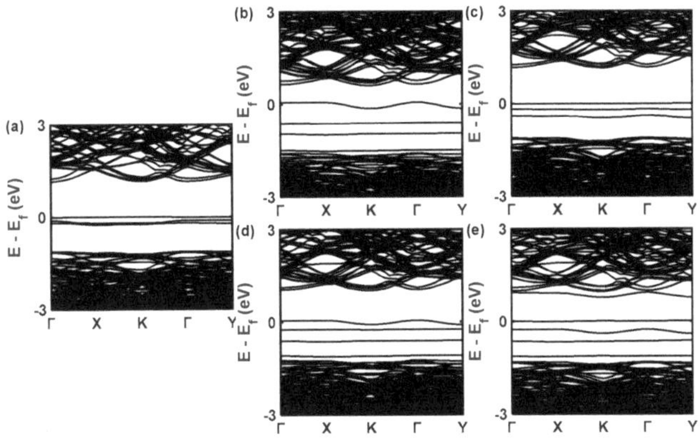

Figure 7. Band structures of 4H-SiC supercells containing 240 atoms and an Al ion. (**a**) Supercell only containing Al_C doping. (**b**–**e**) Supercells containing Al_C–Si_C, C_{Si}, antisite, and V_C coupling defects.

When Al becomes an interstitial atom, the system exhibits distinct n-type doping characteristics, which is defined as the self-compensation effect by the previous literature [37]. From the real space deformation charge densities in Figure 8a below, the electron contribution to the system by the Al_i is clearly observed. The delocalized electrons are injected into the CB and become free carriers. The presence of Si_C, C_{Si}, and C_i defects will not significantly affect the system characteristics, as shown in Figure 8b,c,e. However, when there is an antisite around Al_i, as shown in Figure 8d, a trap state will be generated near the VB, and the free electrons injected into the SiC system by Al_i will be captured, consequently restraining the n-type doping characteristics of the system.

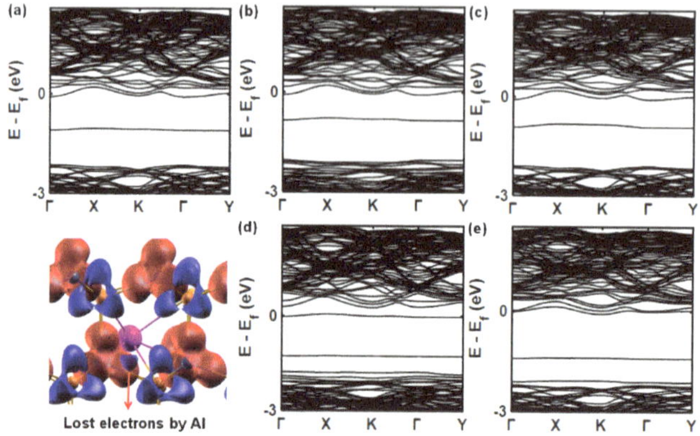

Figure 8. Band structures of 4H-SiC supercells containing 240 atoms and an Al ion. (**a**) Supercell only containing Al_i doping. The real space deformation charge densities near the Al_i are also shown below, with blue and red isosurfaces representing missing and obtaining electrons. (**b–e**) Supercells containing an Al_i and a Si_C, C_{Si}, antisite, and C_i.

3.3. NEB Calculation

When an Al atom replaces a C atom, three deep energy levels will be introduced, which will significantly affect the transport behaviors. However, we find that if an extra C_i is induced near the $Al_{C,}$ as shown in Figure 9a, the Al atom will directly migrate to the interstitial without a barrier, while the C atom will return to the original Al position.

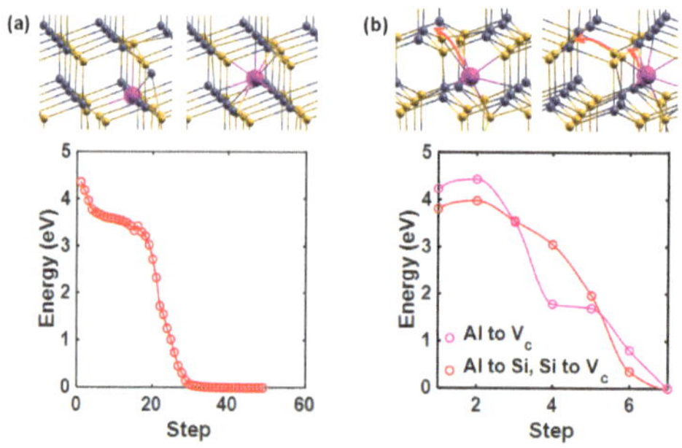

Figure 9. Two metastable Al doping states. (**a**) Energy and structure variations after a C_i move close to an Al_C. The C_i will kick out the Al, resulting in an Al_i without an energy barrier. (**b**) Energy variations during an Al_i compensating a V_C. The left and right local structures correspond to two different paths. Note that the energies of (**b**) are from the NEB calculation, while the energies of (**a**) are from the Broyden–Fletcher–Goldfarb–Shanno (BFGS) quasi-Newton algorithm, adopted in the DFT calculation due to the zero-energy barrier.

However, the single Al_i will also introduce a deep level and even lead to a serious n-type doping phenomenon. Therefore, we continued to study its couplings with other defects. It was found that when an Al_i meets a V_C near it, as shown in Figure 9b, it will occupy the vacancy in two almost barrier-free ways. The first one is that the Al_i directly migrates to the V_C, resulting in an Al_C and encountering an energy barrier of 0.2 eV. The second path is that the Al replaces the Si, and the Si migrates to the V_C, resulting in an Al_{Si}–Si_C coupling and encountering an energy barrier of 0.17 eV. Both results will compensate for the n-type doping effect induced by Al_i, which is beneficial to the p-type doping purpose.

It seems that point defect densities are high near the Al from the MD simulation results. Next, we verified if point defects like V_C and C_i can be attracted or directly induced by

Al in 4H-SiC. First, the NEB calculations on the formation energies of generating a C_i–V_C pair were carried out. As shown in Figure 10a, it was found that the potential barrier of the C_i–V_C pair formation in the presence of Al ion is almost 1 eV lower than that in the perfect 4H-SiC crystal. Moreover, in the presence of Al ion, it provides a metastable state with a repairing barrier of 0.27 eV for the C_i–V_C pair, indicating that C_i–V_C pairs are easily induced by the Al ion implantation. Next, we calculated the total energies of the supercell structures with different Al–V_C and Al–C_i distances, as shown in Figure 10b,c. There is no doubt that V_C can be attracted by the Al, reducing the system energies of nearly 1 eV. For the C_i cases, the energy variations are complicated, and no obvious relationship can be established between the Al–C_i distance and system energy. From the view of energy, the attraction between Al and V_C also tends to separate the C_i–V_C pairs and increase the related defect densities. These microscopic effects all support the contradiction between the electrical activation of the Al ion and the control of the V_C concentration [15].

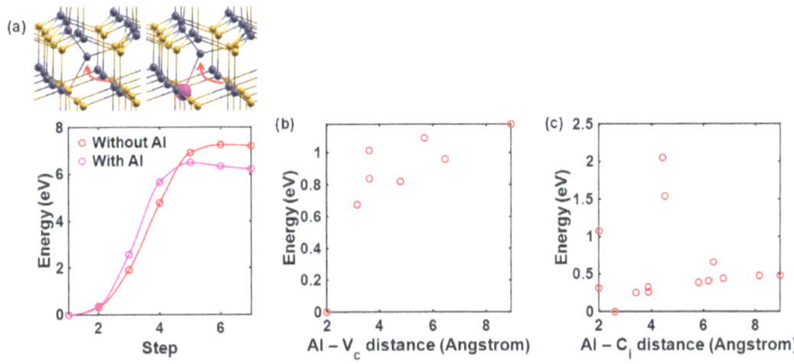

Figure 10. Probabilities of Al_{Si} doping inducing or attracting additional points defects. (**a**) Energy variations to generate a V_C–C_i pair due to the movements of C atoms denoted by the red arrows, with or without an Al_{Si}. (**b**,**c**) Energy variations along the Al_{Si}–V_C and Al_{Si}–C_i distances.

3.4. Impact of Defects on Carrier Transport of 4H-SiC MOSFET Channel

3.4.1. Couplings between Al_{Si} and Other Defects

The transfer characteristics of the 4H-SiC MOSFETs with only Al_{Si} are shown as green lines in Figure 11a,b. Compared with the perfect channel without any defect and Al ion, the local p-type doping effect is clearly observed at the Al_{Si} channel since the carriers are mainly holes at $V_g = 0$ V. Along the increasing V_g, the hole current is reduced rapidly, and then the electron current dominates the transport. To further verify the transport mechanism, the channel density-of-states (DOS) (background color) and current spectrum (pink lines with peaks along the x-axis) of the Al_{Si} channel were calculated and shown in Figure 11c,e. The bright DOS regions represent the VB and CB, while the blue dark regions are bandgaps. Since the 0.3 V V_d is applied by setting the Fermi levels of the source and drain regions to ±0.15 eV, the DOS, located at the ±0.15 eV transport window, will obtain high probabilities to contribute currents. At $V_g = 0$ V, the VB DOS are closer to the transport window, and a hole current peak is clearly observed, which not only verifies the hole current phenomenon but also shows that the p-type doping has moved up the VB of the local 10 nm channel to become a p-type region. At $V_g = 2$ V, both the CB and VB are moved down, and the CB DOSs are closer to the transport window, resulting in the majority of carriers changing from holes to electrons.

In contrast, the Al_{Si} + Si_C and Al_{Si} + antisite channels do not show much threshold voltage (V_{th}) moving behavior, but their hole currents are severely reduced. Taking the Al_{Si} + Si_C channel shown in Figure 11d as an example, the free holes from Al ions are captured by the mid-gap trap levels. Although these levels have already entered the transport window, they can hardly contribute to the currents due to the carrier trapping effect. On the other hand, the Si_C or antisite coupling to Al_{Si} maintains the p-type property of the local channel, and this is the reason that the V_{th} variations are negligible.

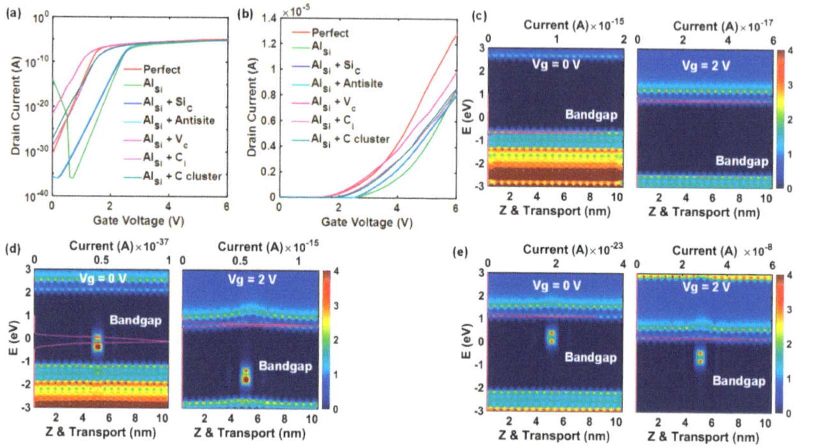

Figure 11. Transfer characteristics of the 4H-SiC MOSEFT local channels when an Al_{Si} couples with another point defect. (**a**,**b**) Transfer characteristic curves of the local channels in log and linear axes. (**c**–**e**) DOS of the channels containing Al_{Si}, Al_{Si} + Si_C, and Al_{Si} + V_C at different V_g.

For the Al_{Si} + V_C, Al_{Si} + C_i, and Al_{Si} + C-cluster channels, the p-type property of the local channels is flipped completely, and the V_{th} shifts towards the negative x-axis are significant. In Figure 11e, the two bright dots correspond to the two trap levels, as shown in Figure 6e, verifying that the trap levels have been successfully incorporated into the channel transport simulation. Moreover, they exhibit obvious resistance to V_g. As shown on the right side, the CB edge (bright line on the top of the two trap states) shows a small peak at the trap position, which will block electron transports, corresponding to the classical Fermi level pinning effect caused by trap states. The pinning effect cannot only cause serious leakage currents at V_g = 0 V but can also impede the CB moving downwards at $V_g > 0$ and form electron barriers.

To exclude the potential influences caused by gate oxide thickness and V_d, the 4H-SiC MOSEFT local channels with a 3 nm oxide layer and 0.1 V V_d are simulated and shown in Figure 12a,b. When the oxide thickness increases, the gate control ability is weakened. But the defect-induced transport behaviors do not change obviously. The results in Figure 12b are similar, such as that reducing the V_d only leads to some current drops. Therefore, the defect impacts on local carrier transports are persistent in SiC MOSEFT channels. Moreover, the coupling effect of the Al_{Si} + V_C + C_i ternary defect is studied in Figure 12c. Clearly, the transport ability is further degraded, indicating that the defect clusters could cause serious local current unbalance, which could be a fatal reason for the device failure.

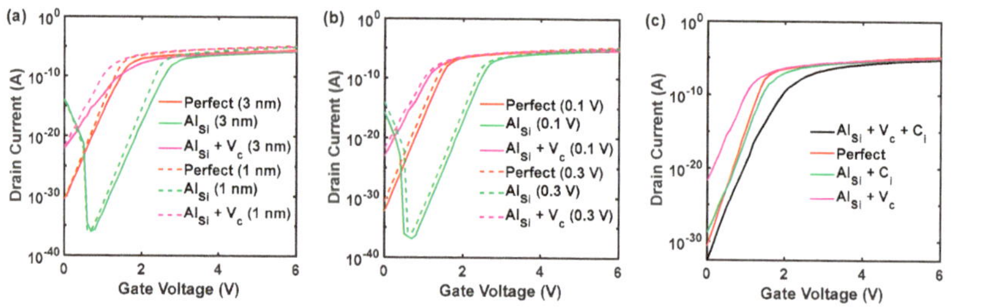

Figure 12. Transfer characteristics of the 4H-SiC MOSEFT local channels. (**a**) Comparisons between 1 nm and 3 nm thickness gate oxide cases. (**b**) Comparisons between V_d = 0.1 V and V_d = 0.3 V cases. (**c**) Transfer characteristics of the channel containing Al_{Si} + V_C + C_i and comparisons with other relevant cases.

3.4.2. Couplings between Al_C and Other Defects

Figure 13a,b show the transfer characteristics of the 4H-SiC MOSFET local channels in the presence of Al_C and are coupled with other defects. From Figure 7, Al_C does not contribute to the doping effect, but it induces three mid-gap trap states. These states seriously impede the carrier transport, as the subthreshold swings (SS) of the Al_C and Al_C + C_{Si} channels are obviously enlarged compared with the perfect channel. These phenomena can also be well explained in Figure 13c, as these trap states induce a non-negligible electron barrier at V_g = 2 V.

When the Si_C appears near the Al_C in Figure 13d, the channel exhibits a n-type characteristic due to the new trap level showing in the upper panel of the bandgap, leading to the increasing leakage current at V_g = 0 V. When an antisite appears near the Al_C in Figure 13e, three of the four trap levels move into the bottom panel of the bandgap, which will not significantly influence the electron transports.

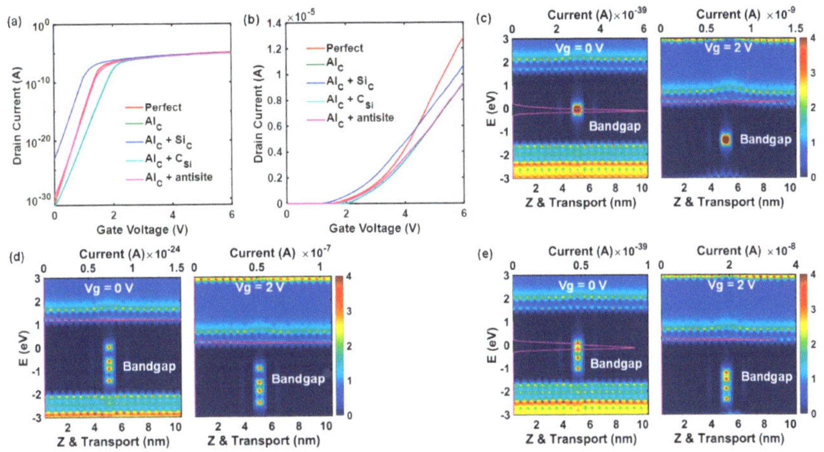

Figure 13. Transfer characteristics of the 4H-SiC MOSEFT local channels when an Al_C coupling with another point defects. (**a**,**b**) Transfer characteristic curves of the local channels in log and linear axes. (**c**–**e**) Calculated density-of-states (DOS) of the channels containing Al_C, Al_C + Si_C, and Al_C + antisite at different V_g.

3.4.3. Couplings between Al_i and Other Defects

When an Al ion replaces C, it has already lost the p-type doping effect, as shown above. In Figure 14 below, the doping effect could further be flipped completely if an Al ion became an interstitial in 4H-SiC. When an Al_i locates at the MOSEFT channel, it simultaneously injects electrons into the CB and induces a mid-gap trap level, as shown in Figure 14c. Therefore, the CB is moved close to the transport window, leading to significant leakage currents at V_g = 0 V. Under this circumstance, an antisite coupling with the Al_i could relieve the n-type doping effect. As shown in Figure 14d, the antisite coupling causes more trap levels, which can not only trap excessive free electrons but can also prevent the downward movement of CB. However, the n-type doping can hardly be eliminated. Due to its strong ability to destroy the p-type doping effect, it is therefore suggested that the Al_i should be avoided to the maximum extent at the ion implantation and post-annealing processes.

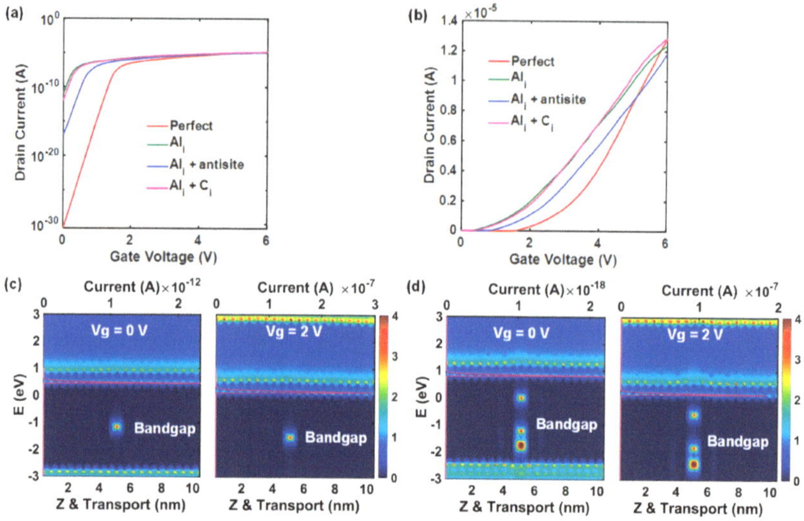

Figure 14. Transfer characteristics of the 4H-SiC MOSEFT local channels when an Al_i coupling with another point defects. (**a**,**b**) Transfer characteristic curves of the local channels in log and linear axes. (**c**,**d**) Calculated density-of-states (DOS) of the channels containing Al_i and Al_i + antisite at different V_g.

4. Conclusions

The Al ion implantation and post-annealing processes were studied using MD simulations in this work. Except for the Al_{Si}, the non-ideal Al_C and Al_i were also observed. Moreover, additional point defects caused by the Al ion bombardment also frequently appeared and exhibited tendencies of moving towards the Al ion. The DFT and quantum simulation results show that the Al_C will cause dense trap levels at the mid-gap of 4H-SiC, and the Al_i even leads to an n-type doping effect. When they couple with the additional point defects nearby, more complicated doping and trap levels will be generated, increasing the leakage current and SS, shifting the V_{th}, and reducing the carrier mobility. This study provides new theoretical evidence for experiments to explain complex doping results.

Author Contributions: Conceptualization and methodology, Y.L.; software, validation, and investigation, Y.W. and H.L.; resources, C.J.; data curation, Q.S.; writing—original draft preparation, Y.L., Y.W. and H.L.; writing—review and editing, C.J. All authors have read and agreed to the published version of the manuscript.

Funding: This work was supported by the National Key R&D Program of China (2023YFB3611600), the Science and Technology Innovation Program of Hunan Province (2023RC3112), the Natural Science Foundation of Hunan Province (2022JJ40094), and the Creative Research Groups Program of the National Natural Science Foundation of China (62321003).

Data Availability Statement: The simulation codes are available from the authors upon request.

Conflicts of Interest: The authors declare no conflicts of interest.

References

1. Kimoto, T.; Watanabe, H. Defect engineering in sic technology for high-voltage power devices. *Appl. Phys. Express* **2020**, *13*, 120101. [CrossRef]
2. Deng, H.; Wei, S.; Li, S. Review of defect physics and doping control in wide-band-gap semiconductors. *Chin. Sci. Bull.* **2023**, *68*, 1753–1761. [CrossRef]
3. Puschkarsky, K.; Grasser, T.; Aichinger, T.; Gustin, W.; Reisinger, H. Review on sic mosfets high-voltage device reliability focusing on threshold voltage instability. *IEEE Trans. Electron Devices* **2019**, *66*, 4604–4616. [CrossRef]
4. Han, L.; Liang, L.; Kang, Y.; Qiu, Y. A review of sic igbt: Models, fabrications, characteristics, and applications. *IEEE Trans. Power Electron.* **2021**, *36*, 2080–2093. [CrossRef]

5. She, X.; Huang, A.Q.; Lucia, O.; Ozpineci, B. Review of silicon carbide power devices and their applications. *IEEE Trans. Ind. Electron.* **2017**, *64*, 8193–8205. [CrossRef]
6. Liu, Y.; Yuan, L.; Sun, C.; Zhang, Y.; Tang, X.; Zhang, Y. Three-dimensional design of a 4h-sic npn lateral phototransistor for micro-pixel in ultraviolet optoelectronic integration. *IEEE Trans. Electron Devices* **2023**, *70*, 6399–6405. [CrossRef]
7. Yang, W.; Sun, Y.; Qi, M.; Tang, Z.; Mao, S.; Yuan, L.; Sun, L.; Zhang, Y.; Zhang, Y. Wide temperature range modeling of implanted resistors based on 4h-sic cmos process. *IEEE Trans. Electron Devices* **2024**, *71*, 3575–3581. [CrossRef]
8. Fan, Y.; Song, Y.; Xu, Z.; Dong, B.; Wu, J.; Rommel, M.; Zhang, K.; Zhao, J.; Zhu, R.; Li, B.; et al. Molecular dynamics simulation of color centers in silicon carbide by helium and dual ion implantation and subsequent annealing. *Ceram. Int.* **2021**, *47*, 24534–24544. [CrossRef]
9. Wachowicz, E.; Ossowski, T.; Kiejna, A. Dft study of stepped 4h-sic{0001} surfaces. *Appl. Surf. Sci.* **2017**, *420*, 129–135. [CrossRef]
10. Wellmann, P.J. Review of sic crystal growth technology. *Semicond. Sci. Technol.* **2018**, *33*, 103001. [CrossRef]
11. Tsuchida, H.; Kamata, I.; Miyazawa, T.; Ito, M.; Zhang, X.; Nagano, M. Recent advances in 4h-sic epitaxy for high-voltage power devices. *Mater. Sci. Semicond. Process.* **2018**, *78*, 2–12. [CrossRef]
12. Huang, Y.; Qian, Y.; Zhang, Y.; Yang, D.; Pi, X. Kick-out diffusion of al in 4h-sic: An ab initio study. *J. Appl. Phys.* **2022**, *132*, 015701. [CrossRef]
13. Nipoti, R.; Carnera, A.; Alfieri, G.; Kranz, L. About the electrical activation of 1×1020 cm-3 ion implanted al in 4h-sic at annealing temperatures in the range 1500–1950 °C. *Mater. Sci. Forum* **2018**, *924*, 333–338. [CrossRef]
14. Poggi, A.; Bergamini, F.; Nipoti, R.; Solmi, S.; Canino, M.; Carnera, A. Effects of heating ramp rates on the characteristics of al implanted 4h–sic junctions. *Appl. Phys. Lett.* **2006**, *88*, 162106. [CrossRef]
15. Nipoti, R.; Ayedh, H.M.; Svensson, B.G. Defects related to electrical doping of 4h-sic by ion implantation. *Mater. Sci. Semicond. Process.* **2018**, *78*, 13–21. [CrossRef]
16. Hornos, T.; Gali, A.; Son, N.T.; Janzén, E. A theoretical study on aluminium-related defects in sic. *Mater. Sci. Forum* **2007**, *556–557*, 445–448. [CrossRef]
17. Wu, J.; Xu, Z.; Liu, L.; Hartmaier, A.; Rommel, M.; Nordlund, K.; Wang, T.; Janisch, R.; Zhao, J. Md simulation study on defect evolution and doping efficiency of p-type doping of 3c-sic by al ion implantation with subsequent annealing. *J. Mater. Chem. C* **2021**, *9*, 2258–2275. [CrossRef]
18. Asada, S.; Murata, K.; Tsuchida, H. Modeling of stacking faults in 4h-sic n-type epilayer for tcad simulation. *IEEE Trans. Electron Devices* **2023**, *70*, 1757–1762. [CrossRef]
19. Bernstein, N.; Gotsis, H.J.; Papaconstantopoulos, D.A.; Mehl, M.J. Tight-binding calculations of the band structure and total energies of the various polytypes of silicon carbide. *Phys. Rev. B* **2005**, *71*, 075203. [CrossRef]
20. Plimpton, S. Fast parallel algorithms for short-range molecular dynamics. *J. Comput. Phys.* **1995**, *117*, 1–19. [CrossRef]
21. Stukowski, A. Visualization and analysis of atomistic simulation data with ovito–the open visualization tool. *Model. Simul. Mater. Sci. Eng.* **2010**, *18*, 015012. [CrossRef]
22. Berendsen, H.J.; Postma, J.v.; Van Gunsteren, W.F.; DiNola, A.; Haak, J.R. Molecular dynamics with coupling to an external bath. *J. Chem. Phys.* **1984**, *81*, 3684–3690. [CrossRef]
23. Ziegler, J.F. Srim-2003. *Nucl. Instrum. Methods Phys. Res. Sect. B: Beam Interact. Mater. At.* **2004**, *219–220*, 1027–1036. [CrossRef]
24. Tersoff, J. Modeling solid-state chemistry: Interatomic potentials for multicomponent systems. *Phys. Rev. B* **1989**, *39*, 5566–5568. [CrossRef]
25. Biersack, J.P.; Ziegler, J.F. The stopping and range of ions in solids. *Ion Implant. Sci. Technol.* **1984**, *1*, 51–108.
26. Jacobsen, K.W.; Norskov, J.K.; Puska, M.J. Interatomic interactions in the effective-medium theory. *Phys. Rev. B* **1987**, *35*, 7423–7442. [CrossRef] [PubMed]
27. Dandekar, C.R.; Shin, Y.C. Molecular dynamics based cohesive zone law for describing al–sic interface mechanics. *Compos. Part A Appl. Sci. Manuf.* **2011**, *42*, 355–363. [CrossRef]
28. Zhao, H.; Chen, N.; Long, Y. Interfacial potentials for al/sic(111). *J. Phys. Condens. Matter* **2009**, *21*, 225002. [CrossRef] [PubMed]
29. Maras, E.; Trushin, O.; Stukowski, A.; Ala-Nissila, T.; Jónsson, H. Global transition path search for dislocation formation in ge on si(001). *Comput. Phys. Commun.* **2016**, *205*, 13–21. [CrossRef]
30. Giannozzi, P.; Andreussi, O.; Brumme, T.; Bunau, O.; Buongiorno Nardelli, M.; Calandra, M.; Car, R.; Cavazzoni, C.; Ceresoli, D.; Cococcioni, M.; et al. Advanced capabilities for materials modelling with quantum espresso. *J. Phys. Condens. Matter.* **2017**, *29*, 465901. [CrossRef]
31. Lv, Y.; Tong, Q.; Liu, Y.; Li, L.; Chang, S.; Zhu, W.; Jiang, C.; Liao, L. Band-offset degradation in van der waals heterojunctions. *Phys. Rev. Appl.* **2019**, *12*, 044064. [CrossRef]
32. Yan, X.; Li, P.; Kang, L.; Wei, S.-H.; Huang, B. First-principles study of electronic and diffusion properties of intrinsic defects in 4h-sic. *J. Appl. Phys.* **2020**, *127*, 085702. [CrossRef]
33. Henkelman, G.; Uberuaga, B.P.; Jónsson, H. A climbing image nudged elastic band method for finding saddle points and minimum energy paths. *J. Chem. Phys.* **2000**, *113*, 9901–9904. [CrossRef]
34. Pizzi, G.; Vitale, V.; Arita, R.; Blügel, S.; Freimuth, F.; Géranton, G.; Gibertini, M.; Gresch, D.; Johnson, C.; Koretsune, T.; et al. Wannier90 as a community code: New features and applications. *J. Phys. Condens. Matter.* **2020**, *32*, 165902. [CrossRef] [PubMed]
35. Datta, S. Nanoscale device modeling: The green's function method. *Superlattices Microstruct.* **2000**, *28*, 253–278. [CrossRef]

36. Marian, D.; Marin, E.G.; Perucchini, M.; Iannaccone, G.; Fiori, G. Multi-scale simulations of two dimensional material based devices: The nanotcad vides suite. *J. Comput. Electron.* **2023**, *22*, 1327–1337. [CrossRef]
37. Huang, Y.; Wang, R.; Zhang, Y.; Yang, D.; Pi, X. Compensation of p-type doping in al-doped 4h-sic. *J. Appl. Phys.* **2022**, *131*, 185703. [CrossRef]

electronics

MDPI

Article

A Geometrically Scalable Lumped Model for Spiral Inductors in Radio Frequency GaN Technology on Silicon

Simone Spataro [1], Giuseppina Sapone [1], Marcello Giuffrida [1] and Egidio Ragonese [2,*]

1 STMicroelectronics, 95121 Catania, Italy; simone.spataro@st.com (S.S.); giuseppina.sapone@st.com (G.S.); marcello.giuffrida@st.com (M.G.)
2 Dipartimento di Ingegneria Elettrica Elettronica e Informatica (DIEEI), University of Catania, 95123 Catania, Italy
* Correspondence: egidio.ragonese@unict.it; Tel.: +39-095-738-2331

Abstract: This paper presents a lumped scalable model for spiral inductors in a radio frequency (RF) gallium nitride (GaN) technology on silicon substrate. The model has been developed by exploiting electromagnetic (EM) simulations of geometrically scaled spiral inductors. To this aim, the technology substrate, i.e., the metal back-end-of-line along with dielectric and semiconductor layers of the adopted GaN process, has been validated by means of experimental data and then used to define the EM simulator set-up for the spiral inductors. The proposed model adopts a simple π-topology with only seven lumped components and predicts inductor performance in terms of inductance, quality factor (*Q*-factor) and self-resonance frequency (*SRF*) for a large range of geometrical parameters of the spiral (i.e., number of turns, metal width, inner diameter).

Keywords: electromagnetic (EM) simulations; gallium nitride (GaN); radio frequency (RF)-integrated circuits; spiral inductors

check for updates

Citation: Spataro, S.; Sapone, G.; Giuffrida, M.; Ragonese, E. A Geometrically Scalable Lumped Model for Spiral Inductors in Radio Frequency GaN Technology on Silicon. *Electronics* **2024**, *13*, 2665. https://doi.org/10.3390/electronics13132665

Academic Editors: Yimeng Zhang, Yuming Zhang and Lejia Sun

Received: 3 June 2024
Revised: 2 July 2024
Accepted: 5 July 2024
Published: 7 July 2024

1. Introduction

Radio frequency (RF) integrated circuits (ICs) extensively employ inductive components (i.e., inductors and transformers) to implement crucial functionality, such as 50-ohm impedance matching at the input/output chip terminals, impedance matching in multi-stage amplifiers, tuned resonant load in the RF amplifiers, LC tank in voltage-controlled oscillators, LC filters, etc. [1]. For this reason, modern RF silicon technologies adopt a proper back-end-of-line (BEOL) with thick metals and thick intermetal oxides to reduce the resistive losses and the parasitic capacitances, respectively [2]. Typically, RF silicon technologies provide general purpose inductor devices into the process design kit (PDK). However, scalable models or stack-up descriptions of electromagnetic (EM) simulations are required for a customized inductor design [3,4].

In the last few years, GaN-based High-Electron-Mobility Transistors (HEMTs) have demonstrated excellent performance for high-power applications thanks to the wide bandgap and the crystal structure of GaN, which allow achieving both high voltage and high currents in a wide temperature operative range. Despite the vast literature available for GaN-based technologies, very few papers deal with inductive components (i.e., inductors and transformers) in RF GaN processes, either on silicon [5,6] or on SiC substrates [7–9]. Moreover, to the authors' knowledge, very scarce literature is available for inductor modeling in GaN technologies since only in [6] has a lumped scalable model been proposed, whereas [9] presents an equivalent circuit extracted for a single geometry. Therefore, unlike silicon-based inductors and transformers [10–18], there is a lack of knowledge on the analysis, design and modeling of inductive devices in GaN technologies. This study aims to partially cover this gap.

This paper is organized as follows. Section 2 briefly describes the adopted RF GaN technology on a Si substrate. Section 3 summarizes the physical phenomena taking place in integrated spiral inductors with the aim of providing the reader with the main design guidelines. The proposed scalable lumped model for spiral inductors in RF GaN technology is detailed in Section 4, while its validation is reported in Section 5. Finally, the main conclusions are summarized in Section 6.

2. RF GaN Technology

The adopted technology is an RF GaN process fabricated on a thin silicon (Si) substrate, which has been developed by STMicroelectronics mainly for sub-6 GHz band power applications [19–22]. The process features 50 V HEMTs with a gate length of 0.5 µm. Since the technology has been specifically addressed to RF applications, transistors have been optimized for normally-on depletion operation, thus exploiting the total two-dimensional electron gas (2DEG) density at the AlGaN/GaN heterointerface. The technology also provides resistors and high-quality-factor (*Q*-factor) metal–insulator–metal (MIM) capacitors for the integration of biasing and matching networks, respectively, while allowing for the fabrication of polygonal/circular spiral inductors and interleaved/stacked transformers that are essential for advanced RF ICs [23–29]. To this aim, the technology BEOL provides three AlCu metals, two thick layers at the top of the stack, which are used for the inductor coil and the underpass, along with a lower thinner one to access the transistor terminals. The upper metal layers, namely MTL3 and MTL2, have equal thickness to avoid *Q*-factor degradation on one side and a bottleneck for the maximum current due to the underpass connection on the other side. Moreover, the two thick metals are suitable for the implementation of high-performance integrated transformers, especially for staked configurations. For the sake of completeness, Figure 1 shows the Scanning Electron Microscope (SEM) microphotograph of the upper metal layers (i.e., MTL3 and MTL2) as they are connected in the underpass region of a spiral inductor.

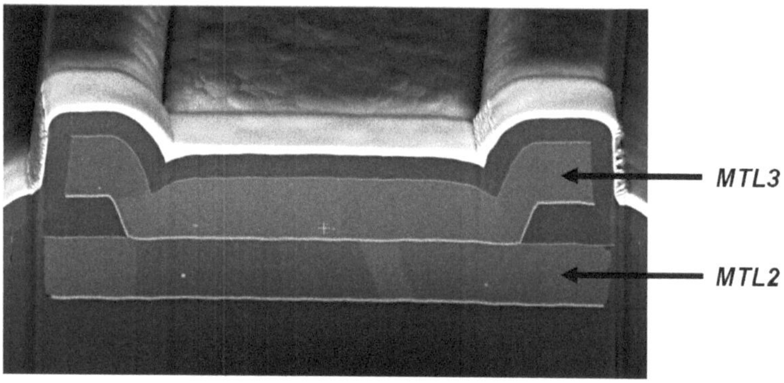

Figure 1. A SEM microphotograph of an inductor in the RF GaN technology (underpass region).

The adopted RF GaN technology enables the design and implementation of watt-level sub-6 GHz power amplifiers (PAs) for base station applications. The final target is replacing the traditional micromodule approach based on discrete components with a fully integrated system, thus achieving a significant reduction in cost and size. This study has been developed in the framework of the development of the above-described RF GaN technology with the aim of driving the implementation of a geometrically scaled library of spiral inductors for RF applications, while providing a reliable design tool for inductor customization into the PDK of the technology.

3. Integrated Inductor Physics: Design Guidelines

The design of integrated inductors is aimed at increasing both the *Q*-factor and the self-resonance frequency (*SRF*) of the component by minimizing the energy losses and the parasitic capacitances, respectively. Energy dissipation occurs into both the metal windings and the underneath layers (i.e., oxides, insulators, diffused semiconductors, etc.). On the other hand, the operative frequency range of the inductor is limited by the interwinding capacitances and especially by the area/perimeter capacitances of the spiral toward the substrate.

Figure 2 depicts the main EM phenomena taking place in an integrated spiral inductor [1]. The losses in the metal layers are mainly due to the current crowding, i.e., the non-uniform distribution of the current in a conductor, which is particularly exacerbated by the peculiar structure of a spiral inductor. Indeed, the current crowding is higher on the internal sides of the spiral where the magnetic field is at its maximum. This phenomenon is the result of the superimposition of the skin and proximity effects, and rises with the increase in the operative frequency, thus resulting in the frequency dependency of the inductor resistance [30]. Therefore, thick and highly conductive metal layers are mandatory to reduce the inductor series resistance, especially in the lower RF bands, since the effectiveness progressively degrades at higher frequencies. On the other hand, the use of "hollow" spirals is a common rule for high-frequency inductive components to avoid the losses in the inner turns [30,31].

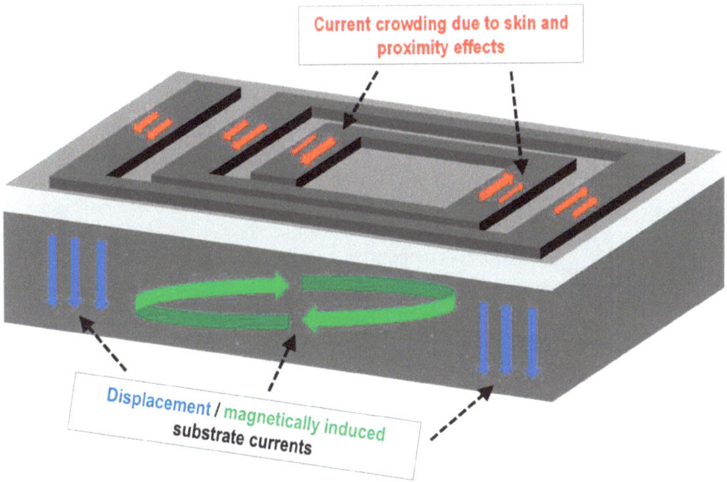

Figure 2. Main EM phenomena in integrated spiral inductor.

Another significant impact on the performance of an integrated inductor is also due to the substrate losses. Indeed, both electric and magnetic fields are significant in the layers underneath the inductor. Specifically, two different mechanisms take place, as represented in Figure 2 using blue and green arrows. The vertical currents (blue arrows) injected into the substrate are the displacement currents (i.e., due to the electrical field and therefore related to the capacitive effects), while the horizontal currents (green arrows) flowing into the conductive substrate layers are magnetically induced currents (i.e., due to the time-varying electrical field according to the Faraday–Lenz law). Both electrically and magnetically induced currents increase at radio frequencies and their effect can be dominant with respect to the ohmic losses into the spiral, especially for large inductors. Specifically, in GaN technology, the bidimensional electron gas in the AlGaN/GaN heterojunction must be properly neutralized below the inductor footprint to avoid the flowing of magnetically induced eddy currents and consequent energy losses [22].

An important role in enlarging the operative frequency range of an integrated inductor or equally in increasing the maximum inductance value that can be used at a given working frequency is played by the minimization of the parasitic capacitances. Typically, the main contributions are due to the electrical coupling with the substrate, especially when its thickness is reduced for thermal reasons, such as in the adopted GaN technology. On the other hand, a thinned substrate helps in reducing the losses due to the displacement currents flowing toward the ground. The main parasitic capacitances are related to the overlap regions between the spiral and the underpass, especially for large-width metal inductors. When the metal spacing is higher than a few microns, the fringing winding capacitances are lower, but their contribution could become significant for inductors with a high perimeter/area ratio (i.e., low w, high n and d_{IN}).

4. Geometrically Scalable Lumped Model Description

The model has been developed to predict the electrical performance of circular inductors for a large range of geometrical parameters to comply with circuit designer demands in RF IC optimization. A circular shape is preferred since it guarantees a better Q-factor performance than squared or polygonal shapes [1]. Although in nanoscale CMOS, the circular shape is forbidden [2,27–29], in the adopted GaN process, circular spirals can be manufactured without any process problems, which do occur in other technologies with wider lithography [16,32,33]. Due to the lack of a sufficiently high number of available fabricated inductors, model parameter extraction has been carried out by using 2D EM simulations of geometrically scaled circular inductors in the adopted GaN technology. To increase the accuracy of the simulations, the process substrate (i.e., the metal back-end-of-line along with dielectric and semiconductor layers of adopted GaN technology) has been validated by means of experimental data of microstrips fabricated in the three metal layers (i.e., MTL1, MTL2 and MTL3).

Figure 3 shows the micrograph of a MTL1 microstrip fabricated on the adopted GaN technology on a 60-μm thick silicon substrate. To achieve an accurate estimation of both the inductance and Q-factor, a frequency dispersity model of the silicon substrate has also been included in the EM simulations. Specifically, the Svensson/Djordjevic model has been used [34]. The comparisons between measured and EM-simulated scattering parameters, S_{21} and S_{11}, are reported in Figure 4 for the sake of completeness. The EM simulated S-parameters are in good agreement with the measured ones over a wide frequency range, especially for S_{21}. The errors are quite low and largely acceptable for our purpose.

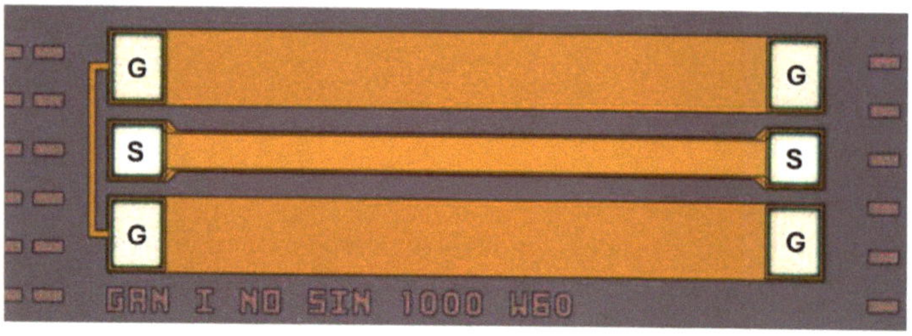

Figure 3. A microphotograph of a MTL1 microstrip fabricated in the adopted RF GaN technology on a 60-μm thick silicon substrate (w = 60 μm).

As an example, a three-turn circular spiral inductor with the main geometrical parameters (i.e., the metal width, w, the inner diameter, d_{IN}, and the metal spacing, s) is shown in Figure 5.

The proposed model adopts a simple π-topology, shown in Figure 6, which employs only geometrically scalable circuit parameters. The ideal inductor, L, is used to model the spiral and underpass inductances, as well as the magnetic coupling within the substrate. Two capacitors, C_{P1} and C_{P2} (whose values are different due to asymmetric inductor layout), are adopted considering the capacitive effects toward the substrate. Moreover, a further capacitor, C_S, is exploited to account for other capacitive effects. Specifically, parasitic capacitances throughout the inductor turns along with the overlap between the spiral and the underpass are considered. Finally, ohmic and substrate losses are modeled by means of frequency-variable resistances, R_S and R_P, respectively.

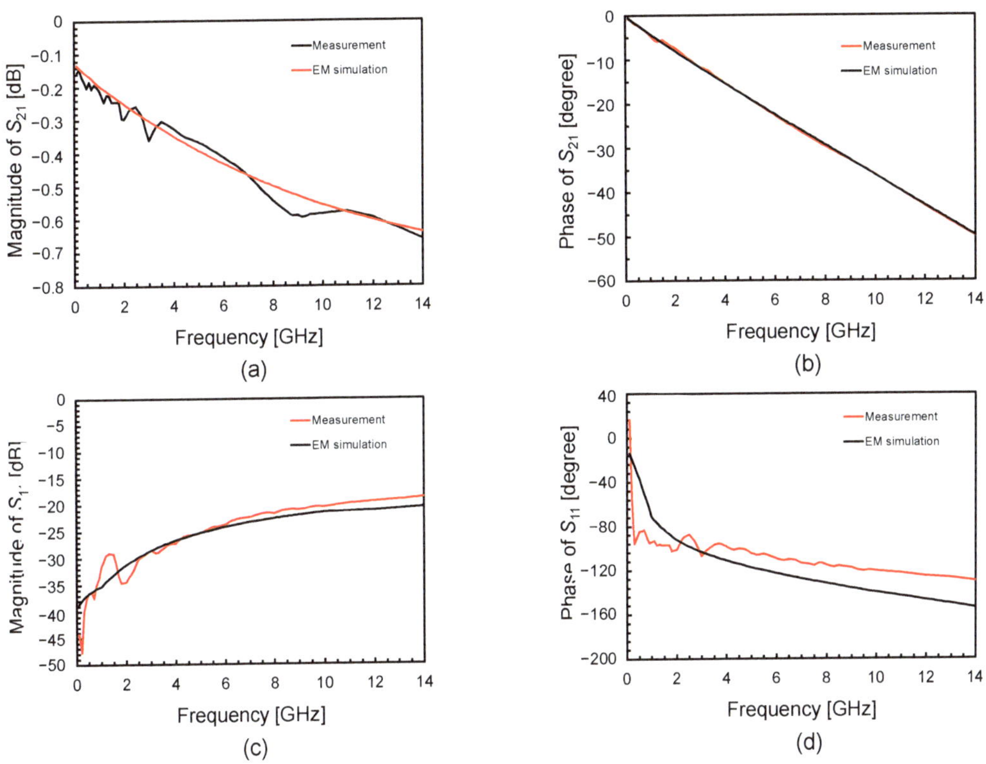

Figure 4. A comparison between the measured and EM-simulated S-parameters of the microstrip in Figure 3: the magnitude of S_{21} (**a**), the phase of S_{21} (**b**), the magnitude of S_{11} (**c**) and the phase of S_{11} (**d**).

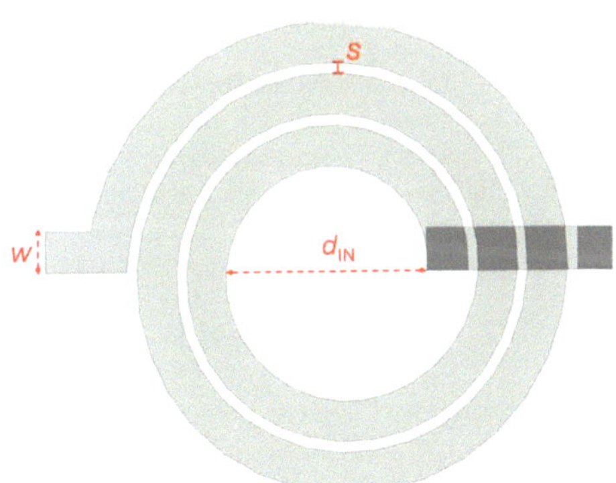

Figure 5. Layout and geometrical parameters of typical circular RF inductor.

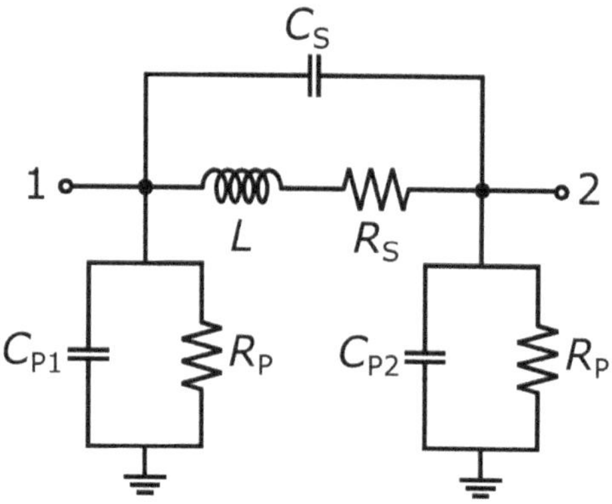

Figure 6. Proposed lumped model for spiral inductors in GaN technology on Si (Terminal 2 is the underpass of the inductor).

The inductance value, L, is the sum of three contributions,

$$L = L_{COIL} + L_{UND} + L_{SUB}, \tag{1}$$

where L_{COIL} and L_{UND} are the inductances of the spiral and the underpass, respectively, while L_{SUB} is used to account for the induced magnetic field in the substrate in a simple way [35–37]. The first two terms of (1) can be calculated by using well-known closed form expressions available in the literature [38,39], whereas L_{SUB} has been evaluated by using the following monomial formula:

$$L_{SUB} = -4\pi\alpha n^2 d_{AVG}^2 w \tag{2}$$

where d_{AVG} is the average inductor diameter and α is a fitting coefficient of around 0.92.

It should be noted that L_{SUB} is a negative value since a reduction in the total inductance is expected due to the induced magnetic field within the substrate.

Capacitive contributions (i.e., C_{S} and $C_{\text{P1,2}}$) determine the inductor *SRF* and can be extracted from EM data. The following scalable equation has been used for C_{S}:

$$C_S = \frac{1}{N_{OVER}} \frac{\varepsilon_M w^2}{t_M} p(n) \tag{3}$$

where N_{OVER} is the number of overlaps between the spiral and the underpass (i.e., N_{OVER} = n + 1.5), and ε_{M} and t_{M} are the electric permittivity and the thickness of the intermetal (i.e., MTL3-MTL2) dielectric layer, respectively. It is worth noting that $\varepsilon_M w^2 / t_M$ is the capacitance of a single overlap between the spiral and the underpass. To improve the geometrical scalability of (3), the fitting function $p(n)$ has also been included, which adopts the second-order polynomial expression reported in (4).

$$p(n) = 0.9\left(n^2 - 3n + 4\alpha\right) \tag{4}$$

Substrate capacitances, $C_{\text{P1,2}}$, are calculated by means of the following equation:

$$C_{P1,2} = \frac{C_{SUB} A_{COIL}}{2} \left[1 + \left(1 + \alpha\right) e^{-\beta w}\right] \tag{5}$$

where C_{SUB} is the substrate capacitance per unit area, A_{COIL} is the coil area and β is an experimental coefficient estimated to be 3.53×10^4. The coil area, A_{COIL}, is given by the product of the width, w, and the spiral length, l_{COIL}, which is calculated as follows:

$$l_{COIL} = \pi n[d_{in} + w + (n - 0.5)(w + s)] \quad (6)$$

Since spirals are not symmetric, the capacitances C_{P1} and C_{P2} must be slightly different to properly model the two different values of the self-resonance frequencies of the inductor, SRF_1 and SRF_2, when terminal 1 or 2 is grounded, respectively. To account for this asymmetry, l_{COIL} is calculated with Equation (6) for C_{P1}, while it is adjusted according to (7) for C_{P2}.

$$l_{COIL2} = \pi n \left[d_{in} + w + \left(\frac{7}{12}n - 0.5\right)(w + s)\right] \quad (7)$$

It is worth noting that when the metal width, w, increases, the capacitance value given by (5) reduces to its ideal value since the fringing effects tend to be negligible.

The series resistance, R_S, is modeled by using a roughly parabolic law according to Equation (8):

$$R_s = R_{DC}\left[1 + \left(\frac{freq}{f_{RO}} - \frac{freq^2}{2f_{RO}f_{RM}}\right)\left(\frac{f_{RO}}{2f_{RM}} - 1\right)^{-1}\right] \quad (8)$$

where R_{DC} is the overall dc resistance of the inductor, which can easily be calculated by using the metal sheet resistance for the number of squares (i.e., l_{COIL}/w), while f_{RM} and f_{RO} are the frequencies for which R_S assumes its maximum and zero, respectively. Since both f_{RM} and f_{RO} cannot be easily predicted from the inductor geometrical parameters, the following empirical formulas have been used:

$$f_{RO} \approx \left(1 - 2200\sqrt{w \cdot d_{in}}\right) \cdot SRF \quad (9)$$

$$f_{RM} \approx \frac{0.04 \cdot SRF}{w^{0.1} \cdot n^{0.23} \cdot d_{in}^{0.12}} \quad (10)$$

Finally, substrate losses are modeled by means of R_P that is calculated by using the following expression:

$$R_P = \frac{1}{6} \frac{\rho_{Si} t_{Si}}{\pi \left(\frac{d_{OUT}}{2}\right)^2} \frac{f_0}{freq} \quad (11)$$

where ρ_{Si} and t_{Si} are the resistivity and the thickness of the silicon substrate, d_{OUT} is the inductor outer diameter and f_0 is equal to 3.5 GHz.

5. Model Validation

The proposed lumped model has been validated by comparison with the EM data of geometrically scaled inductors whose parameters are listed in Table 1. The model can predict the performance of circular inductors with 10-µm metal spacing, a number of turns from 1.5 to 6.5, a metal width from 10 µm to 100 µm, and inner diameters from 60 µm to 300 µm. Specifically, the large variability in the metal width is essential to demonstrate the soundness of the model not only for low-current inductors but also for high-current inductors required in the power amplifiers. The errors in low-frequency inductance, peak Q-factor, Q_{MAX}, and SRF are summarized in Table 2.

Table 1. Layout parameters of spiral inductors for model validation.

Inductors	1	2	3	4	5	6	7	8	9	10
w [µm]	10	10	20	20	40	40	60	60	100	100
n	3.5	6.5	1.5	3.5	1.5	2.5	1.5	2.5	1.5	1.5
d_{in} [µm]	60	60	220	150	120	200	290	240	100	300

Table 2. Model errors of for low-frequency inductance, Q_{MAX}, and *SRF*.

Errors	Maximum	Minimum
L @ 100 MHz	9.2%	1.7%
Q_{MAX}	16.4%	3.6%
SRF	12.3%	0.3%

As an example of the very good estimation of inductor performance, comparisons for inductors 2, 6 and 10 are reported in Figures 7a, 7b and 7c, respectively. It is worth noting that since the model covers a very wide range of geometrical parameters in terms of n, w and d_{IN}, the inductors selected for the comparison in Figure 7 represent very different geometries for a fair validation of the model. The prediction of the *SRF* for inductor 2 is less accurate than for the other inductors due to the fringing capacitance of the spiral that has a higher perimeter/area ratio. Generally, the model is more accurate with larger values of w. Indeed, the main aim is to guarantee very high accuracy for large-width inductors mainly used in PAs.

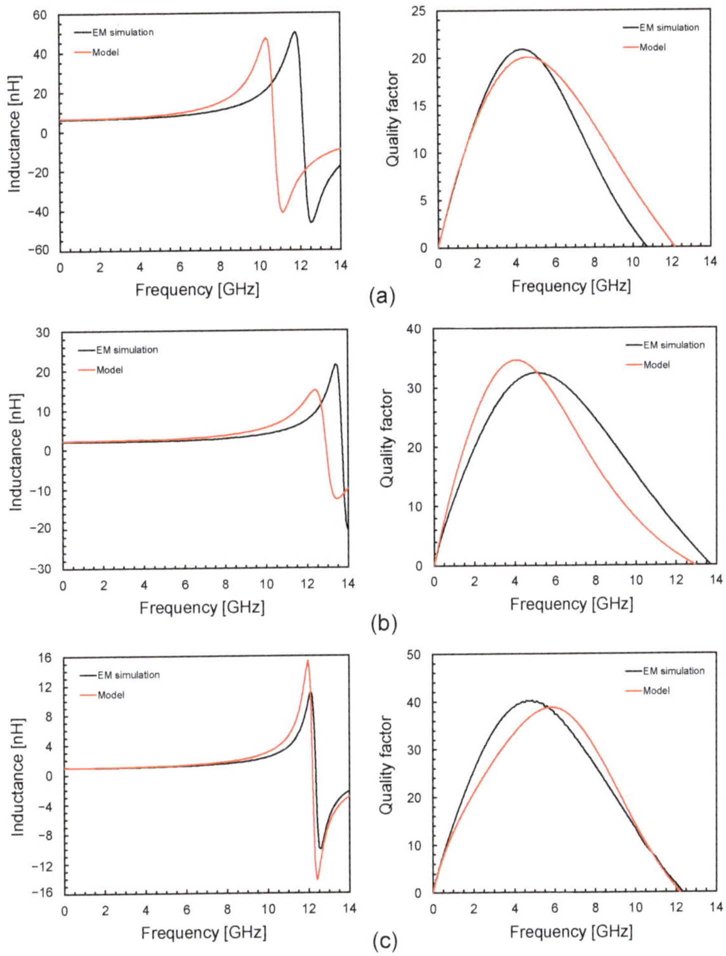

Figure 7. Comparison between proposed geometrically scaled lumped model and EM simulations for three circular inductors: (**a**) inductor 2, (**b**) inductor 6 and (**c**) inductor 10.

Figure 8 depicts the microphotograph of inductor 7 fabricated in the adopted RF GaN technology for on-wafer characterization [15]. Figure 9 compares the proposed model, the EM-simulated data and the experimental data for inductor 7. The agreement between measurements and EM simulations further confirms the soundness of the EM simulation set-up adopted to develop the inductor model, as defined in Section 4. Moreover, the comparison highlights the accuracy of the proposed lumped model with respect to the experimental measurements.

Figure 8. Microphotograph of inductor 7 test structure in adopted RF GaN technology.

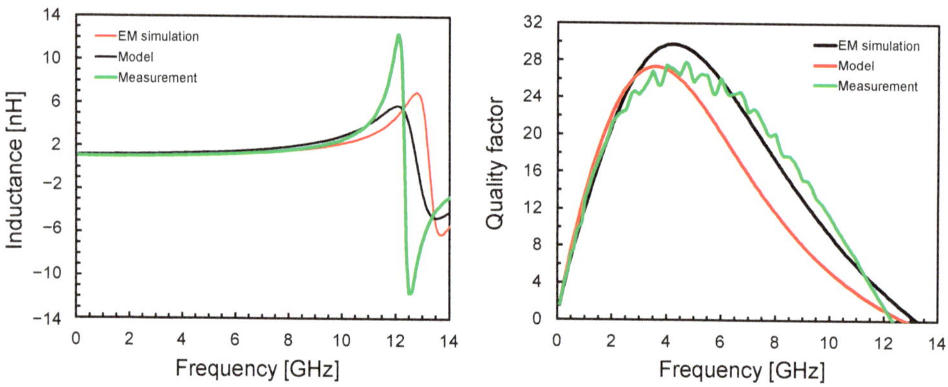

Figure 9. Comparison between proposed geometrically scaled lumped model, EM simulation and on-wafer measurement for inductor 7.

It is interesting to highlight some remarkable differences between this study and a recently published study [6] on a similar topic (i.e., modeling of GaN on Si inductors). Indeed, although a larger number of inductors with different shapes have been analyzed in [6], the model proposed here has been developed to cover a much wider range of the metal trace width. This circumstance is reflected in the explicit dependency of the model equations, specifically (2) and (5), on the layout parameter, w. Moreover, a considerable difference between the two studies concerns the modeling methodology itself. Indeed, a pure analytical approach has been followed in [6], using a lot of circuit elements to account for the several physical phenomena occurring in the metal and substrate layers. On the other hand, the proposed scalable model adopts a much easier equivalent circuit, consisting of only seven lumped components and starting from some theoretical considerations, exploits experimental coefficients for fitting.

It is worth briefly comparing this study with [6,9]. Despite its simplicity and wider geometrical scalability, the proposed model turns out to be broadly comparable to the model in [6] in terms of percentage errors in the main performance parameters. On the other hand, the model in [9] is more complex and accurate, exploiting customized parameter extraction from single inductor measurements, but it is not geometrically scalable. As already mentioned, the key feature of the proposed model is to predict inductor performance for a wide range of geometrical parameters with very good accuracy.

6. Conclusions

A simple yet effective lumped model for circular inductors has been developed in a 0.5-μm RF GaN technology for sub-6 GHz power applications. The model adopts only seven lumped components and is fully geometrically scalable. It demonstrates very good accuracy in the prediction of inductor performance (i.e., inductance, *Q*-factor and *SRF*) for a wide range of geometrical parameters. Both the geometrical scalability and network simplicity of the proposed model allow for its use in actual RF IC design. Thanks to its simplicity, the model can be used in other RF GaN technologies provided that proper tuning for the fitting parameters is performed to adapt it to a different BEOL. To the authors' knowledge, this is the very first model for spiral inductors in RF GaN technology guaranteeing state-of-the-art accuracy for such a wide range of geometrical parameters.

Author Contributions: Conceptualization, S.S., G.S. and E.R.; validation, S.S. and G.S.; formal analysis, S.S.; methodology, S.S., G.S. and E.R.; project administration, E.R. and M.G.; supervision, E.R., G.S. and M.G.; writing—original draft, E.R.; writing—review and editing, E.R., S.S. and M.G. All authors have read and agreed to the published version of the manuscript.

Funding: This study was partially supported by STMicroelectronics under the Research Contract with the DIEEI titled "Progettazione e ottimizzazione di componenti passivi reattivi integrati per applicazioni 5G in tecnologia GaN on Si e relativa attività propedeutica per rilascio di un futuro design kit".

Data Availability Statement: The data presented in this study are available on request from the corresponding author.

Conflicts of Interest: Authors Simone Spataro, Giuseppina Sapone and Marcello Giuffrida are employed by the company STMicroelectronics. The remaining authors declare that the research was conducted in the absence of any commercial or financial relationships that could be construed as a potential conflict of interest.

References

1. Scuderi, A.; Ragonese, E.; Biondi, T.; Palmisano, G. *Integrated Inductors and Transformers: Characterization, Design and Modeling for RF and MM-Wave Applications*; CRC Press: Boca Raton, FL, USA; Taylor & Francis Group: Boca Raton, FL, USA, 2010.
2. Ragonese, E.; Nocera, C.; Cavarra, A.; Papotto, G.; Spataro, S.; Palmisano, G. A Comparative analysis between standard and mm-wave optimized BEOL in a nanoscale CMOS technology. *Electronics* **2020**, *9*, 2124. [CrossRef]
3. Salnikov, A.S.; Goryainov, A.E.; Dobush, I.M.; Kalentyev, A.A.; Garays, D.V. Approach to scalable modeling for planar inductor using EM simulation and a few samples measurement. In Proceedings of the 2017 IEEE MTT-S International Conference on Numerical Electromagnetic and Multiphysics Modeling and Optimization for RF, Microwave, and Terahertz Applications (NEMO), Seville, Spain, 17–19 May 2017; pp. 55–57.
4. Dobush, I.M.; Vasil'evskii, I.S.; Zykov, D.D.; Bragin, D.S.; Salnikov, A.S.; Popov, A.A.; Gorelov, A.A.; Kargin, N.I. Development of a 0.15 μm GaAs pHEMT process design kit for low-noise applications. *Electronics* **2021**, *10*, 2775. [CrossRef]
5. Eblabla, A.; Li, X.; Wallis, D.J.; Guiney, I.; Elgaid, K. High-performance MMIC inductors for GaN-on-low-resistivity silicon for microwave applications. *IEEE Microw. Wirel. Compon. Lett.* **2018**, *28*, 99–101. [CrossRef]
6. Montesdeoca, M.S.M.; Angulo, S.M.; Duarte, D.M.; Del Pino, J.; García, J.A.G.Y.; Khemchandani, S.L. An analytical scalable lumped-element model for GaN on Si inductors. *IEEE Access* **2020**, *8*, 52863–52871. [CrossRef]
7. Chander, S.; Bansal, K.; Gupta, S.; Gupta, M. Design and analysis of high performance air-bridge spiral circular inductors for GaN MMICs up to Ku band. In Proceedings of the 2017 Devices for Integrated Circuit (DevIC), Kalyani, India, 23–24 March 2017; pp. 734–736.
8. Chander, S.; Bansal, K.; Gupta, S.; Gupta, M. Design and analysis of spiral circular inductors for GaN based low noise amplifier (MMICs). In Proceedings of the 2015 International Conference on Microwave, Optical and Communication Engineering (ICMOCE), Bhubaneswar, India, 18–20 December 2015; pp. 292–294.

9. Wang, H.-S.; He, W.-L.; Wang, R.-D.; Zhang, M.-H. A double-π equivalent circuit model for GaN on-chip inductors. In Proceedings of the IEEE International Conference on Solid-State and Integrated Circuit Technology, Hangzhou, China, 25–28 October 2016; pp. 811–815.
10. Long, J.R.; Copeland, M.A. The modeling, characterization, and design of monolithic inductors for silicon RF IC's. *IEEE J. Solid-State Circuits* **1997**, *32*, 357–369. [CrossRef]
11. Kapur, S.; Long, D.E. Modeling of integrated RF passive devices. In Proceedings of the IEEE Custom Integrated Circuits Conference, San Jose, CA, USA, 19–22 September 2010; pp. 1–8.
12. Cavarra, A.; Nocera, C.; Papotto, G.; Ragonese, E.; Palmisano, G. Transformer design for 77-GHz down-converter in 28-nm FD-SOI CMOS technology. In *Applications in Electronics Pervading Industry, Environment and Society, ApplePies 2018*; Saponara, S., De Gloria, A., Eds.; Lecture Notes in Electrical Engineering; Springer: Cham, Switzerland, 2019; Volume 573.
13. Huo, X.; Chan, P.C.H.; Chen, K.J.; Luong, H.C. A physical model for on-chip spiral inductors with accurate substrate modeling. *IEEE Trans. Electron Devices* **2006**, *53*, 2942–2949. [CrossRef]
14. Sia, C.B.; Ong, B.H.; Chan, K.W.; Yeo, K.S.; Ma, J.G.; Do, M.A. Physical layout design optimization of integrated spiral inductors for silicon-based RFIC applications. *IEEE Trans. Electron Devices* **2005**, *52*, 2559–2567. [CrossRef]
15. Biondi, T.; Scuderi, A.; Ragonese, E.; Palmisano, G. Characterization and modeling of silicon integrated spiral inductors for high-frequency applications. *Analog. Integr. Circuits Signal Process.* **2007**, *51*, 89–100. [CrossRef]
16. Ragonese, E.; Scuderi, A.; Biondi, T.; Palmisano, G. Scalable lumped modeling of single-ended and differential inductors for RF IC Design. *Wiley Int. J. RF Microw. Comput.-Aided Eng.* **2009**, *19*, 110–119.
17. Biondi, T.; Scuderi, A.; Ragonese, E.; Palmisano, G. Analysis and modeling of layout scaling in silicon integrated stacked transformers. *IEEE Trans. Microw. Theory Tech.* **2006**, *54*, 2203–2210. [CrossRef]
18. El-Gharniti, O.; Kerherve, E.; Begueret, J.-B. Modeling and characterization of on-chip transformers for silicon RFIC. *IEEE Trans. Microw. Theory Tech.* **2007**, *55*, 607–615. [CrossRef]
19. Cerantonio, V.; Giuffrida, M.; Miccoli, C.; Chini, A.; Iucolano, F. From T-CAD simulations to large signal model for GaN RF device. In Proceedings of the 2020 AEIT International Conference of Electrical and Electronic Technologies for Automotive (AEIT AUTOMOTIVE), Turin, Italy, 8–10 November 2020; pp. 1–6.
20. Giorgino, G.; Cioni, M.; Miccoli, C.; Gervasi, L.; Giuffrida, M.F.S.; Ruvolo, M.; Castagna, M.E.; Cappellini, G.; Luongo, G.; Moschetti, M.; et al. Study of 100V GaN power devices in dynamic condition and GaN RF device performances in sub-6GHz frequencies. *Elsevier E-Prime Adv. Electr. Eng. Electron. Energy* **2023**, *6*. [CrossRef]
21. Vandendaele, W.; Leurquin, C.; Lavieville, R.; Jaud, M.A.; Viey, A.G.; Gwoziecki, R.; Mohamad, B.; Nowak, E.; Constant, A.; Iucolano, F. Reliability of GaN MOSc-HEMTs: From TDDB to threshold voltage instabilities (Invited). In Proceedings of the IEEE International Reliability Physics Symposium (IRPS), Monterey, CA, USA, 26–30 March 2023; pp. 1–8.
22. Scandurra, A.; Testa, M.; Franzò, G.; Greco, G.; Roccaforte, F.; Castagna, M.E.; Calabretta, C.; Severino, A.; Iucolano, F.; Bruno, E.; et al. Isolation of bidimensional electron gas in AlGaN/GaN heterojunction using Ar ion implantation. *Mater. Sci. Semicond. Process.* **2023**, *168*, 107871. [CrossRef]
23. Santhakumar, R.; Pei, Y.; Mishra, U.K.; York, R.A. Monolithic millimeter-wave distributed amplifiers using AlGaN/GaN HEMTs. In Proceedings of the IEEE MTT-S Int. Microwave Symposium Digest, Atlanta, GA, USA, 15–20 June 2008; pp. 1063–1066.
24. Korndorfer, F.; Muhlhaus, V. Lumped modeling of integrated MIM capacitors for RF applications. In Proceedings of the Microwave Measurement Conference (ARFTG), Austin, TX, USA, 8–9 December 2016; pp. 1–4.
25. Bevilacqua, A. Fundamentals of integrated transformers: From principles to applications. *IEEE Solid-State Circuits Mag.* **2020**, *12*, 86–100. [CrossRef]
26. Long, J.R. Monolithic transformers for silicon RF IC design. *IEEE J. Solid-State Circuits* **2000**, *35*, 1368–1382. [CrossRef]
27. Giammello, V.; Ragonese, E.; Palmisano, G. Transmitter chipset for 24/77-GHz automotive radar sensors. In Proceedings of the IEEE Radio Frequency Integrated Circuits Symposium (RFIC) Technical Digest, Anaheim, CA, USA, 23–25 May 2010; pp. 75–78.
28. Nocera, C.; Papotto, G.; Cavarra, A.; Ragonese, E.; Palmisano, G. A 13.5-dBm 1-V power amplifier for W-band automotive radar applications in 28-nm FD-SOI CMOS technology. *IEEE Trans. Microw. Theory Tech.* **2021**, *69*, 1654–1660. [CrossRef]
29. Papotto, G.; Nocera, C.; Finocchiaro, A.; Parisi, A.; Cavarra, A.; Castorina, A.; Ragonese, E.; Palmisano, G. A 27-mW W-band radar receiver with effective TX leakage suppression in 28-nm FD-SOI CMOS. *IEEE Trans. Microw. Theory Tech.* **2021**, *69*, 4132–4141. [CrossRef]
30. Kuhn, W.B.; Ibrahim, N.M. Approximate analytical modeling of current crowding effects in multi-turn spiral inductors. In Proceedings of the 2000 IEEE MTT-S International Microwave Symposium Digest, Boston, MA, USA, 11–16 June 2000; pp. 405–408.
31. Craninckx, J.; Steyaert, M.S.J. A 1.8-GHz low-phase-noise CMOS VCO using optimized hollow spiral inductors. *IEEE J. Solid State Circuits* **1997**, *32*, 736–744. [CrossRef]
32. Greco, N.; Parisi, A.; Lombardo, P.; Spina, N.; Ragonese, E.; Palmisano, G. A double-isolated dc–dc converter based on integrated LC resonant barriers. *IEEE Trans. Circuits Syst. I Reg. Pap.* **2018**, *65*, 4423–4433. [CrossRef]
33. Greco, N.; Parisi, A.; Spina, N.; Ragonese, E.; Palmisano, G. Scalable lumped models of integrated transformers for galvanically isolated power transfer systems. *Integration* **2018**, *63*, 323–331. [CrossRef]
34. Keysight Helpfiles. PCB Material Characterization. Available online: https://helpfiles.keysight.com/csg/N1930xB/ToolsAndUtilities/PCB_Material_Characterization.htm#The_Svensson-Djordjevic_Model (accessed on 3 June 2024).

35. Kuhn, W.B.; Ibrahim, N.M. Analysis of current crowding effects in multiturn spiral inductors. *IEEE Trans. Microw. Theory Tech.* **2001**, *49*, 31–38. [CrossRef]
36. Kuhn, W.B. Loss mechanisms and quality factor improvement for inductors in high-resistivity SOI processes. In Proceedings of the IEEE 12th Topical Meeting on Silicon Monolithic Integrated Circuits in RF Systems, Santa Clara, CA, USA, 16–18 January 2012; pp. 29–32.
37. Kuhn, W.B.; He, S.; Mojarrad, M. Modeling spiral inductors in SOS processes. *IEEE Trans. Electron Devices* **2004**, *51*, 677–683. [CrossRef]
38. Mohan, S.S.; del Mar Hershenson, M.; Boyd, S.P.; Lee, T.H. Simple accurate expressions for planar spiral inductances. *IEEE J. Solid-State Circuits* **1999**, *34*, 1419–1424. [CrossRef]
39. Mohan, S.S. The Design, Modeling and Optimization of on Chip Inductor and Transformer Circuit. Ph.D. Thesis, Stanford University, Stanford, CA, USA, 1999.

electronics

MDPI

Article

Mobility Extraction Using Improved Resistance Partitioning Methodology for Normally-OFF Fully Vertical GaN Trench MOSFETs

Valentin Ackermann [1,2,*], Blend Mohamad [1,*], Hala El Rammouz [1], Vishwajeet Maurya [1], Eric Frayssinet [3], Yvon Cordier [3], Matthew Charles [1], Gauthier Lefevre [2,4], Julien Buckley [1,*] and Bassem Salem [2,*]

1 Univ. Grenoble Alpes, CEA, LETI, F-38000 Grenoble, France; hala.elrammouz@cea.f (H.E.R.); vishwajeet.maurya@cea.fr (V.M.); matthew.charles@cea.fr (M.C.)
2 Univ. Grenoble Alpes, CNRS, CEA/LETI Minatec, Grenoble INP, LTM, F-38054 Grenoble, France; gauthier.lefevre@cea.fr
3 Univ. Côte d'Azur, CNRS, CRHEA, rue Bernard Grégory, 06560 Valbonne, France; eric.frayssinet@crhea.cnrs.fr (E.F.); yvon.cordier@crhea.cnrs.fr (Y.C.)
4 Univ. Grenoble Alpes CEA, LITEN, F-38000 Grenoble, France
* Correspondence: valentin.ackermann@cea.fr (V.A.); blend.mohamad@cea.fr (B.M.); julien.buckley@cea.fr (J.B.); bassem.salem@cea.fr (B.S.)

Abstract: In this work, fully vertical GaN trench MOSFETs were fabricated and characterized to evaluate their electrical performances. Transistors show a normally-OFF behavior with a high I_{ON}/I_{OFF} (~10^9) ratio and a significantly small gate leakage current (10^{-11} A/mm). Thanks to an improved resistance partitioning method, the resistances of the trench bottom and trench channel were extracted accurately by taking into account different charging conditions. This methodology enabled an estimation of the effective channel and bottom mobility of 11.1 $cm^2/V{\cdot}s$ and 15.1 $cm^2/V{\cdot}s$, respectively.

Keywords: GaN; MOSFET; vertical device; resistance partitioning

check for updates

Citation: Ackermann, V.; Mohamad, B.; El Rammouz, H.; Maurya, V.; Frayssinet, E.; Cordier, Y.; Charles, M.; Lefevre, G.; Buckley, J.; Salem, B. Mobility Extraction Using Improved Resistance Partitioning Methodology for Normally-OFF Fully Vertical GaN Trench MOSFETs. *Electronics* **2024**, *13*, 2350. https://doi.org/10.3390/electronics13122350

Academic Editors: Yimeng Zhang, Yuming Zhang and Lejia Sun

Received: 19 May 2024
Revised: 11 June 2024
Accepted: 12 June 2024
Published: 15 June 2024

1. Introduction

Over recent years, vertical gallium nitride (GaN) power transistors have demonstrated increasing potential for efficient power switching applications. The wide-bandgap GaN material is known to have superior intrinsic material properties over its silicon (Si) and silicon carbide (SiC) material counterpart. Indeed, its larger critical electric field, higher electron mobility, and higher saturation velocity favor the technological development of high breakdown voltage and low on-state resistance devices. In addition, the vertical topology is expected to be more adapted for high-power switching applications compared to the lateral one (like the HEMT), due to its capability of reaching higher breakdown voltages, its robustness to high electric fields, and its potentially higher power capability.

Nowadays, different vertical technology concepts have been under development, such as the CAVET [1,2], the semipolar gate structure [3], the fin-FET [4], or the well-known vertical power trench MOSFET [5–7]. Among these architectures, the vertical power trench MOSFET shows even higher potential regarding the benefits granted by MOS gate technology (high switching capability, low gate leakage current, gate robustness to voltage overshoot...).

To improve its quality (and also its control), different research groups have tried to fabricate vertical GaN trench MOSFETs and study the effects of MOS gate module variation on the devices' electrical performance (p-GaN doping [8], GaN trench surface optimization [9,10], GaN trench orientation [7,11], gate stack comparison [12–14]). Some studies have demonstrated, for instance, groundbreaking results with devices showing adequate normally-OFF operation (V_{th} ~ 3–8 V [5,7]), very low $R_{ON,sp}$ (<10 $m\Omega.cm^2$ [6,7]), a high I_{ON}/I_{OFF} ratio (~10^9), a significantly low gate leakage current (<10^{-11} $A.mm^{-1}$

for instance), as well as encouraging channel mobility results (~10–130 $cm^2/V{\cdot}s$ [5,7]). However, for the latter key parameter, the common methodology used in the literature to estimate it is based on the calculation of the transconductance, extracted from the transfer characteristic in the MOSFET linear stage of operation [7,10]. While this method allows us to easily extract the effective field-effect mobility on a single device, its main drawback is that it overestimates the channel mobility value considering the drift layer mobility contribution in the output current value during the mobility calculation.

On the other hand, mobility extraction methods related to the more mature lateral recessed MOS-HEMT use either the transconductance method [15,16], the Y-function method [17], or the resistance partitioning one [18,19]. For the latter method, the resistance and effective mobility are evaluated for the trench bottom and trench sidewalls separately. To do so, sheet resistance and carrier density of the different trench regions must be evaluated under the assumption of a uniform carrier density and identical electrostatic behavior between the trench bottom and trench sidewall. While this methodology can be used in the case of a lateral recessed MOS-HEMT, it is not the case for a vertical GaN MOSFET since it does not take into account the difference in electrostatic behavior between the n^- GaN trench bottom and the p-GaN trench sidewall is this technology.

In this study, we present an improved methodology that aims to separate the resistance contributions in the trench region and then extracts the effective mobilities of the trench channel and trench bottom. Firstly, the fabrication and electrical characterization of fully vertical GaN trench MOSFETs are reported. The transistors' key parameters are extracted by means of vertical I-V and C-V measurements, showing devices with a threshold voltage V_{th} of ~1 V, a significantly small gate leakage I_G (10^{-11} A/mm), and adequate switching capability (I_{ON}/I_{OFF} ~ 10^9). Then, using an improved resistance partitioning methodology that takes into account the difference in electrostatic behavior between the trench bottom and trench channel, we evaluate the effective mobilities of both of these areas as being 15.1 $cm^2/V{\cdot}s$ and 11.1 $cm^2/V{\cdot}s$, respectively.

2. Materials and Methods

The epitaxy layers were grown on a free-standing (FS) 2-in n-type GaN wafer using metal–organic vapor phase epitaxy (MOVPE). From bottom to top, the different doped GaN layers were grown as follows: 100 nm n^+ GaN drain layer (Si, 1×10^{19} cm^{-3}), 10 µm n^- GaN drift layer (Si, 1×10^{16} cm^{-3}), 700 nm p-GaN (Mg, $N_A - N_D = 3.5 \times 10^{18}$ cm^{-3}), and 200 nm n^+ GaN source layer (Si, 6×10^{18} cm^{-3}).

The process flow starts with the deposition of a 1 µm SiO_2 hardmask by PECVD. Then, the drain contact is fabricated with a Ti/Al/Ni/Au metal stack deposited on the wafer backside by e-beam evaporation. The deposition is followed by rapid thermal annealing (RTA) at 750 °C in N_2 atmosphere for 3 min 30 s.

The process continues with the patterning through photolithography of the GaN gate trench and a mesa structure that will terminate the n-p-n heterostructure. Firstly, the resist pattern is transferred into the hardmask by performing CF4 dry etching in an inductive couple plasma reactor (ICP-RIE). Secondly, the GaN trenches and the mesa are etched to a depth of 1.1 µm through the epi layers by ICP-RIE using a Cl_2 dry-etch process. Afterwards, 10 min RTA at 600 °C in O_2 atmosphere is performed for p-GaN layer activation.

To reduce the etch-induced damages and remove the etch residues on the trench sidewalls, a HCl pre-deposition wet surface treatment is applied to the sample for 4 min at an ambient temperature. Immediately after this step, 20 nm of thermal Al_2O_3 is deposited by ALD at 300 °C using a trimethylaluminium (TMA) precursor and H_2O vapor oxidant for the deposition. A total of 40 nm of TiN is then deposited as the gate metal through sputtering. To finalize gate fabrication, the metal gate is patterned and etched on top of the mesa structures by ion beam etching (IBE). Finally, 3 min RTA at 400 °C in N_2 is performed on the sample.

The process flow is followed by the fabrication of the source contacts. A dry-etch fluorocarbide process is used to open the gate oxide as well as the SiO_2 hardmask in the source contact region. The source contact is made of a Ti/Al metal stack defined by a lift-off step to finalize MOSFET processing.

A schematic diagram of a processed vertical MOSFET along with its top-view SEM image is shown in Figure 1a. As can be seen, the resulting MOS gate is located inside the trench as well as all around the mesas, as a consequence of the deposition and etching steps of the gate dielectric and gate metal. The dimension of the expected gate trench width (W_{tr}) varies from 1 to 6 μm depending on the device studied. In addition to these test structures, planar capacitors located on the drift layer are also included in the initial layout, as illustrated in Figure 1.

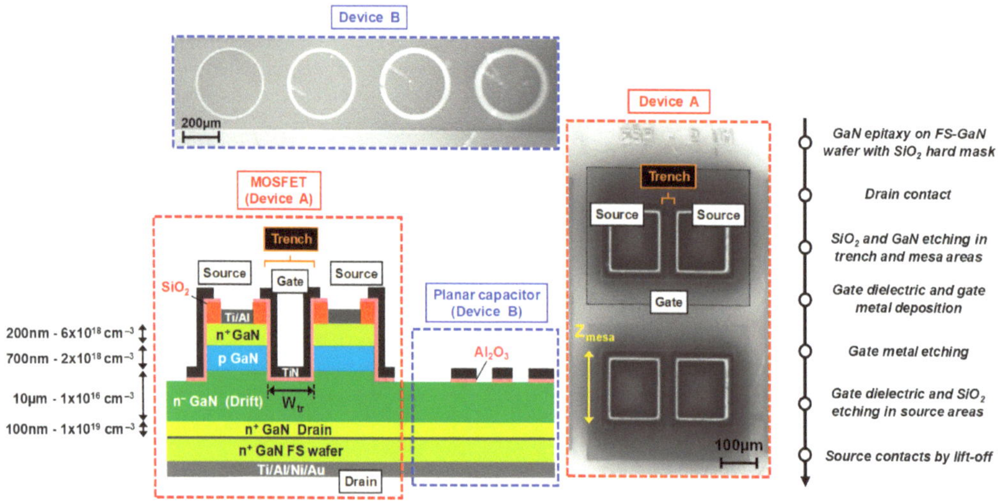

(a) (b)

Figure 1. (**a**) A cross-sectional schematic of a given processed vertical MOSFET (device A) and planar capacitor (device B) test structures, each device being linked to their top-view SEM image (**b**). The main process flow steps to fabricate the vertical MOSFETs and planar capacitor test structures.

The high-angle annular dark field (HAADF)-STEM image in Figure 2a shows the gate of a given processed trench MOSFET. While the dielectric and gate metal deposited seems sufficiently conformal with the GaN surface, etch-induced non-uniformities can still be observed along the trench sidewalls, as observed in Figure 2b.

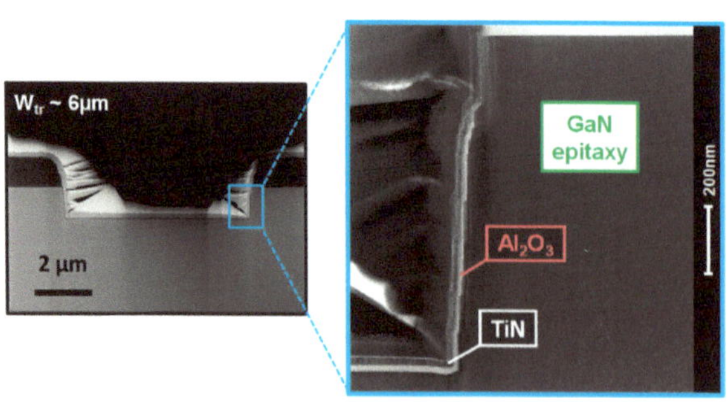

(a) (b)

Figure 2. HAADF-STEM images of (**a**) the cross-section of a given vertical MOSFET and (**b**) the right trench sidewall of the device.

3. Results and Discussion

The transfer characteristics of a given fully vertical MOSFET are shown in Figure 3, with the current values normalized to the mesa width (Z_{mesa}, cf. Figure 1a). The transfer characteristic on a linear scale (Figure 3a) confirms the transistor behavior and demonstrates the normally-OFF switching operation of the device with a threshold voltage (V_{th}) of ~1 V, determined through extrapolation of the linear region of the characteristic at V_D = 2 V (correlation factor *r* of the fitting as being around ~1). As shown in Figure 3b, the device demonstrates a good ON/OFF current ratio of 10^9 and a significantly low gate leakage current of 10^{-11} A/mm, indicating the benefits of both the n-p-n heterostructure and the MOS gate building blocks. The value of the subthreshold slope calculated is around ~139 mV/dec.

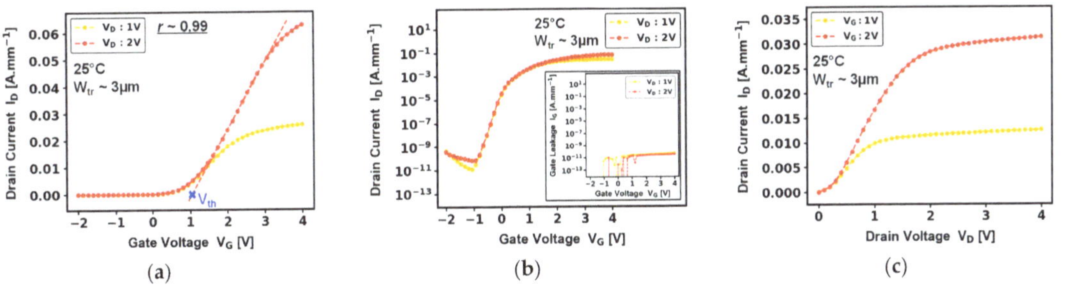

Figure 3. (**a**) Linear and (**b**) semi-log transfer characteristics: gate leakage current versus applied gate voltage; (**c**) output characteristics of given vertical GaN-on-GaN MOSFET.

Focusing on the V_{th} value, different hypotheses can be made to explain its origins. In our case, this V_{th} result could be related to the thinner gate dielectric (~20 nm) compared to what can be found in previous works (80–100 nm [6,7,20]). Also, the presence of a positive charge density trapped at the dielectric/GaN interface or inside the dielectric bulk could have a detrimental effect on the V_{th} by shifting it toward negative values [21–23], thus reducing the V_{th} of the device. Finally, the issue of the insufficient electrically activated Mg doping concentration is still a major process concern nowadays, since a low Mg doping concentration drastically reduces the V_{th} value [6,8,24].

The normalized C-V curves related to both the MOSFET and planar capacitor are illustrated in Figure 4a. Firstly, a different capacitance behavior is clearly noticed as device A is based on charge inversion and device B on charge accumulation. Consequently, a higher V_{th} is observed for device A compared to the V_{FB} for device B. This result can be explained due to the p-type GaN layer integrated into the MOSFET heterostructure that should enhance the work function of the GaN and thus enhance the V_{FB}. A small hysteresis of ~100 and ~47 mV is shown for device A and B, respectively, suggesting a dielectric/GaN interface of good quality, despite the sidewall macroscopic roughness observed in Figure 2b.

Since the capacitance is by definition related to a variation in charge with the applied voltage V_S, one can calculate the charge density (Q_c) simply by combining the capacitance from the onset voltage value V_{onset} (defined either by V_{FB} (device B) or V_{th} (device A)) with the maximum voltage value V_{max} of the applied voltage sweep, as described in Equation (1):

$$Q_c = \int_{V_{onset}}^{V_{max}} C.dV_S \quad (1)$$

Thus, Figure 4b represents the calculated charge density as a function of the applied voltage. Since the onset of the charge inversion (V_{th} ~ 0.14 V) in the channel regions for device A occurs at a different gate voltage than the charge accumulation in the bottom region (V_{FB} ~ −2.64 V), a difference in charge densities is clearly noticed at a given voltage value.

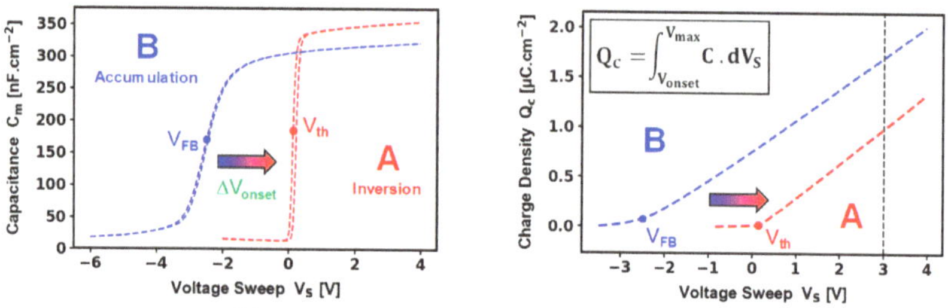

(a) (b)

Figure 4. (**a**) The capacitance–voltage characteristics measured at 10 kHz for device A and device B (cf. Figure 1a for the schematic diagram of the test structures' cross-section). (**b**) The evolution of the calculated characteristics with applied voltage, calculated at the V_{th} (A) or V_{FB} (B) for the same set of devices.

To extract the mobility contributions in the trench region, based on our test structures, we used an improved resistance partitioning method that takes into account the evolution of the output current with the gate trench width [17,18], as well as the different electrostatic conditions of device A and B illustrated in Figure 4. To do so, lateral I_D-V_G characteristics based on the planar MOSFET configuration were measured. In this measurement configuration, one of the source electrodes was considered the drain on our MOSFET devices, which means the current flows laterally from one source to the other.

Figure 5a shows the lateral $I_D - V_G$ characteristics as a function of the gate trench width at $V_D = 0.5$ V. A clear reduction in output current is visible when increasing the gate trench width (i.e., the trench bottom), meaning the total lateral resistance R_{tot} should increase as well with this parameter. Thus, in Figure 5b, a visible linear dependence of the total lateral resistance on the gate trench width can be observed (for I_D values selected at $V_G = 3$ V, symbolized by the dashed line in Figure 5a). Since the current flows laterally (Figure 5c), R_{tot} can be divided into different resistance contributions, defined as follows:

$$R_{tot} = 2R_S + 2R_{ch} + R_{bot}.W_{tr} = 2(R_{acc} + R_{con}) + 2R_{ch} + R_{bot}.W_{tr} \quad (2)$$

where R_S is the source resistance, R_{acc} is the access resistance, R_{con} is the contact resistance, R_{ch} is the channel resistance, and R_{bot} is the trench bottom resistance.

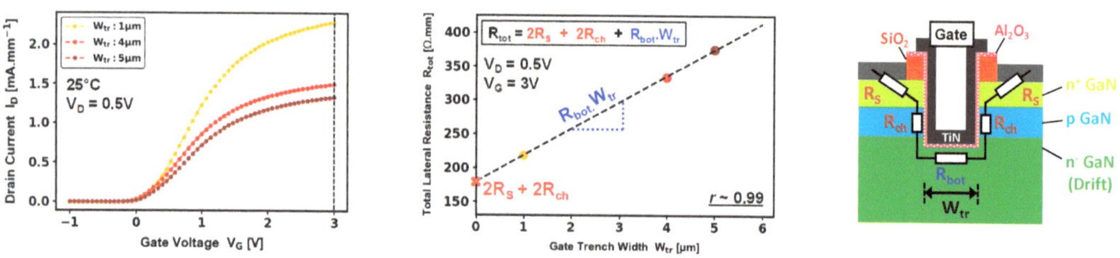

(a) (b) (c)

Figure 5. (**a**) The transfer characteristics of a lateral GaN-on-GaN MOSFET with different gate trench widths, when one of the source contacts of the MOSFET is considered the drain. (**b**) The evolution of the total lateral resistance with the expected gate trench width, calculated for I_D values at $V_G = 3$ V. (**c**) A schematic diagram of resistance partitioning around the trench area in the planar MOSFET configuration.

The R_{bot} term can be directly determined by the slope of the fitted curve shown in Figure 5b and is evaluated as ~39.0 Ω·mm. The sheet resistance $R_{sheet,bot}$ is then easily deduced as being around ~39.0 kΩ·sq. In addition, one can estimate through extrapolation the resistance value for which W_{tr} tends to 0 (i.e., a very thin gate), symbolized by the red cross on the *y*-axis. In that case, the associated resistance value is estimated at around ~179.4 Ω·mm and takes into consideration the source and channel resistance components from the left and right trench sidewalls. The source contribution R_S can then be measured with transfer length measurements (TLMs), as shown in Figure 6. From these measurements, for one source electrode, the contact and access resistance contributions can be calculated, estimated as being around ~7.1 Ω·mm and ~5.9 Ω·mm, respectively.

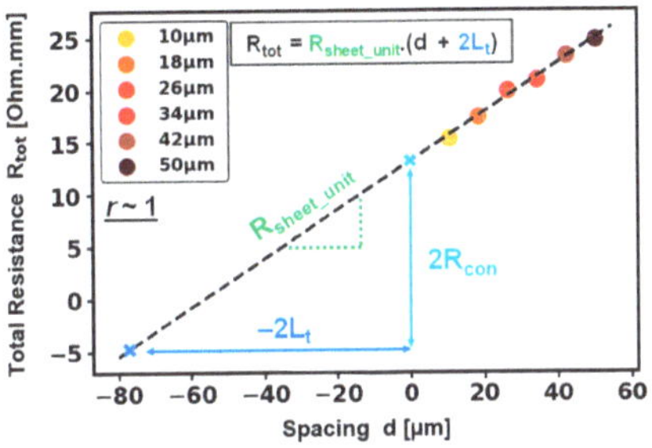

Figure 6. Transfer length measurements on n^+ GaN source layer of GaN-on-GaN epitaxy.

After removing the R_S contribution (measured with TLMs as being around ~13 Ω·mm for one source electrode) from the R_{tot}, and by considering the thickness of the p-GaN layer, we estimated a channel sheet resistance $R_{sheet,ch}$ of ~91.0 kΩ·sq.

At this point, since the resistance contributions and the charge densities are known, one can extract the mobilities (μ) from the trench bottom and trench channel areas with Equation (3):

$$\mu = \frac{1}{QR_{sheet}} = \frac{1}{qNR_{sheet}} \quad (3)$$

where ρ is the material resistivity, Q is the charge density, q is the elemental charge, N is the charge carrier density, and R_{sheet} is the sheet resistance. The charge density and the charge carrier density values chosen for devices A and B correspond to those selected at V_S = 3 V by the black dashed line in Figure 4b (i.e., the same gate voltage used for the previous R_{tot} calculation in Figure 5b). Finally, synthesis of the main parameters extracted from resistance partitioning (cf. Figure 5) and the C-V measurements (cf. Figure 4) is exposed in Table 1.

Table 1. Synthesis of parameters extracted from improved resistance partitioning method for trench bottom and trench sidewall areas.

Trench Area	Sheet Resistance (kΩ·sq)	Q (C·cm^{-2})	N (cm^{-2})	μ (cm^2/V·s)
Channel	~91.0	9.86×10^{-7} (A)	6.15×10^{12} (A)	11.1
Bottom	39.0	1.70×10^{-6} (B)	1.06×10^{13} (B)	15.1

Consequently, applying Equation (3) with the parameters in Table 1 leads to the extraction of the effective trench bottom and channel mobilities of 15.1 and 11.1 cm^2/V·s, respectively. As explained in [25], this poor channel mobility value could be mainly related to the damaged trench sidewalls following the GaN trench etching step, which is a critical process step for the fabrication of vertical GaN trench MOSFETs [7,26,27]. This should result, on the one hand, in carrier scattering coming from surface roughness (as seen in Figure 2b), and on the other hand, in oxide interface traps at the dielectric/GaN interface, significantly reducing channel mobility.

Finally, a way higher channel mobility of 30 $cm^2/V{\cdot}s$ is obtained when using the transconductance method defined as follows:

$$\mu_{ch} = g_m \cdot \frac{L}{Z} \cdot \frac{1}{C_{ox}} \cdot \frac{1}{V_D} \tag{4}$$

where g_m is the transconductance, Z is the channel width of 200 μm, L is the channel length of 0.7 μm, C_{ox} is the gate oxide conductance of 354.8 $nF.cm^{-2}$, and V_D is the drain voltage defined at 1 V. This low channel mobility is in agreement with the values reported in the literature for state-of-the-art vertical GaN MOSFETs. Indeed, among the research groups that extracted the mobility using the transconductance method, studies from Khadar et al. [7], Ishida et al. [10], and Zhu et al. [28] have fabricated vertical GaN MOSFETs that demonstrate channel mobilities ranging from 15 to 41 $cm^2/V{\cdot}s$, while the current record has been reached by Otake et al. [5], with a channel mobility of ~131 $cm^2/V{\cdot}s$.

Consequently, the mobility results estimated in this study pave the way for further improvements, especially by applying a finer post-etch GaN surface treatment to mend the trench sidewalls from etching damages [10], or by improving the critical dielectric/GaN interface with a higher-quality gate dielectric [13].

4. Conclusions

This study aimed to evaluate the electrical performances of a fully vertical GaN trench by means of I-V and C-V measurements. We first proved the normally-OFF behavior (V_{th} ~ 1 V), the adequate switching operation (I_{ON}/I_{OFF} ~ 10^9), as well as the significantly small gate leakage current (10^{-11} A/mm) of our devices. Then, the resistances of the trench bottom and trench channel were also extracted using an improved resistance partitioning method. Subsequently, by considering the difference in electrostatic behavior between the devices linked to these trench areas, we estimated the effective channel and bottom mobilities to be 11.1 $cm^2/V{\cdot}s$ and 15.1 $cm^2/V{\cdot}s$, respectively. These mobility values are promising and act as a starting point to be improved upon thanks to the optimization of the gate module, with, for instance, a well-controlled trench etching process, a better pre-deposition GaN surface treatment, or the integration of an alternative alumina dielectric into the MOS gate.

Author Contributions: Conceptualization, V.A., H.E.R., V.M., E.F., Y.C., M.C., J.B., and B.S.; methodology, V.A. and B.M.; investigation, V.A., B.M., B.S., and J.B.; writing—original draft preparation, V.A.; writing—review and editing, V.A., B.M., H.E.R., V.M., E.F., Y.C., M.C., G.L., J.B., and B.S.; visualization, V.A.; supervision, B.S., B.M., and J.B. All authors have read and agreed to the published version of the manuscript.

Funding: This study was partially supported by the Labex GANEXT (ANR-11-LABX-0014), the VERTIGO project supported by the French Agence Nationale de la Recherche as part of France 2030 with reference ANR-22-PEEL-0004, and the French RENATECH network through the PTA technological platforms.

Data Availability Statement: The original contributions presented in the study are included in the article, further inquiries can be directed to the corresponding authors.

Conflicts of Interest: The authors declare no conflicts of interest.

References

1. Ji, D.; Laurent, M.A.; Agarwal, A.; Li, W.; Mandal, S.; Keller, S.; Chowdhury, S. Normally OFF Trench CAVET With Active Mg-Doped GaN as Current Blocking Layer. *IEEE Trans. Electron Devices* **2017**, *64*, 805–808. [CrossRef]
2. Nie, H.; Diduck, Q.; Brian, A.; Edwards, A.P.; Kayes, B.M.; Zhang, M.; Ye, G.; Prunty, T.; Bour, D.; Kizilyalli, I.C. 1.5-kV and 2.2-mΩ.Cm^2 Vertical GaN Transistors on Bulk-GaN Substrates. *IEEE Electron Device Lett.* **2014**, *35*, 939–941. [CrossRef]
3. Shibata, D.; Kajitani, R.; Ogawa, M.; Tanaka, K.; Tamura, S.; Hatsuda, T.; Ishida, M.; Ueda, T. 1.7 KV/1.0 MΩcm 2 Normally-off Vertical GaN Transistor on GaN Substrate with Regrown p-GaN/AlGaN/GaN Semipolar Gate Structure. In Proceedings of the 2016 IEEE International Electron Devices Meeting (IEDM), San Francisco, CA, USA, 3–7 December 2016; IEEE: Piscataway, NJ, USA, 2016. [CrossRef]

4. Zhang, Y.; Sun, M.; Perozek, J.; Liu, Z.; Zubair, A.; Piedra, D.; Chowdhury, N.; Gao, X.; Shepard, K.; Palacios, T. Large Area 1.2 KV GaN Vertical Power FinFETs with a Record Switching Figure-of-Merit. *IEEE Electron Device Lett.* **2018**, *40*, 1. [CrossRef]
5. Otake, H.; Chikamatsu, K.; Yamaguchi, A.; Fujishima, T.; Ohta, H. Vertical GaN-Based Trench Gate Metal Oxide Semiconductor Field-Effect Transistors on GaN Bulk Substrates. *Appl. Phys. Express* **2018**, *1*, 011105. [CrossRef]
6. Oka, T.; Ina, T.; Ueno, Y.; Nishii, J. 1.8 MΩ·cm 2 Vertical GaN-Based Trench Metal–Oxide–Semiconductor Field-Effect Transistors on a Free-Standing GaN Substrate for 1.2-KV-Class Operation. *Appl. Phys. Express* **2015**, *8*, 054101. [CrossRef]
7. Khadar, R.A.; Liu, C.; Soleimanzadeh, R.; Matioli, E. Fully Vertical GaN-on-Si Power MOSFETs. *IEEE Electron Device Lett.* **2019**, *40*, 443–446. [CrossRef]
8. Zhu, R.; Jiang, H.; Tang, C.W.; Lau, K.M. Effects of P-GaN Body Doping Concentration on the ON-State Performance of Vertical GaN Trench MOSFETs. *IEEE Electron Device Lett.* **2021**, *42*, 970–973. [CrossRef]
9. Kodama, M.; Sugimoto, M.; Hayashi, E.; Soejima, N.; Ishiguro, O.; Kanechika, M.; Itoh, K.; Ueda, H.; Uesugi, T.; Kachi, T. GaN-Based Trench Gate Metal Oxide Semiconductor Field-Effect Transistor Fabricated with Novel Wet Etching. *Appl. Phys. Express* **2008**, *1*, 021104. [CrossRef]
10. Ishida, T.; Nam, K.P.; Matys, M.; Uesugi, T.; Suda, J.; Kachi, T. Improvement of Channel Property of GaN Vertical Trench MOSFET by Compensating Nitrogen Vacancies with Nitrogen Plasma Treatment. *Appl. Phys. Express* **2020**, *13*, 124003. [CrossRef]
11. Gupta, C.; Chan, S.H.; Lund, C.; Agarwal, A.; Koksaldi, O.S.; Liu, J.; Enatsu, Y.; Keller, S.; Mishra, U.K. Comparing Electrical Performance of GaN Trench-Gate MOSFETs with a-Plane and m-Plane Sidewall Channels. *Appl. Phys. Express* **2016**, *9*, 121001. [CrossRef]
12. Mukherjee, K.; De Santi, C.; Borga, M.; You, S.; Geens, K.; Bakeroot, B.; Decoutere, S.; Meneghesso, G.; Zanoni, E.; Meneghini, M. Use of Bilayer Gate Insulator in GaN-on-Si Vertical Trench MOSFETs: Impact on Performance and Reliability. *Materials* **2020**, *13*, 4740. [CrossRef] [PubMed]
13. Goncalez Filho, W.; Borga, M.; Geens, K.; Cingu, D.; Chatterjee, U.; You, S.; Bakeroot, B.; Decoutere, S.; Knaepen, W.; Arnou, P.; et al. AlON gate dielectric and gate trench cleaning for improved reliability of vertical GaN MOSFETs. In Proceedings of the CIPS 2022; 12th International Conference on Integrated Power Electronics Systems, Berlin, Germany, 15–17 March 2022; pp. 1–5.
14. Zhu, R.; Jiang, H.; Tang, C.W.; Lau, K.M. Enhancing ON- and OFF-State Performance of Quasi-Vertical GaN Trench MOSFETs on Sapphire With Reduced Interface Charges and a Thick Bottom Dielectric. *IEEE Electron Device Lett.* **2022**, *43*, 346–349. [CrossRef]
15. Wang, Y.; Wang, M.; Xie, B.; Wen, C.P.; Wang, J.; Hao, Y.; Wu, W.; Chen, K.J.; Shen, B. High-Performance Normally-Off Al_2O_3/GaN MOSFET Using a Wet Etching-Based Gate Recess Technique. *IEEE Electron Device Lett.* **2013**, *34*, 1370–1372. [CrossRef]
16. Hu, Q.; Hu, B.; Gu, C.; Li, T.; Li, S.; Li, S.; Li, X.; Wu, Y. Improved Current Collapse in Recessed AlGaN/GaN MOS-HEMTs by Interface and Structure Engineering. *IEEE Trans. Electron Devices* **2019**, *66*, 4591–4596. [CrossRef]
17. Le Royer, C.; Mohamad, B.; Biscarrat, J.; Vauche, L.; Escoffier, R.; Buckley, J.; Becu, S.; Riat, R.; Gillot, C.; Charles, M.; et al. Normally-OFF 650V GaN-on-Si MOSc-HEMT Transistor: Benefits of the Fully Recessed Gate Architecture. In Proceedings of the 2022 IEEE 34th International Symposium on Power Semiconductor Devices and ICs (ISPSD), Vancouver, BC, Canada, 22–25 May 2022; IEEE: Piscataway, NJ, USA, 2022; pp. 49–52. [CrossRef]
18. Kammeugne, R.K.; Leroux, C.; Cluzel, J.; Vauche, L.; Le Royer, C.; Krakovinsky, A.; Gwoziecki, R.; Biscarrat, J.; Gaillard, F.; Charles, M.; et al. Accurate Statistical Extraction of AlGaN/GaN HEMT Device Parameters Using the Y-Function. *Solid-State Electron.* **2021**, *184*, 108078. [CrossRef]
19. Piotrowic, C.; Mohamad, B.; Fernandes Paes Pinto Rocha, P.; Malbert, N.; Ruel, S.; Pimenta-Barros, P.; Jaud, M.-A.; Vauche, L.; Le Royer, C. Impact of Gate Morphology on Electrical Performances of Recessed GaN-on Si MOS Channel-HEMT for Different Channel Orientations. In Proceedings of the 2023 35th International Symposium on Power Semiconductor Devices and ICs (ISPSD), Hong Kong, 28 May–1 June 2023; IEEE: Piscataway, NJ, USA, 2023; pp. 382–385. [CrossRef]
20. Ruzzarin, M.; Geens, K.; Borga, M.; Liang, H.; You, S.; Bakeroot, B.; Decoutere, S.; De Santi, C.; Neviani, A.; Meneghini, M.; et al. Exploration of Gate Trench Module for Vertical GaN Devices. *Microelectron. Reliab.* **2020**, *114*, 113828. [CrossRef]
21. Esposto, M.; Krishnamoorthy, S.; Nath, D.N.; Bajaj, S.; Hung, T.-H.; Rajan, S. Electrical Properties of Atomic Layer Deposited Aluminum Oxide on Gallium Nitride. *Appl. Phys. Lett.* **2011**, *99*, 133503. [CrossRef]
22. Bisi, D.; Chan, S.H.; Liu, X.; Yeluri, R.; Keller, S.; Meneghini, M.; Meneghesso, G.; Zanoni, E.; Mishra, U.K. On Trapping Mechanisms at Oxide-Traps in Al2O3/GaN Metal-Oxide-Semiconductor Capacitors. *Appl. Phys. Lett.* **2016**, *108*, 112104. [CrossRef]
23. Rrustemi, B.; Piotrowicz, C.; Jaud, M.-A.; Triozon, F.; Vandendaele, W.; Mohamad, B.; Gwoziecki, R.; Ghibaudo, G. Effect of Doping on Al2O3/GaN MOS Capacitance. *Solid-State Electron.* **2022**, *194*, 108356. [CrossRef]
24. Götz, W.; Johnson, N.M.; Walker, J.; Bour, D.P.; Street, R.A. Activation of Acceptors in Mg-Doped GaN Grown by Metalorganic Chemical Vapor Deposition. *Appl. Phys. Lett.* **1996**, *68*, 667–669. [CrossRef]
25. Treidel, E.B.; Hilt, O.; Hoffmann, V.; Brunner, F.; Bickel, N.; Thies, A.; Tetzner, K. On the Conduction Properties of Vertical GaN N-Channel Trench MISFETs. *IEEE J. Electron Devices Soc.* **2021**, *9*, 215–228. [CrossRef]
26. Zhang, Y.; Sun, M.; Liu, Z.; Piedra, D.; Hu, J.; Gao, X.; Palacios, T. Trench Formation and Corner Rounding in Vertical GaN Power Devices. *Appl. Phys. Lett.* **2017**, *110*, 193506. [CrossRef]

27. Yamada, S.; Sakurai, H.; Osada, Y.; Furuta, K.; Nakamura, T.; Kamimura, R.; Narita, T.; Suda, J.; Kachi, T. Formation of Highly Vertical Trenches with Rounded Corners via Inductively Coupled Plasma Reactive Ion Etching for Vertical GaN Power Devices. *Appl. Phys. Lett.* **2021**, *118*, 102101. [CrossRef]
28. Zhu, R.; Jiang, H.; Tang, C.W.; Lau, K.M. GaN quasi-vertical trench MOSFETs grown on Si substrate with ON-current exceeding 1A. *Appl. Phys. Express* **2022**, *15*, 121004. [CrossRef]

electronics

MDPI

Article

Collaborative Design of Pulsed-Power Generator Based on SiC Drift Step Recovery Diode

Jingkai Guo, Yahui Chen, Yu Zhang, Lejia Sun *, Yu Zhou, Qingwen Song, Xiaoyan Tang and Yuming Zhang

School of Microelectronics, Xidian University, Xi'an 710071, China; jk_guo@stu.xidian.edu.cn (J.G.); yahuichen@stu.xidian.edu.cn (Y.C.); yz_99@stu.xidian.edu.cn (Y.Z.); yzhou_6083@stu.xidian.edu.cn (Y.Z.); qwsong@xidian.edu.cn (Q.S.); xytang@mail.xidian.edu.cn (X.T.); zhangym@xidian.edu.cn (Y.Z.)
* Correspondence: ljsun@xidian.edu.cn

Abstract: Despite the extensively researched physical principles, numerous published simulations on SiC drift step recovery diodes (SiC DSRD) and the practical implementation of SiC DSRD-based pulses, there are few kinds of research focusing on collaborative design between a SiC DSRD and its driving circuit. In this paper, a collaborative design method of a SiC DSRD and its driving circuit are presented. In addition, a detailed simulation is conducted to verify design considerations and to analyze the impact of driving parameter changes on the output pulse waveform. A pulse generator prototype with a self-developed SiC DSRD is implemented. The experimental results show that the circuit can output a peak voltage of 790 V on a matching load of 50 Ω, with a rise time of 520 ps (20%~80%), and can work at a 1 MHz repetition frequency rate with good stability.

Keywords: SiC; pulsed power circuit; drift step recovery diodes; sub-nanosecond

check for updates

Citation: Guo, J.; Chen, Y.; Zhang, Y.; Sun, L.; Zhou, Y.; Song, Q.; Tang, X.; Zhang, Y. Collaborative Design of Pulsed-Power Generator Based on SiC Drift Step Recovery Diode. *Electronics* **2024**, *13*, 2152. https://doi.org/10.3390/electronics13112152

Academic Editor: Francesco Giuseppe Della Corte

Received: 26 March 2024
Revised: 20 April 2024
Accepted: 1 May 2024
Published: 31 May 2024

1. Introduction

Pulsed power technology is a method that compresses energy from milli- or microseconds to nano- or picoseconds on the time scale and brings several orders of magnitude increase in the power level. This technology finds extensive applications across various industries such as food processing, nitric oxide production, water and exhaust gas treatment, and ion implantation, among others [1]. Numerous experiments have demonstrated that nanosecond pulses are particularly advantageous for these applications [2–4]. While conventional semiconductor switches like IGBTs, MOSFETs, and thyristors can effectively generate microsecond power pulses in such scenarios, they are not suitable for generating high-voltage nanosecond pulses. To address this issue, drift step recovery diodes (DSRDs) can be employed [5].

Drift step recovery diodes (DSRDs), as a new type of semiconductor opening switch, can generate a nanosecond pulse on the load through a matching external trigger circuit. However, the DSRD has strict requirements for circuit matching. The unmatched driving circuits will lead to a decrease in the output peak, the slowdown of the pulse front, and an increase in power loss owing to the appearance of pre-pulse.

Based on DSRD devices, a number of scholars have reported extensive research. The Russian Ioffe Physical-Technical Institute developed a pulsed generator with a front edge of 600 ps and an amplitude of 2 kV [6]. Israel's Soreq Nuclear Laboratory presented a solid-state pulsed generator with a front edge of less than 1 ns and an amplitude of 6 kV [7]. The American SLAC National Accelerator Laboratory reported a pulsed power supply. The rise time was less than 2 ns and the amplitude was 15 kV based on the DSRD front-end pump circuit [8,9]. The Northwest Institute of Nuclear Technology in China developed a pulsed generator with a pulse front of 700 ps and an amplitude of 2 kV [10]. In 2016, V.A. Ilyin et al. reported the circuit test results of an eight-diode-stacked 4H-SiC DSRD device with the P+PN+ structure. The rise time of the output pulse was 900 ps with an amplitude of 10.5 kV and verified the possibility of stacking 4H-SiC DSRD devices for the first time [11].

In the same year, B.V. Ivanov et al. reported the results of the repetition frequency operation test of a single diode P+PN+ structure. The rise time of the output pulse was 500 ps with an amplitude of 1.8 kV and a maximum repetition frequency of 500 kHz [12]. In 2018, T Goto et al. from Tsukuba University of Japan prepared a SiC DSRD with a P+PNN+ structure. It is different from the conventional P+PNN+ structure. In conventional P+PNN+ structure, the P-type base region stores plasma, and the N-type base region bears pressure. However, based on the result of the simulation, T Goto et al. found that the 9 μm thick P-type base region matches the 8 μm thick N-type base region when the pulse generator output voltage is the largest and the rise time is shorter. After stacking five diodes, the rise time of the output pulse was 2.3 ns with an amplitude of 11 kV [13]. In addition, V.A. Ilyin et al. reported the test results of 15 diodes stacked with a 4H-SiC DSRD, with an output pulse rise time of 1.6 ns and an amplitude of 30.5 kV [14]. In the same year, Wang Yajie et al. reported a pulse generator with an output pulse amplitude of 2 kV and a pulse front of 680 ps (20%~90%). Based on the equivalent model of DSRD, the simulation model of the forward-and-reverse pump circuit is established. According to the requirements of output voltage, the values of energy storage inductance are simulated and analyzed. Finally, optimal values of the component parameters in the circuit are obtained [15]. In 2020, Sun Ruize et al. reported a SiC DSRD device with the P+PNN+ structure, in which the breakdown voltage was above 10.9 kV. The measured output pulse rise time was 1.75 ns with an amplitude of 10.56 kV, and the maximum repetition frequency was 1 MHz [16]. In 2021, Liang Lin et al., from Huazhong University of Science and Technology of China, reported the testing results of a SiC DSRD circuit using four-way parallel triggering. The measured rise time of the output pulse was 1.011 ns, and the amplitude was 2.302 kV. By adjusting the DC bias voltage of the input circuit, they found that the amplitude and rise time of the output voltage of the device increase with the increase in the DC bias voltage [17].

Although the match between a DSRD and its driving circuits is an important issue, little research pays attention to the collaborative design of DSRD and its driving circuits. In this paper, a collaborative design method of a SiC DSRD and its driving circuit are presented. In addition, a detailed simulation is conducted to verify design considerations and to analyze the impact of driving parameter changes on the output pulse waveform. A pulse generator prototype with self-developed SiC DSRD is implemented.

2. Design of SiC DSRD Device

2.1. Operation Principle of SiC DSRD

A simplified operating circuit is shown in Figure 1a. The DSRD is connected in parallel to the load. The waveform of input current I_{in} is shown in Figure 1b. I_d and I_{out} are the currents flowing through the DSRD and the load, respectively. When the DSRD is turned off, I_{in} is switched from the the DSRD loop to the load loop, then forms a high-voltage fast pulse on the load (R_L).

Figure 1c shows the carrier concentration and electric field distributions at different times. Initially, the DSRD is in the blocking state due to the existence of the space charge region (SCR). Before t_0, the hole concentration and doping concentration are basically the same in the base region, whereas the electron concentration can be ignored. Then, I_{in} provides a forward current to the DSRD. Electrons and holes, from the N^+ region and P^+ region, are injected into the base region. It decreases the resistance of the base region. During the t_1-t_2 process, the current I_{in} changes direction into reverse to extract the plasma in the base area. The resistance of the base region is still low due to the existence of high-concentration plasma. At time t_2, the plasma is completely extracted, and the reverse current reaches the maximum. At this moment, the majority of carriers are extracted by the reverse current I_{in} and move with the saturated velocity. Meanwhile, the space charge region forms at the PN^+ junction and expands with the speed of saturated velocity. As a result, the resistance of the base region sharply increases, and the current switches from the

DSRD to the parallel connected load with the output pulse. At time t_3, the space charge region is completely established, and the output pulse reaches its maximum voltage.

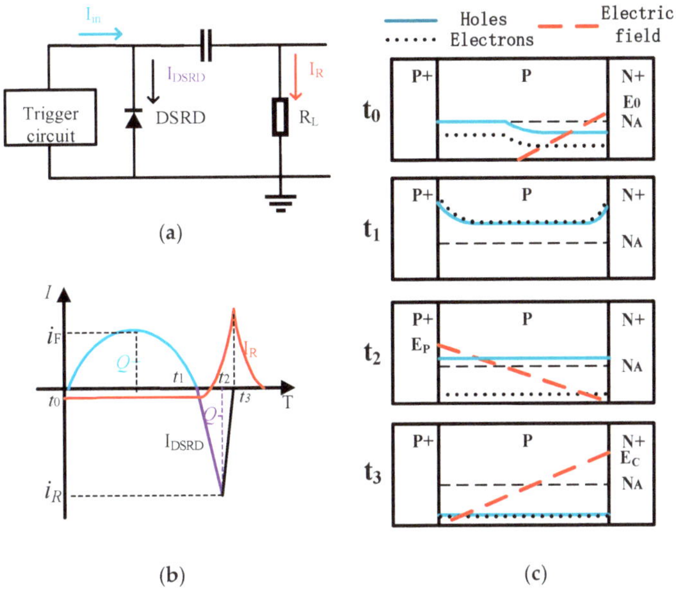

Figure 1. The working principle of DSRD: (**a**) the simplified working circuit of DSRD; (**b**) switching characteristics of DSRD; (**c**) the carrier concentration and electric field distributions at different times.

In addition, for SiC DSRDs, vast loss of injected charge is a general issue [18,19] which may lead to a decrease in DSRD energy efficiency. The charge loss rate is defined as Q-/Q+. The extracted non-equilibrium charge Q- and injected charge Q+ are defined by the areas under current transient response at different stages of the recovery process of the DSRD [20].

2.2. Parameter Selection and Optimization of SiC DSRD

It is worth noting that, in the process of plasma extraction, if the carrier concentration at the base region cannot sustain the reverse current, major carriers will accumulate at the border of the P+P junction [21]. In this case, the voltage drop on the base will occur before the cutoff, resulting in the appearance of pre-pulse. It increases the power loss of the DSRD. To avoid the appearance of the pre-pulse, in terms of the DSRD design, the maximum reverse current (I_{max}) must satisfy the following:

$$I_{max} \leq qN_A v_s S \quad (1)$$

In Equation (1), q is the elementary charge quantity; N_A is the base doping concentration; v_s is the carrier-saturated drift velocity of the base region; and S is the active area of the DSRD.

In addition, the breakdown voltage (V_{BV}) of the DSRD is the main parameter that restricts the output peak voltage (V_{max}) of the pulse generator. In other words, V_{BV} puts an upper limitation on the output peak current. The DSRD breakdown voltage V_{BV} can be expressed as

$$V_{BV} \propto N_A^{-1} W \quad (2)$$

The output peak voltage V_{max} is

$$V_{max} = I_{max}R_L \tag{3}$$

The relation between V_{BV} and V_{max} is

$$V_{max} \approx V_{BV} \tag{4}$$

Therefore, the I_{max} is limited by N_A and W, as shown in Equation (6):

$$I_{max} = \frac{V_{max}}{R_L} \tag{5}$$

$$I_{max} \propto {N_A}^{-1}W \tag{6}$$

In Equations (2) and (3), W is the base width; R_L is the load resistance which is usually 50 Ω.

Then, the suitable structure parameter of DSRD can be determined. At first, the I_{max} value or V_{max} value will be given in advance according to practical requirements. Next, the value of the W, N_A, and S can be chosen based on Equations (1) and (6). That is to say, when one of the three parameters is given, the other two parameters will also be obtained. For example, in this paper, the given V_{max} value is 800 V. Based on the given design method, a SiC DSRD is designed. The thicknesses of the p+ and p-base layers are 2 μm and 6 μm, with accepted concentrations of 1×10^{19} and 2×10^{16} cm^{-3}, respectively; the n+ layer has a thickness of 5 μm and a doping concentration of 1×10^{19} cm^{-3}. The active area S is 0.28 mm^2.

3. Collaborative Design of DSRD-Based Pulse Generator

3.1. Working Principle of Driving Circuit

The schematic diagram of the driving circuit of the DSRD is illustrated in Figure 2. which consists of two DC power supplies with voltage V_1 and V_2, the primary switch of SiC MOSFET(MOS), energy storage inductances L_1 and L_2, the decoupling capacitance C_1, the resonant capacitance C_2, the dc-blocking capacitance C_3, the decoupling capacitance C_4, the DSRD device, and the load R_L.

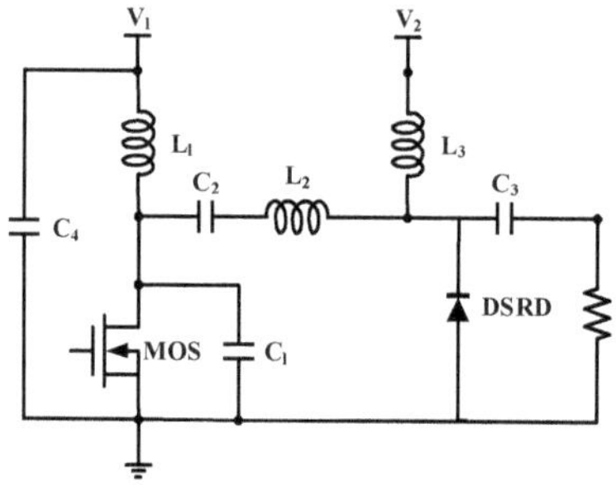

Figure 2. Schematic diagram of DSRD-based pulse generator.

The working process can be divided into four intervals, which are shown in Figure 3. Figure 4 shows the current waveforms during the operation.

Interval 1: Initially, the MOS is off-state, and the resonant capacitance C_2 is charged by V_1 and V_2 to the initial voltage $\Delta V = V_1 - V_2$.

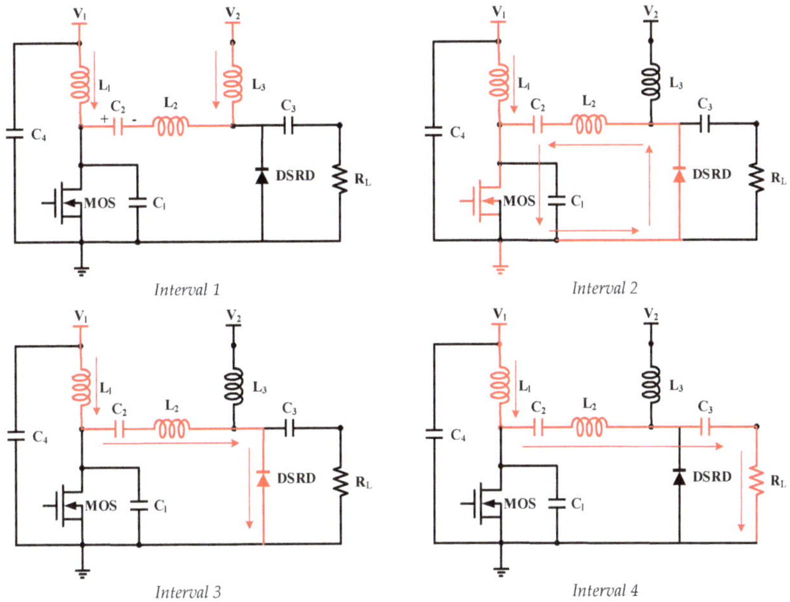

Figure 3. Equivalent circuit of pulse generator.

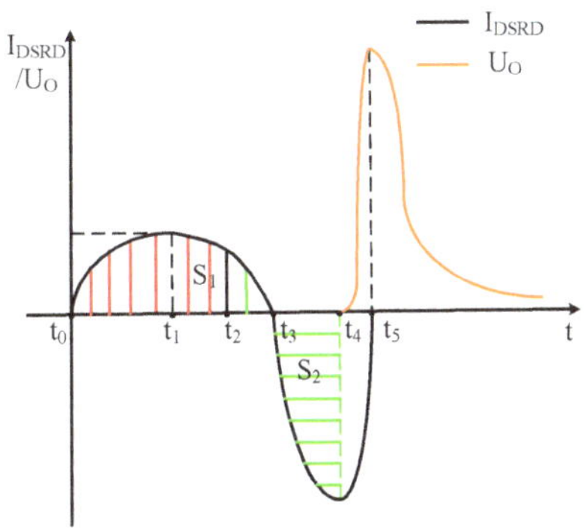

Figure 4. Theoretical waveforms in the circuit shown in Figure 3.

Interval 2: At t_0, the MOSFET is turned on; then, the circuit can be divided into two loops, which are V_1-L_1-*MOSFET(loop1)* and C_2-*MOSFET-DSRD*-L_2*(loop2)*. In loop1, the inductor L_1 stores the energy from V_1, and the current of L_1 can be expressed as

$$i_{L_1} = \frac{V_1}{L_1}t \tag{7}$$

In loop2, the inductor L_2 and capacitor C_2 begin to resonate, and at t_1 the resonant current reaches its peak value, then it begins to decrease. The resonant current which can be considered as the forward current accumulates charge carriers in the DSRD; the resonant current can be expressed as

$$i_{L_2} = \frac{\Delta V}{\omega_2 L_2}\sin(\omega_2 t),\ \omega_2 = 1/\sqrt{L_2 C_2} \tag{8}$$

Interval 3: At t_2, the MOSFET turns off. The energy stored in L_1 needs to be released through C_2, L_2, and DSRD. Then, the resonant current decreases rapidly. At t_3, the resonant current reverses. During the interval, the resonant current extracts charged plasma which has been accumulated in interval 2.

Interval 4: At t_4, the plasma in the DSRD base region is depleted by the extracting process, and the DSRD is rapidly turned off. Then, the high current through the DSRD is swiftly switched to the load R_L, generating a high-voltage pulse.

3.2. Conduction Time of the MOSFET

In order to weaken the effect of carrier recombination, the charge carrier's accumulation process in DSRD should satisfy the following equation:

$$\Delta T + (t_4 - t_2) \leq \tau \tag{9}$$

where ΔT represents the conduction time of the MOSFET, τ represents the carrier lifetime of the SiC DSRD, which equals 40 ns. Therefore, the conduction time ΔT is selected as 25 ns.

3.3. Inductance Design

The inductors used in power pulse technology often require a wide frequency response. Due to the easy saturation and large inductance deviation under different frequencies of the magnetic core inductor, the hollow inductor is selected in this design.

Based on the working principle illustrated above, the high-voltage pulse is generated by the high current through the DSRD swiftly switched to the load R_L. During the voltage pulse output, the current of L_2 can be expressed as

$$I_{L_2} = I_{peak} e^{\left(-\frac{t}{L_2/R_L}\right)} \tag{10}$$

The energy stored in the inductor L_2 can be expressed as

$$W_{\max} = \frac{1}{2} L_2 I_{L_2}^2 = \frac{1}{2} L_2 I_{\text{peak}}^2 e^{\left(\frac{2t}{L_2/R_L}\right)} \tag{11}$$

Based on (11), the relationship between the value of L_2 and its stored energy is shown in Figure 5.

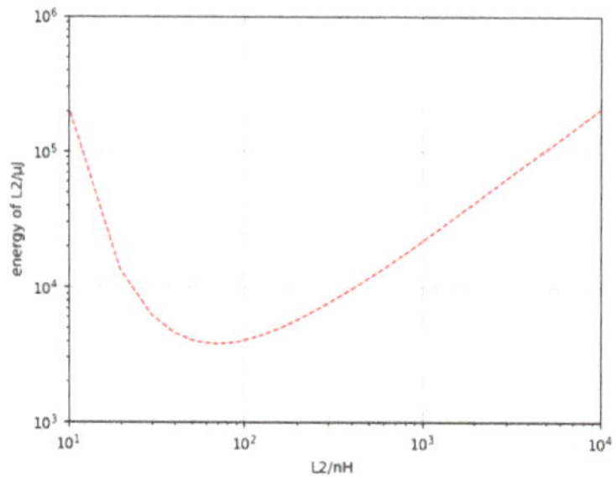

Figure 5. Energy varies with the value of L2.

It can be seen that when the L_2 equals 75 nH, the energy stored in L_2 reaches its maximum value. In practice, the L_2 is selected as 100 nH.

In order to ensure that the designed inductance works in the linear region, the maximum current of L_1 is selected as $i_{L1\text{-max}}$ = 9.5 A. Therefore, the inductance of L_1 can be calculated as

$$L_1 = \frac{V_1}{i_{L1-MAX}} \Delta T \tag{12}$$

Based on (12), when $\Delta T = 25$ ns, $V_1 = 120$ V, and $i_{L1\text{-max}} = 9.5$ A, $L_1 = 320$ nH is selected. Table 1 shows the selected circuit parameters.

Table 1. Pulse generator parameters.

Parameters	Values
L_1	320 nH
L_2	100 nH
C_1	1 nF
C_2	15 nF
ΔT	25 ns

3.4. Capacitance Design

During the interval2, in circuit loop2, the on-resistance of MOSFET (R_{DSon}), the on-resistance of the DSRD (R_{DSRD}), the capacitance C_2, and the inductance L_2 begin to resonate and the equation can be achieved as

$$L_2 C_2 \frac{d^2 u_{C_2}}{dt^2} + RC_2 \frac{du_{C_2}}{dt} + u_{C_2} = 0 \tag{13}$$

Assuming that the MOSFET is turned off when the resonant current of L_2 and C_2 reaches its peak value, the following equation can be achieved as

$$C_2 = \frac{4\Delta T^2}{\pi^2 L_2} \tag{14}$$

where the R represents the sum of the $R_{DS(on)}$ and R_{DSRD}. Based on (13) and (14), when $\Delta T = 25$ ns and L_2=100 nH, $C_2 = 15$ nF is achieved.

3.5. MOSFET and Driver Design

In the pulse generator, the MOSFET is used to switch between forward-carrier injection and reverse-carrier extraction. In order to achieve an ultra-fast edge of the output pulse, the MOSFET needs to have a fast turn-off speed. Therefore, the RF MOSFET (DE475-102N21A) with 8 ns turn-off time is selected. To match the RF MOSFET (DE475-102N21A), the driver chip (IXRFD630) which is recommended by the device manufacturer is selected. The device parameters are listed in Table 2.

Table 2. MOSFET and driver parameters.

DE475-102N21A	IXRFD630
$V_{DS} = 1000$ V	$V_{cc} = 30$ V
$I_{D25} = 24$ A	$I_{in} = 10$ μA
$I_{DM} = 144$ A	$I_{peak} = 28$ A
$R_{DS(on)} < 0.45\ \Omega$	$V_{OH} = 30$ V
$T_{on} = 5$ ns	$T_r = 4$ ns
$T_{off} = 8$ ns	$T_f = 4$ ns

4. Simulation Analysis

In order to analyze the impact of the parameter changes on the waveform of the output pulse, the simulation based on the software Pspice (https://www.pspice.com/ (accessed on 26 January 2024)) is conducted. Figure 6a shows the simulation output pulse waveform with a different conduction time ΔT, which is changed from 10 ns to 90 ns with an interval of 20 ns. It can be seen that when $\Delta T = 10$ ns, the output pulse is relatively low due to the MOS being turned off at a small current in circuit loop1. When the ΔT changes from 30 ns to 70 ns, the pulse amplitude increases slightly, and the energy of each power pulse also improves. As energy increases, electromagnetic interference becomes more serious, which brings a larger secondary pulse. When ΔT = 90 ns, the output pulse disappears due to the large recombination of carriers in the DSRD base region. Figure 6b shows the simulation output pulse waveform with a different conductor, L_2, which is changed from 40 nH to 130 nH with an interval of 30 nH. It can be seen that when L_2 = 70 nH, the pulse amplitude reaches its peak value. As the L_2 increases, both the pulse delay and pulse discharge time are prolonged. Figure 6c shows the simulation output pulse waveform with a different conductor, L_1, which is changed from 40 nH to 160 nH with an interval of 30 nH. It can be seen that a smaller L_1 should make the output pulse appear unexpectedly flat-topped, and a larger L_1 should make the pulse amplitude reduce. Figure 6d shows the simulation output pulse waveform with a different capacitor C_1, which is changed from 40 pF to 1640 pF with an interval of 400 pF. It can be seen that when C_1 = 440 pF, the pulse amplitude reaches its highest value. Figure 6e shows the simulation output pulse waveform with a different ΔV by adjusting V_1, which is changed from 30 V to 150 V with an interval of 30 V. It can be seen that the pulse amplitude does not change significantly, but the pulse width increases which means the pulse energy is related to the ΔV. Figure 6f shows the simulation output pulse waveform with different V_1 and V_2 while maintaining a constant ΔT. When V_1 increases to 150 V, the pulse amplitude no longer changes. However, considering that the energy storage of the inductor L_1 is small when the V_1 is low, the pulse amplitude is decreased due to insufficient energy storage. For trade-offs, V_1 = 210 V is optimal.

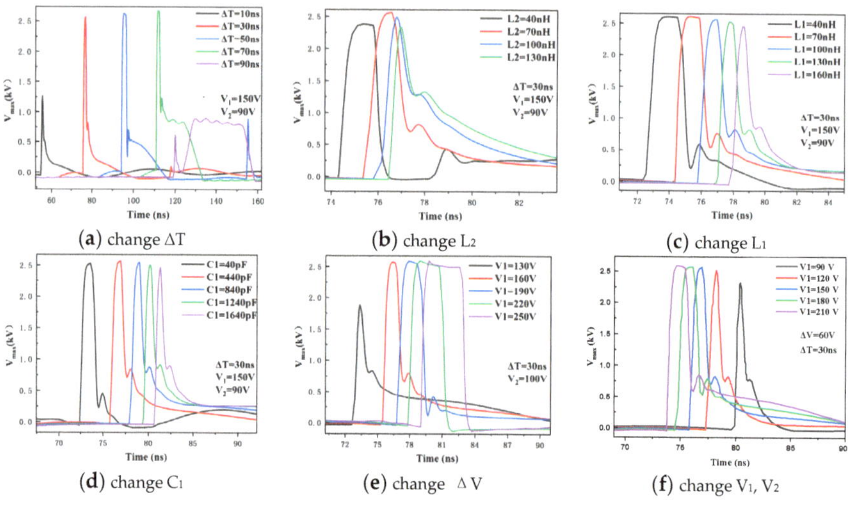

Figure 6. Changing circuit parameters; the output peak V_{max} change curve.

5. Experimental Result

According to the parameters obtained in Table 1, a pulse generator prototype was built based on a self-developed SiC DSRD, as shown in Figure 7a. We selected the IXRFD630 driver from IXYS company and the DE475-102N21A RF MOSFET as a switch. Figure 7b shows that the Tektronix MSO64B Oscilloscope captures the output waveform. The output load uses two 30 dB attenuators connected to the oscilloscope via coaxial lines. The SDN-414-05 current divider from T&M Germany was used to obtain the current flowing through the DSRD.

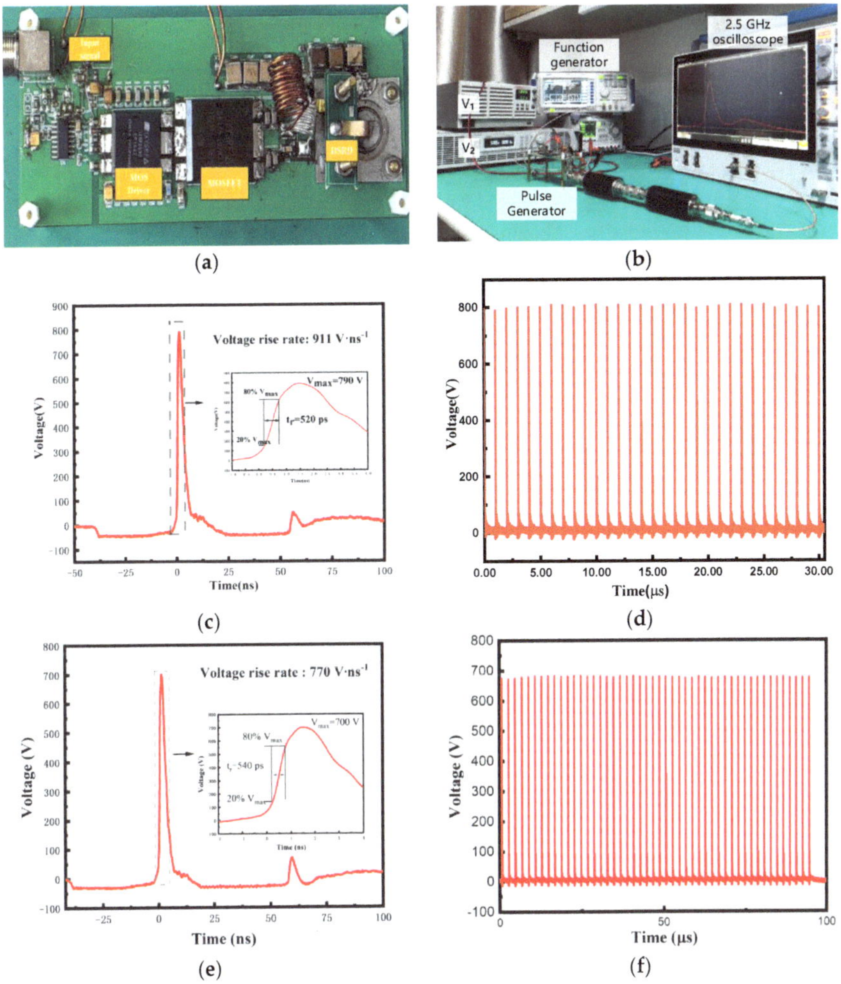

Figure 7. The pulse generator and output waveforms of the pulse generator are based on SiC DSRD. (**a**) Pulse generator based on self-developed SiC DSRD; (**b**) the test platform; (**c**) output voltage waveform of pulse generator with 790V; (**d**) output voltage waveform of pulse generator with peak amplitude 790 V and 1 MHz repetition rate under V_1 = 120 V, V_2 = 40 V; (**e**) output voltage waveform of pulse generator with peak amplitude 700 V under V_1 = 90 V, V_2 = 30 V; (**f**) output voltage waveform of pulse generator with peak amplitude 700 V and 500kHz repetition rate under V_1 = 90 V, V_2 = 30 V.

Figure 7c shows the waveform of a single power pulse. It can be seen that there is no pre-pulse in the output waveform. The peak voltage is 790 V, and the pulse rise time is 520 ps, which meets the design requirements (820 V) of the pulse generator. In addition, there is a slight secondary pulse observed in the waveform. Figure 7d shows the test result of the pulsed generator with a repetition rate of 1 MHz under V_1 = 120 V, V_2 = 40 V. Figure 7e,f show the output voltage waveform of the pulse generator with 700 V and 500 kHz repetition rate under V_1 = 90 V, V_2 = 30 V. The stable output pulse indicates the excellent stability of the pulsed generator.

6. Conclusions

According to the operating principle of DSRDs, this paper presents a collaborative design method of a SiC DSRD and the pulse generator based on it. An optimal method of DSRD parameters is proposed from the viewpoint of driving circuit requirements. In addition, a detailed simulation is conducted to verify design considerations and to analyze the impact of driving parameter changes on the output pulse waveform. A pulse generator based on a SiC DSRD was developed. The experimental results show that the generator can output a peak voltage of 790 V on a matching load of 50 Ω, with a pulse rise time of 520 ps. The pulse front is in the sub-nanosecond level with good stability and can be applied in miniaturized ultra-wideband electromagnetic pulse generators.

Author Contributions: Formal analysis, J.G.; investigation, J.G., Y.Z. (Yu Zhang) and Y.Z. (Yu Zhou); resources, L.S.; data curation, J.G.; writing—original draft preparation, J.G. and Y.C.; writing—review and editing, Y.C.; visualization, Y.C.; supervision, L.S.; project administration, Q.S., X.T. and Y.Z. (Yuming Zhang). All authors have read and agreed to the published version of the manuscript.

Funding: This research was funded by the National Natural Science Foundation of China (Program No. 62174123).

Data Availability Statement: The authors confirm that the data supporting the findings of this study are available within this article.

Acknowledgments: The authors thank all the anonymous reviewers for their very helpful comments in improving the paper.

Conflicts of Interest: The authors declare no conflicts of interest.

References

1. Jiang, W.H. Repetition rate pulsed power technology and its applications. *High Power Laser Part. Beams* **2012**, *24*, 10–15. [CrossRef]
2. van Heesch, E.; Pemen, A.; Huijbrechts, P.; van der Laan, P.; Prasinski, K.; Zanstra, G.; de Jong, P. A fast pulsed power source applied to treatment of conducting liquids and air. *IEEE Trans. Plasma Sci.* **2000**, *28*, 137–143. [CrossRef]
3. Lyublinsky, A.G.; Korotkov, S.V.; Aristov, Y.V.; Korotkov, D.A. Pulse power nanosecond-range DSRD-based generators for electric discharge technologies. *IEEE Trans. Plasma Sci.* **2013**, *41*, 2625–2629. [CrossRef]
4. Matsumoto, T.; Wang, D.; Namihira, T.; Katsuki, S.; Akiyama, H. Performances of nanosecond pulsed discharge. *Acta Phys. Pol. A* **2009**, *115*, 1101–1103. [CrossRef]
5. Grekhov, I.V.; Mesyats, G.A. Nanosecond semiconductor diodes for pulsed power switching. *Phys.-Uspekhi* **2005**, *48*, 703. [CrossRef]
6. Efanov, V.M.; Kardo-Sysoev, A.F.; Larionov, M.A.; Tchashnikov, I.G.; Yarin, P.M.; Kriklenko, A.V. Powerful semiconductor 80 kV nanosecond pulser. In Proceedings of the 11th IEEE International Pulsed Power Conference, Baltimore, MD, USA, 29 June–2 July 1997; pp. 985–987.
7. Kesar, A.S.; Merensky, L.M.; Ogranovich, M.; Kardo-Sysoev, A.F.; Shmilovitz, D. 6-kV, 130-ps rise-time pulsed-power circuit featuring cascaded compression by fast recovery and avalanche diodes. *Electron. Lett.* **2013**, *49*, 1539–1540. [CrossRef]
8. Krasnykh, A. Overview of driver technologies for nanosecond TEM kickers. In Proceedings of the 7th International Particle Accelerator Conference (IPAC 2016), Busan, Republic of Korea, 3 May 2016; pp. 3645–3647.
9. Benwell, A.; Burkhart, C.; Krasnykh, A.; Tang, T.; Kardo-Sysoev, A. A 5 kV, 3 MHz solid-state modulator based on the DSRD switch for an ultra-fast beam kicker. In Proceedings of the IEEE International Power Modulator and High Voltage Conference (IPMHVC), San Diego, CA, USA, 23 May 2013; pp. 328–331.
10. Fang, X.; Ding, Z.; Hao, Q.; Yuan, X.; Hu, L.; Fan, Y. Study on generation of drift step recovery diodes for sub-nanosecond switching and high repetition rate operation. In Proceedings of the 4th China Pulsed Power Conference, Hohhot, China, 25 August 2015; pp. 351–354.
11. Ilyin, V.A.; Afanasyev, A.V.; Ivanov, B.V.; Kardo-Sysoev, A.F.; Luchinin, V.V.; Reshanov, S.A. High-voltage ultra-fast pulse diode stack based on 4H-SiC. *Mater. Sci. Forum* **2016**, *858*, 786–789. [CrossRef]
12. Ivanov, B.V.; Smirnov, A.A.; Shevchenko, S.A.; Afanasyev, A.V.; Ilyin, V.A. High voltage sub-nanosecond silicon carbide opening switch. In Proceedings of the 57th International Scientific Conference on Power and Electrical Engineering of Riga Technical University (RTUCON), Riga, Latvia, 13–14 October 2016; pp. 1–4.
13. Goto, T.; Shirai, T.; Tokuchi, A.; Naito, T.; Fukuda, K.; Iwamuro, N. Experimental demonstration on ultra-high voltage and high speed 4H-SiC DSRD with smaller numbers of die stacks for pulse power. *Mater. Sci. Forum* **2018**, *924*, 858–861. [CrossRef]
14. Ilyin, V.A.; Afanasyev, A.V.; Demin, Y.S.; Ivanov, B.V.; Kardo-Sysoev, A.F.; Luchinin, V.V. 30 kV pulse diode stack based on 4H-SiC. *Mater. Sci. Forum* **2018**, *924*, 841–844. [CrossRef]

15. Wang, Y.; He, P.; Jing, X.; Tie, W.; Xie, J.; Zhao, C. Simulation and calculation of pulsed power source based on drift step recovery diode switching. *High Power Laser Part. Beams* **2018**, *30*, 095005.
16. Sun, R.; Zhang, K.; Chen, W.; Xia, Y.; Tan, J.; Chen, Y. 10-kV 4H-SiC Drift Step Recovery Diodes (DSRDs) for Compact High-repetition Rate Nanosecond HV Pulse Generator. In Proceedings of the 32nd International Symposium on Power Semiconductor Devices and ICs (ISPSD), Vienna, Austria, 18 August 2020; pp. 62–65.
17. Yang, Z.W.; Liang, L.; Yan, X.X. Dynamic Electrical Characteristics of 4H-SiC Drift Step Recovery Diodes of High Voltage. *IEEE Trans. Plasma Sci.* **2022**, *50*, 1276–1281. [CrossRef]
18. Kyuregyan, A.S. Theory of Drift Step Recovery Diodes. *Solid State Electron.* **2004**, *49*, 720–727. [CrossRef]
19. Ivanov, B.V.; Smirnov, A.A.; Shevchenko, S.A. A Study of Charge Losses in 4H-SiC Drift Step Recovery Diodes (DSRD). In Proceedings of the of EIConRusNW, St. Petersburg, Russia, 3 February 2016; pp. 51–52.
20. Ivanov, P.A.; Konkov, O.I.; Samsonova, T.P.; Potapov, A.S.; Grekhov, I.V. Dynamic Characteristics of 4H-SiC Drift Step Recovery Diodes. *Semiconductors* **2015**, *49*, 1511–1515. [CrossRef]
21. Galeckas, A.; Linnros, J. Investigation of surface recombination and carrier lifetime in 4H/6H-SiC. *Mater. Sci. Eng.* **1999**, *61*, 239–243. [CrossRef]

electronics

Article

Study of 1500 V AlGaN/GaN High-Electron-Mobility Transistors Grown on Engineered Substrates

An-Chen Liu [1], Pei-Tien Chen [1], Chia-Hao Chuang [2], Yan-Chieh Chen [3], Yan-Lin Chen [4], Hsin-Chu Chen [3,*], Shu-Tong Chang [5], I-Yu Huang [6] and Hao-Chung Kuo [1,7,*]

1 Department of Photonics and Institute of Electro-Optical Engineering, National Yang Ming Chiao Tung University, Hsinchu 30010, Taiwan; arsen.liou@gmail.com (A.-C.L.); dy1998416@gmail.com (P.-T.C.)
2 Industry Academia Innovation School, National Yang Ming Chiao Tung University, Hsinchu 30010, Taiwan; charleschc.ee06@nctu.edu.tw
3 Institute of Advanced Semiconductor Packaging and Testing, National Sun Yat-sen University, Kaohsiung 804201, Taiwan; jschen20000415@gmail.com
4 Master Program in Semiconductor and Green Technology, National Chung Hsing University, Taichung 402202, Taiwan; g111002610@mail.nchu.edu.tw
5 Department of Electrical Engineering, National Chung Hsing University, Taichung 402202, Taiwan; stchang@dragon.nchu.edu.tw
6 College of Semiconductor and Advanced Technology Research, National Sun Yat-sen University, Kaohsiung 804201, Taiwan; iyuhuang@mail.nsysu.edu.tw
7 Semiconductor Research Center, Hon Hai Research Institute, Taipei 114699, Taiwan
* Correspondence: chenhc@mail.nsysu.edu.tw (H.-C.C.); hckuo0206@nycu.edu.tw (H.-C.K.)

Abstract: In this study, we demonstrate breakdown voltage at 1500 V of GaN on a QST power device. The high breakdown voltage and low current collapse performance can be attributed to the higher quality of GaN buffer layers grown on QST substrates. This is primarily due to the matched coefficient of thermal expansion (CTE) with GaN and the enhanced mechanical strength. Based on computer-aided design (TCAD) simulations, the strong electric-field-induced trap-assisted thermionic field emissions (TA-TFEs) in the GaN on QST could be eliminated in the GaN buffer. This demonstration showed the potential of GaN on QST, and promises well-controlled performance and reliability under high-power operation conditions.

Keywords: GaN on engineered poly-AlN substrates; QST substrate; GaN on Si substrate; HEMT; high breakdown voltage

Citation: Liu, A.-C.; Chen, P.-T.; Chuang, C.-H.; Chen, Y.-C.; Chen, Y.-L.; Chen, H.-C.; Chang, S.-T.; Huang, I.-Y.; Kuo, H.-C. Study of 1500 V AlGaN/GaN High-Electron-Mobility Transistors Grown on Engineered Substrates. *Electronics* **2024**, *13*, 2143. https://doi.org/10.3390/electronics13112143

Academic Editor: Yahya M. Meziani

Received: 10 May 2024
Revised: 22 May 2024
Accepted: 29 May 2024
Published: 30 May 2024

1. Introduction

Gallium nitride (GaN) has emerged as a crucial material for the next generation of high-frequency and high-power devices due to its exceptional properties, including a high concentration of two-dimensional electron gas (2DEG), superior carrier mobility, low on-resistance, and high breakdown voltage [1–3]. These attributes have propelled GaN to the forefront of semiconductor materials, enabling its utilization across a broad spectrum of high-performance electronic and optoelectronic devices. Traditionally, sapphire and silicon substrates have been favored for GaN device fabrication due to their availability and cost-effectiveness. However, their lower thermal conductivity presents a significant challenge for GaN epitaxy, particularly in achieving thick epi-layer stacks for high breakdown voltages exceeding 1200 V [4].

This challenge has been addressed through the incorporation of carbon-doped GaN (GaN:C) buffer layers with HEMT devices, which primarily improves drain leakage currents and increases the breakdown voltage [5]. To tackle these challenges effectively, enhancing the growth of high-resistivity GaN buffers is crucial for achieving robust electrical insulation from silicon substrates, characterized by minimal leakage currents and a high breakdown voltage.

In our study, we introduce superlattice buffer layers on Qromis Substrate Technology (QS) substrates to reduce the difference in thermal expansion coefficients. This approach enhances the interfacial quality and electronic properties of GaN-based devices. The superior epitaxial quality of GaN layers on QST substrates, compared to those on Si substrates, is expected to significantly enhance device performance [6,7]. Our analysis revealed the superior performance of AlGaN/GaN high-electron-mobility transistors (HEMTs) with QST substrates, achieving a hard breakdown of up to 1500 V. This improvement can be attributed to the superior thermal expansion coefficient between QST substrates and GaN epitaxy, contributing to the enhanced effectiveness of GaN:C thickness to reduce drain leakage currents compared to Si substrates. It is believed that the improved GaN on QST epitaxy process leads to a higher breakdown value.

2. Materials and Methods

The epitaxial layers of the $Al_{0.24}Ga_{0.76}N$/GaN-power high-electron-mobility transistors (HEMTs) were grown on high-thermal-conductivity QST substrates by metal–organic chemical vapor deposition (MOCVD). Prior to the preparation of the buffer and active layers, an AlN nucleation layer (NL) was grown to compensate for lattice mismatch and reduce dislocation density within the fabricated devices. This involved the initial growth of a 60 nm AlN nucleation layer to elevate the conduction band energy within the buffer layer and mitigate leakage currents. The AlN NL served to establish a robust foundation for epitaxial growth, enhancing the electrical isolation and thermal conductivity of the device. Following the nucleation layer, a 2 µm thick AlN/GaN superlattice buffer layer was prepared. Subsequently, buffer layer thicknesses ranging from 1 µm to 3 µm were prepared for a comparison of breakdown voltages. Following this, a 300 nm thick undoped GaN channel layer was grown. The structural diagram is depicted in Figure 1a. Subsequently, the AlGaN barrier layer and GaN cap layer were grown on the channel layer, facilitating the formation of high-electron-mobility and two-dimensional electron gas (2DEG) at the interface with the AlGaN barrier layer. The quality of this layer is crucial for the device's overall performance, particularly in terms of on-resistance (R_{ON}).

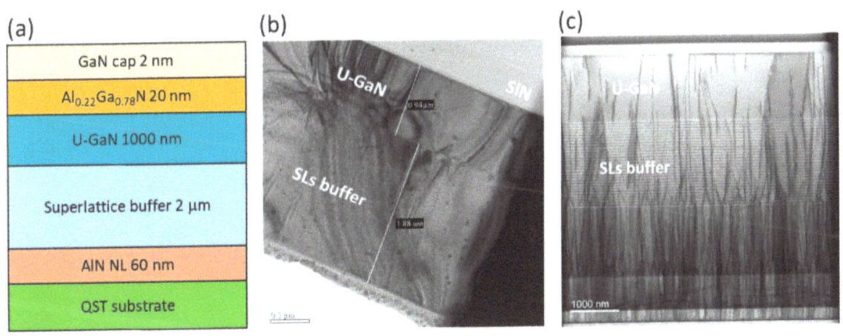

Figure 1. The epitaxial structure of (**a**) cross-section schematic view. TEM images of (**b**) GaN on QST and (**c**) GaN on Si substrate.

The composition and thickness of these layers are meticulously calculated to optimize the device's electrical characteristics. The epitaxial layer structure of the GaN on QST and Si substrate and its cross-sectional transmission electron microscopy (TEM) image are illustrated in Figure 1b,c, showcasing the QST substrate high-quality epitaxial growth and the interfaces between the various layers. HEMTs fabricated on QST substrates demonstrate high breakdown voltages and mechanical strength, highlighting the advantages of employing QST for the development of robust and efficient HEMT power devices. This fabrication approach underscores the significance of layer engineering in creating devices capable of withstanding high voltages and thermal stresses, rendering them suitable for a broad range of high-power applications.

For the fabrication of MIS-HEMTs, the device process commenced with mesa isolation via Ar implantation. Subsequently, source/drain (S/D) ohmic contact formation was carried out using Ti/Al/Ni/Au stacking layers, followed by annealing at 875 °C. After this high-temperature process, ALD-grown 2/20 nm thick AlN/Al_2O_3 layers were deposited to serve as both the gate dielectric layer and the first passivation layer [8]. Next, the gate metal was deposited and patterned. Following this, a thick SiO_2 inter-layer dielectric (ILD) was applied, followed by the deposition metal 1 (M1), thick SiO_2 inter-metal-layer dielectric (IMD), and metal 2 (M2). The M1 layer was 1 mm thick Al and the M2 layer was 2.5 mm thick Al. The backend process followed the CMOS BEOL rule. Finally, a thick SiNx passivation layer was deposited onto the patterned device structure. The device dimensions, denoted by $L_G/L_{GS}/L_{GD}$, were 2/3.5/22 µm, with these geometric parameters playing a critical role in determining the overall device performance. Figure 2 presents a cross-sectional schematic view of the fabricated devices.

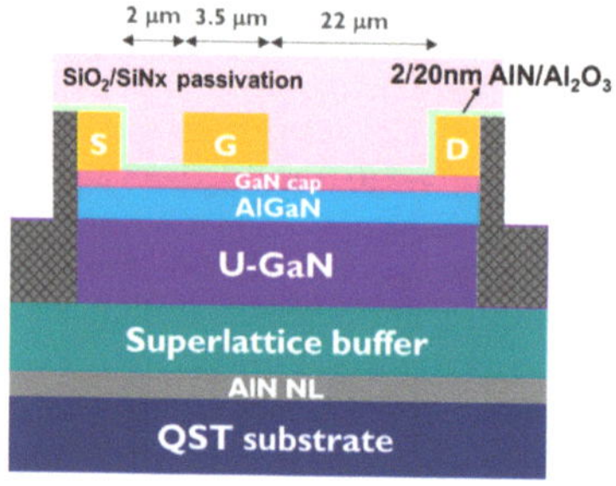

Figure 2. Cross-sectional schematic view of the AlGaN/GaN HEMT.

3. Result and Discussion

To investigate the impact of QST substrates on device performance, we measured the transfer device characteristics of the fabricated devices, as shown in Figure 3. The threshold voltage was about −12.7 V, and the subthreshold swing was about 98.1 mV/dec, which implies that the device is a well-controlled metal–insulator–semiconductor (MIS) depletion-mode (D-mode) HEMT.

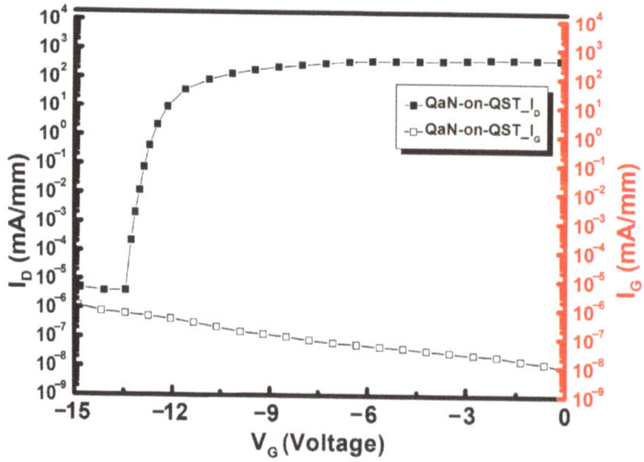

Figure 3. Transfer characteristic of AlGaN/GaN HEMT with GaN on QST substrate in semi-log scale: gate leakage current (red axis) and drain current (black axis).

To demonstrate the superiority of QST substrates over Si substrates, the investigation extended into the off-state behavior of devices. The off-state leakage current was defined as V_D for I_D achieved at 1 µA, as shown in Figure 4a. The variation in GaN:C thickness

(1~3 μm) of the $I_{D,off}$ leakage current was analyzed when fixing the SL thickness at 2 μm. It was demonstrated that a thicker GaN:C layer of 3 μm more effectively improved the $I_{D,off}$ leakage current. The obtained values of BV showed a positive correlation to the thickness of GaN:C, which is consistent with a previous report of GaN on Si epitaxy [4]. The breakdown voltages were investigated as a function of the total thickness of the epitaxial structure, as depicted in Figure 4b. The maximum thickness of the GaN buffer layer (SLs+GaN:C) in the QST substrate was 5 μm, whereas the maximum thickness of the GaN buffer layer in the Si substrate was 5.5 μm. A breakdown voltage capability of 1500 V was achieved with GaN on the QST substrate, compared to only 1200 V for GaN on the Si substrate. A linear trend was observed between high breakdown voltage and thickness in GaN on the QST substrate. Therefore, it is believed that a GaN buffer layer thicker than 5 μm on the QST substrate could achieve high performance. A thick GaN buffer layer (>10 μm) will be pursued in future work. Devices grown on highly resistive substrates experience a limited supply of carriers from the depleted region of the substrate, primarily from thermally generated carriers. This leads to substrate depletion [9,10], a phenomenon observed only with highly resistive substrates. As substrate depletion occurs, the electric field on the substrate intensifies, triggering different carrier generation processes such as Shockley–Read–Hall (SRH) generation and/or impact ionization. The QST substrate, composed of materials with a coefficient of thermal expansion (CTE) as a handling layer of the Si(111) layer, exhibits higher resistivity [11]. Consequently, it can withstand a higher electric field, resulting in a higher breakdown voltage than the GaN on Si substrate.

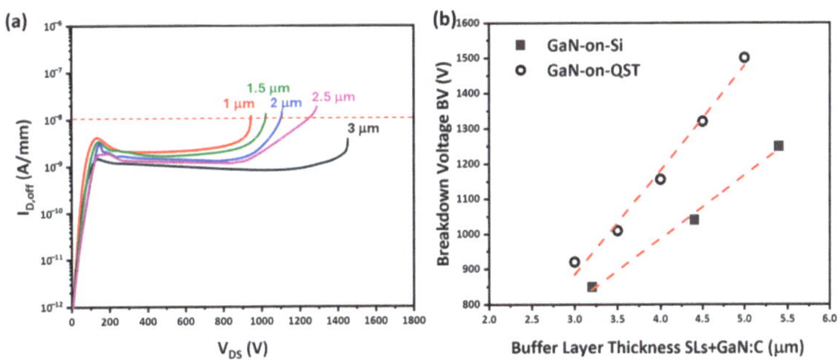

Figure 4. (**a**) Off-state I_D–V_D characteristics with GaN:C thickness; (**b**) breakdown voltage of epitaxial structures on Si substrate for various epitaxial thickness.

The bowing of epitaxial structures on Si or QST substrates is depicted as a function of the full width at half maximum (FWHM) of X-ray rocking curves from GaN (10–12) diffraction, as shown in Figure 5. Moreover, bowing corresponds to a concave shape; the definition of bowing is illustrated in the inset of this figure. This indicates that threading dislocations alleviate the compressive stress induced in the superlattice (SL) structure and following the thick GaN:C layer during growth, resulting in larger concave bowing with tensile stress due to thermal expansion mismatches during the cooling process after growth. Dislocation dynamics significantly influence the bowing of semiconductor layers. This can be referenced by studies that explore the impact of dislocation density on material bowing and mechanical properties [12,13]. It has been observed that the peak FWHM of the XRD rocking curve remains consistent, while the bowing varies. This indicates that although lattice mismatching is compatible, the resultant thermal coefficient mismatching differs from that observed in GaN on Si structures. Considering QST substrates, it is noted that while the top layer remains Si(111), a thick core layer with a matched CTE is introduced into the QST substrate. Therefore, the thermal-induced lattice mismatch is mitigated [11]. The wafer bowing of GaN on QST is smaller than that of GaN on Si, despite similar dislocation densities. This is because the CTE layer of the QST substrate enables the release of substrate bowing during the cooling process after the growth of GaN in MOCVD. GaN on QST not only ensures precise stress management in the epitaxial layer structure, but also necessitates improvement in the crystallinity of epitaxial films to grow GaN on Si with reduced bowing. The obtained bowing ranges from 10 to 30 μm for 200 mm wafers, which is sufficiently small for processing in a conventional fabrication line for Si devices.

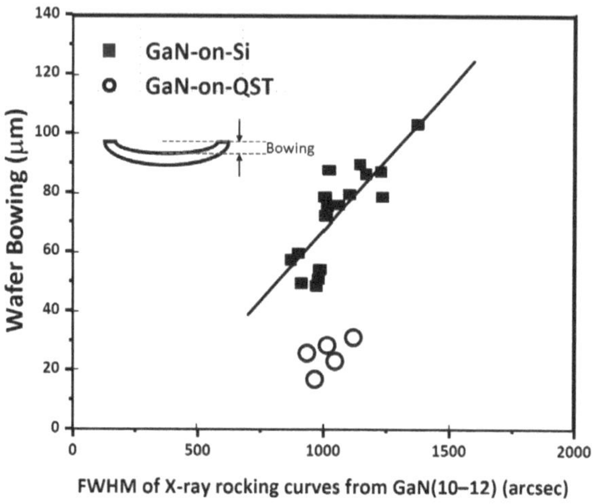

Figure 5. Bowing of epitaxial structures as a function of the FWHM of X-ray rocking curves from GaN(10–12).

The pulsed output characteristics obtained at bias points of (V_{GS}, V_{DS}) = (0 V, 0 V), (−12 V, 0 V), and (−12 V, 10 V). The results indicate the presence of trap states that significantly affect the devices' current response to voltage changes. Figure 6 presents the gate and drain lag measurements for GaN HEMTs grown on QST, highlighting their response under varied bias conditions. Specifically, for the GaN on QST sample, the gate lag percentages were 13.3% at V_D = 6 V and 10.6% at V_D = 10 V; drain lag percentages were 31.9% at V_D = 6 V and 30.2% at V_D = 10 V. These findings are crucial for understanding the dynamic behavior of the devices under a range of bias conditions. This behavior suggests that trap-induced lag is a critical factor in device performance, particularly affecting the reliability and operational efficiency under varying electronic loads. The I_D versus the V_{DS} characteristics of GaN HEMTs on QST, when subjected to the aforementioned bias conditions, reveal significant insights, indicating superior buffer layer quality with fewer trap states and a more efficient trap release mechanism. Hence, the advanced buffer layer in the GaN on QST HEMTs correlates with enhanced device performance, characterized by reduced gate/drain lag effects under the specified measurement conditions. This underscores that the buffer layer of GaN on QST substrates providing a more formidable barrier against trap-related degradation, thereby ensuring greater charge carrier mobility and augmented device reliability [14,15].

QST substrates enhance GaN HEMTs by providing a stable, thermal expansion coefficient platform for growth, reducing defects that cause carrier trapping. Optimized epitaxial growth on QST leads to uniform, high-quality buffer layers, crucial for minimizing trapping and the resulting drain lag.

The low-temperature (5 K) photoluminescence (PL) spectrum served as a diagnostic tool to assess the material quality, carrier concentration, and trap states in GaN HEMT devices. In contrast, we carried out previous work on an optimal GaN on Si device as a reference [16]. Regarding material quality, the sharpness and the positioning of the near-band-edge emission peak were crucial indicators. A notably narrow and intense peak, typically at approximately 360 nm for GaN, denoted a high crystalline quality with minimal defects. For GaN HEMT devices, the spectrum revealed that the PL peak, as shown in

Figure 7, for devices on QST substrates was sharper and more pronounced than that on Si substrates, signifying a superior material quality of the QST substrate.

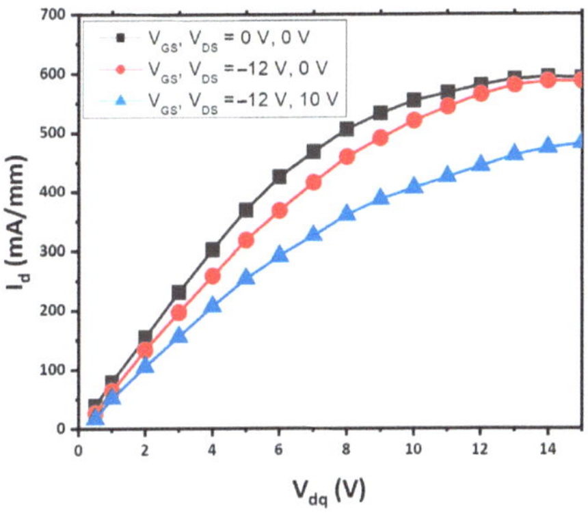

Figure 6. Measurement pulsed output characteristics under the bias point: (V_{GS}, V_{DS}) = (0 V, 0 V), (−12 V, 0 V), (−12 V, 10 V).

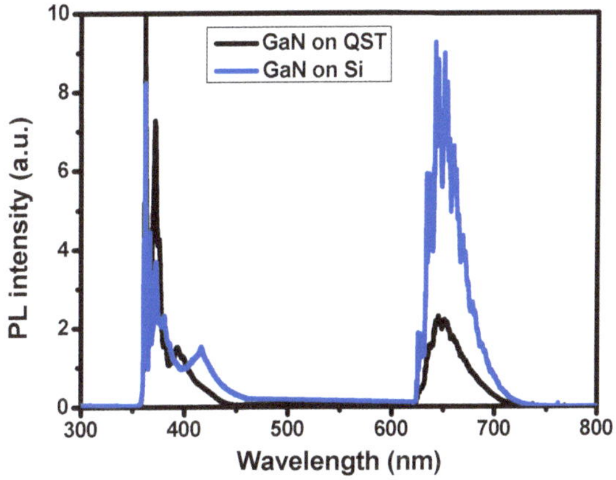

Figure 7. PL spectra from the GaN on QST/Si substrate.

As for carrier concentration, the PL peak intensity correlated directly with the radiative recombination rate, which could be linked to the carrier concentration. It was found that higher-intensity peaks were suggestive of elevated carrier concentrations within the material, provided that non-radiative processes were not predominant. Moreover, trap states were inferred from additional peaks in the longer wavelengths, typically in the visible range, which were indicative of deep-level or defect-related emissions. The PL spectrum for GaN on Si exhibited a broader peak with additional features in comparison to GaN on QST, potentially indicating a higher density of trap states in the Si substrate material. Analyzing the spectral data, the GaN on QST peak was discerned to be narrower and more intense at the band edge, coupled with fewer long-wavelength emissions, thus implying an improved material quality, enhanced carrier concentration, and a reduced number of trap states in comparison to GaN on Si [17,18].

To gain a deeper understanding of the physical phenomenon of TA-TFE in AlGaN/GaN HEMTs of GaN on Si substrate, TCAD simulations were performed under high-bias operation. The simulated device structure of the GaN on Si substrate is the same as that in this study. Simulation models include the drift–diffusion model, polarization, SRH, Auger recombination model, and doping–electric field dependence mobility. The measured breakdown voltage on the device of GaN on Si substrate provides a good calibration of the breakdown dependence TA-TFE model. Y.-H. Li et al. studied the mechanisms of GaN MISHEMT degradation, various negative bias voltages, and various temperatures, as well as dc negative gate bias stress (dc-NGBS) and ac negative gate bias stress (ac-NGBS). The dynamic R_{ON} is higher at higher temperatures, indicating the extracted trap energy levels in the GaN layer due to TA-TFE dominating the degradation of dc-NGBS [19].

The TA-TFE phenomenon occurs due to the increased negative bias applied to the gate. This causes the lateral energy band to elevate beneath the gate region within the GaN layer. Consequently, there is significant bending of the energy band at the channel edge, as illustrated by the lateral energy band depicted in Figure 8c. The cut line referred to is indicated in the x-direction from −4 µm to 14 µm in Figure 8a,b. These findings also contribute to a deeper understanding of semiconductor device physics, particularly emphasizing electron transport mechanisms and leakage paths under strong electric fields [20]. Through TCAD modeling, it was observed that the energy band diagram analysis when V_G was −15 V and V_D was 1200 V showed that the strong electric field caused significant band bending, thereby exacerbating the TA-TFE in the AlGaN/GaN HEMT of the GaN on Si substrate, as shown in Figure 8a. This phenomenon leads to the generation of electron–hole pairs, where electrons are extracted from the drain and holes accumulate in the GaN buffer, especially with the assistance of dislocation traps under high-electric-field conditions. In comparison, in a comparative analysis of AlGaN/GaN HEMT with the GaN on QST substrate, it was found that fewer holes accumulated in the GaN buffer under high electric field conditions, as shown in Figure 8b.

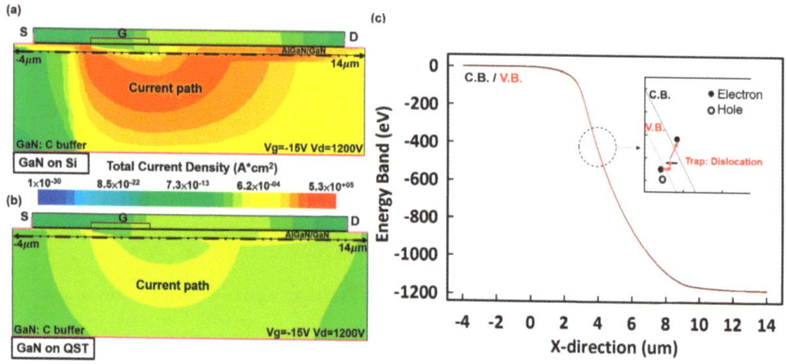

Figure 8. The total current density in the GaN buffer of (**a**) the GaN on Si substrate and (**b**) the GaN on QST substrate at V_G = −15 V, V_D = 1200 V. (**c**) Schematic of the energy band across the GaN channel along the horizontal axis and the TA-TFE mechanism.

In order to further study the breakdown mechanism of AlGaN/GaN HEMT on the GaN on Si substrate, especially under the high electric field, the electrical fitting indicated that as the drain voltage increases, the generation of GaN buffer holes also increases directly. This result shows a significant leakage current from the source to the drain. When the level of hole generation in the GaN buffer is low, the increase in leakage current between the source and drain is only slight (shown by the black line), a situation very similar to that of the GaN on QST substrate. These holes are a consequence of recombination, whereby the electrons trapped by acceptor sites are recombined, creating leakage pathways within the buffer, as shown in Figure 9a. Figure 9b displays the fitting parameters as a ratio of h+/[C]. As the ratio approaches 1, the effect of carbon in mitigating the leakage current becomes negligible.

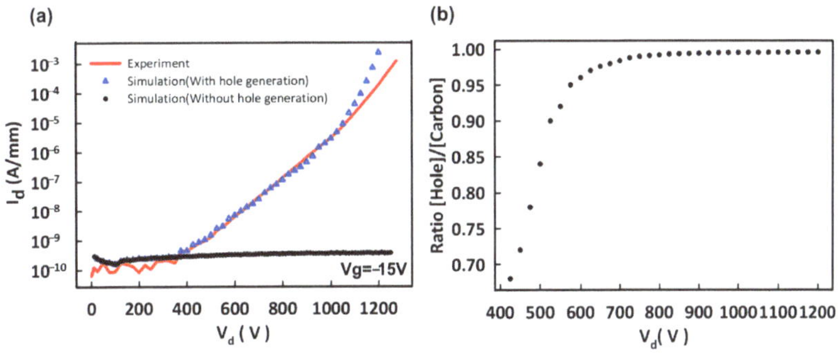

Figure 9. (**a**) GaN on Si, GaN on QST (black line) I_D-V_D curve, and (**b**) the ratio [hole]/[C] vs. V_D of GaN on Si.

GaN on Si and GaN on QST substrates may exhibit differences in the density and characteristics of traps for electrons and holes caused by dislocations and carbon doping. It has been observed that when the level of hole generation in the GaN buffer is low, the increase in leakage current between the source and drain is insignificant, similar to the behavior observed for GaN on QST substrate, and conversely shown for GaN on Si substrate. However, Figure 9 also demonstrates the efficacy of carbon in reducing the leakage current [20,21]. Therefore, while similarities in trap behavior may exist, differences in the effectiveness of carbon doping in reducing leakage current could contribute to distinct breakdown mechanisms between GaN on Si and GaN on QST substrates under high-electric-field conditions. Due to the superior lattice match and thermal expansion coefficient between QST substrates and GaN epitaxy compared to Si substrates, there may be differences in the effectiveness of carbon doping in reducing leakage current. The improved lattice matching and similar thermal expansion coefficient between QST substrates and GaN may lead to a more structurally intact GaN crystal during epitaxial growth on QST substrates, reducing the formation of dislocations and thereby decreasing the trap density for electrons and holes. In contrast, GaN grown on Si substrates may be more susceptible to dislocation formation, resulting in higher trap densities. Consequently, the effectiveness of carbon doping in reducing leakage current differed due to the distinct characteristics of QST and Si substrates, thereby influencing distinct breakdown mechanisms under high electric field conditions.

4. Conclusions

Our research into GaN-based power HEMT devices on QST compared to conventional Si substrates highlights critical insights into device performance, particularly focusing on breakdown voltage capabilities. The GaN on QST devices show a superior high-voltage off-state performance, achieving excellent breakdown voltages of up to VDS = 1500 V at room temperature. Measurements of gate/drain lag and PL have demonstrated the exceptional buffer layer quality of QST substrates. These evaluations revealed a lower prevalence of trap states and a more effective trap release mechanism compared to other substrates. The significant advantage in breakdown voltage for GaN on QST highlights the potential of QST substrates to enhance the performance and reliability of GaN-based power devices. This superior breakdown voltage capability positions GaN on QST as a promising candidate for next-generation high-efficiency power devices, underscoring the importance of substrate technology in advancing semiconductor device performance. Through detailed simulations, we effectively demonstrated how optimized device structures and doping profiles on QST substrates could significantly mitigate substrate leakage and minimize electron injection under high-bias conditions.

Author Contributions: Data analysis and structure designing, A.-C.L. and P.-T.C.; fabrication, C.-H.C., Y.-C.C. and Y.-L.C.; writing—review and editing, A.-C.L. and H.-C.C.; supervision, H.-C.K. and S.-T.C.; project administration, H.-C.C., I.-Y.H. and H.-C.K. All authors have read and agreed to the published version of the manuscript.

Funding: National Science and Technology Council, Taiwan (112-2218-E-008-007- and 112-2222-E-110-008-).

Data Availability Statement: The original contributions presented in the study are included in the article; further inquiries can be directed to the corresponding authors.

Acknowledgments: The authors would like to thank Chang Gung University, Taoyuan, along with Hsien-Chin Chiu for the process support; the Taiwan Semiconductor Research Institute (TSRI) for semiconductor-related equipment support; and the Hon Hai Research Institute for their helpful discussion.

Conflicts of Interest: The authors declare no conflicts of interest.

References

1. Haziq, M.; Falina, S.; Manaf, A.A.; Kawarada, H.; Syamsul, M. Challenges and Opportunities for High-Power and High-Frequency AlGaN/GaN High-Electron-Mobility Transistor (HEMT) Applications: A Review. *Micromachines* **2022**, *13*, 2133. [CrossRef] [PubMed]
2. Hsu, L.-H.; Lai, Y.-Y.; Tu, P.-T.; Langpoklakpam, C.; Chang, Y.-T.; Huang, Y.-W.; Lee, W.-C.; Tzou, A.-J.; Cheng, Y.-J.; Lin, C.-H.; et al. Development of GaN HEMTs Fabricated on Silicon, Silicon-on-Insulator, and Engineered Substrates and the Heterogeneous Integration. *Micromachines* **2021**, *12*, 1159. [CrossRef] [PubMed]
3. Kozak, J.P.; Zhang, R.; Porter, M.; Song, Q.; Liu, J.; Wang, B.; Wang, R.; Saito, W.; Zhang, Y. Stability, Reliability, and Robustness of GaN Power Devices: A Review. *IEEE Trans. Power Electron.* **2023**, *38*, 8442–8471. [CrossRef]
4. Liu, A.C.; Huang, Y.W.; Lin, C.H.; Dong, Y.J.; Lai, Y.Y.; Ting, C.C.; Tu, P.T.; Yeh, P.C.; Chen, H.C.; Kuo, H.C. Effect of O_2 plasma surface treatment on gate leakage current in AlGaN/GaN HEMT. In Proceedings of the 2023 International VLSI Symposium on Technology, Systems and Applications (VLSI-TSA/VLSI-DAT), Hsinchu, Taiwan, 17–20 April 2023; pp. 1–2.
5. Gamarra, P.; Lacam, C.; Tordjman, M.; Splettstösser, J.; Schauwecker, B.; di Forte-Poisson, M.A. Optimization of a carbon doped buffer layer for AlGaN/GaN HEMT devices. *J. Cryst. Growth* **2015**, *2015*, 232–236. [CrossRef]
6. Keshmiri, N.; Wang, D.; Agrawal, B.; Hou, R.; Emadi, A. Current Status and Future Trends of GaN HEMTs in Electrified Transportation. *IEEE Access* **2020**, *8*, 70553–70571. [CrossRef]
7. Richter, E.; Beyer, F.C.; Zimmermann, F.; Gärtner, G.; Irmscher, K.; Gamov, I.; Heitmann, J.; Weyers, M.; Tränkle, G. Growth and Properties of Intentionally Carbon-Doped GaN Layers. *Cryst. Res. Technol.* **2020**, *55*, 1900129. [CrossRef]
8. Tzou, A.J.; Chu, K.H.; Lin, I.F.; Østreng, E.; Fang, Y.S.; Wu, X.P.; Wu, B.W.; Shen, C.H.; Shieh, J.M.; Yeh, W.K.; et al. AlN Surface Passivation of GaN-Based High Electron Mobility Transistors by Plasma-Enhanced Atomic Layer Deposition. *Nanoscale Res. Lett.* **2017**, *12*, 315. [CrossRef] [PubMed]
9. Sayadi, L.; Iannaccone, G.; Haeberlen, O.; Fiori, G.; Tomberger, M.; Knuuttila, L.O.; Curatola, G. The role of silicon substrate on the leakage current through GaN-on-Si epitaxial layers. *IEEE Trans. Electron Devices* **2018**, *65*, 51–58. [CrossRef]
10. Borga, M.; Meneghini, M.; Stoffels, S.; Li, X.; Posthuma, N.; Van Hove, M.; Decoutere, S.; Meneghesso, G.; Zanoni, E. Impact of substrate resistivity on the vertical leakage, breakdown, and trapping in GaN-on-Si E-mode HEMTs. *IEEE Trans. Electron Devices* **2018**, *65*, 2765–2770. [CrossRef]
11. Odnoblyudov, V.; Aktas, O.; Basceri, C. 12-Fundamentals of CTE-matched QST® substrate technology. In *Woodhead Publishing Series in Electronic and Optical Materials, Thermal Management of Gallium Nitride Electronics*; Woodhead Publishing: Sawston, UK, 2022; pp. 251–274.
12. Benito, J.A.; Jorba, J.; Manero, J.M.; Roca, A. Change of Young's modulus of cold-deformed pure iron in a tensile test. *Met. Mater. Trans. A* **2005**, *36*, 3317–3324. [CrossRef]
13. Boughrara, N.; Benzarti, Z.; Khalfallah, A.; Evaristo, M.; Cavaleiro, A. Comparative study on the nanomechanical behavior and physical properties influenced by the epitaxial growth mechanisms of GaN thin films. *Appl. Surf. Sci.* **2022**, *579*, 152188. [CrossRef]
14. Beleniotis, P.; Schnieder, F.; Krause, S.; Haque, S.; Rudolph, M. An efficient drain-lag model for microwave GaN HEMTs based on ASM-HEMT. *Int. J. Microw. Wirel. Technol.* **2022**, *14*, 134–142. [CrossRef]
15. Nazir, M.S.; Pampori, A.; Dangi, R.; Kushwaha, P.; Yadav, E.; Sinha, S.; Chauhan, Y.S. Characterization and modeling of drain lag using a modified RC network in the ASM-HEMT framework. *Solid-State Electron.* **2023**, *199*, 108490. [CrossRef]
16. Liu, A.-C.; Tu, P.-T.; Chen, H.-C.; Lai, Y.-Y.; Yeh, P.-C.; Kuo, H.-C. Improving Performance and Breakdown Voltage in Nor-mally-Off GaN Recessed Gate MIS-HEMTs Using Atomic Layer Etching and Gate Field Plate for High-Power Device Applica-tions. *Micromachines* **2023**, *14*, 1582. [CrossRef] [PubMed]

17. Wang, H.; Lin, Y.; Jiang, J.; Dong, D.; Ji, F.; Zhang, M.; Jiang, M.; Gan, W.; Li, H.; Wang, M.; et al. Investigation of Thermally Induced Threshold Voltage Shift in Normally-OFF p-GaN Gate HEMTs. *IEEE Trans. Electron Devices* **2022**, *69*, 2287–2292. [CrossRef]
18. Pavlidis, G.; Yates, L.; Kendig, D.; Lo, C.F.; Marchand, H.; Barabadi, B.; Graham, S. Thermal Performance of GaN/Si HEMTs Using Near-Bandgap Thermoreflectance Imaging. *IEEE Trans. Electron Devices* **2020**, *67*, 822–827. [CrossRef]
19. Lee, Y.H.; Chang, K.C.; Tai, M.C.; Wang, Y.X.; Lin, H.N.; Yeh, Y.H.; Kuo, H.M.; Tsai, X.Y.; Lee, J.; Huang, I.; et al. The Transition of Threshold Voltage Shift of Al_2O_3/Si_3N_4 AlGaN/GaN MIS-HEMTs Under Negative Gate Bias Stress From DC to AC. *IEEE Trans. Electron Devices* **2024**, *71*, 1792–1797. [CrossRef]
20. Ma, M.; Cao, Y.; Lv, H.; Wang, Z.; Zhang, X.; Chen, C.; Wu, L.; Lv, L.; Zheng, X.; Tian, W.; et al. Effect of Acceptor Traps in GaN Buffer Layer on Breakdown Performance of AlGaN/GaN HEMTs. *Micromachines* **2023**, *14*, 79. [CrossRef] [PubMed]
21. Kurbanbaeva, D.M.; Lashkov, A.V.; Tsarik, K.A. Investigation of the Influence of the Buffer Layer Design in a GaN HEMT Transistor on the Breakdown Characteristics. *Russ. Microelectron.* **2023**, *52* (Suppl. S1), S14–S19. [CrossRef]

electronics

Article

Highly Reliable Short-Circuit Protection Circuits for Gallium Nitride High-Electron-Mobility Transistors

Chul-Min Kim, Hyun-Soo Yoon, Jong-Soo Kim and Nam-Joon Kim *

Department of Electrical Engineering, Daejin University, Pocheon 11159, Republic of Korea; cmkim@daejin.ac.kr (C.-M.K.); ab223015@daejin.ac.kr (H.-S.Y.); jskim2@daejin.ac.kr (J.-S.K.)
* Correspondence: njkim@daejin.ac.kr

Abstract: This paper presents a circuit for detecting and protecting against short circuits in E-mode gallium nitride high-electron-mobility transistors (GaN HEMTs) and analyzes the protection performance of the circuit. GaN HEMTs possess fast switching characteristics that enable high efficiency and power density in power conversion devices. However, these characteristics also pose challenges in protecting against short circuits and overcurrent situations. The proposed method detects short-circuit events by monitoring an instantaneous drop in the DC bus voltage of a circuit with GaN HEMTs applied and uses a bandpass filter to prevent the malfunction of the short-circuit protection circuit during normal switching and ensure highly reliable operation. Using this method, the short-circuit detection time of E-mode GaN HEMTs can be reduced to 257 ns, successfully protecting the device without malfunctions even in severe short-circuit situations occurring at high DC link voltages.

Keywords: GaN HEMT; short-circuit protection; reliability

Citation: Kim, C.-M.; Yoon, H.-S.; Kim, J.-S.; Kim, N.-J. Highly Reliable Short-Circuit Protection Circuits for Gallium Nitride High-Electron-Mobility Transistors. *Electronics* **2024**, *13*, 1203. https://doi.org/10.3390/electronics13071203

Academic Editor: Francesco Giuseppe Della Corte

Received: 19 February 2024
Revised: 17 March 2024
Accepted: 22 March 2024
Published: 25 March 2024

1. Introduction

Power semiconductors have been extensively studied as essential components of various power conversion devices for a long time. Among them, research has been focused on silicon (Si)-based power semiconductors, which are applied in various fields to enhance the efficiency and power density of power conversion devices. However, due to the physical limitations of Si-based power semiconductors, research efforts are actively pursuing new components that can overcome these limitations. Recently, there has been an increasing demand for GaN HEMTs, which possess wide-bandgap (WBG) characteristics, among power semiconductors [1].

GaN HEMT generates a two-dimensional electron gas (2DEG) structure based on the heterojunction of AlGaN and GaN. This results in high charge mobility and density, as well as reduced parasitic capacitance within the device, leading to advantages such as high-speed switching and low switching losses compared to Si power semiconductors. Therefore, the application of GaN HEMT in power conversion devices effectively enhances device efficiency and power density. However, the high-speed switching characteristics of GaN HEMTs can pose significant drawbacks in terms of systems with short-circuit and overcurrent aspects. GaN HEMTs typically have very low gate–source threshold voltages, making them susceptible to faulty turn-on due to high dv/dt and di/dt resulting from high-speed switching, thereby increasing the risk of short-circuit occurrence. The short-circuit performance of GaN HEMTs has been evaluated in [2], and although the duration varies depending on the applied voltage magnitude, the short-circuit withstand time of GaN HEMTs is very low on the order of hundreds of nanoseconds. Additionally, GaN-based power semiconductors have much smaller package sizes compared to conventional Si power semiconductors, which provides an advantage in terms of power density. However, due to the difficulty in dissipating heat through the package in the event of a short circuit, it is necessary to protect the device from short-circuit situations before critical failures or permanent performance degradation can occur.

Various power semiconductors with short-circuit withstand times can be seen in Figure 1. Examining the short-circuit withstand times tested up to a DC bus voltage of 400 V, E-mode GaN HEMTs can withstand durations of 520 ns and 400 ns at 350 V and 400 V conditions, respectively.

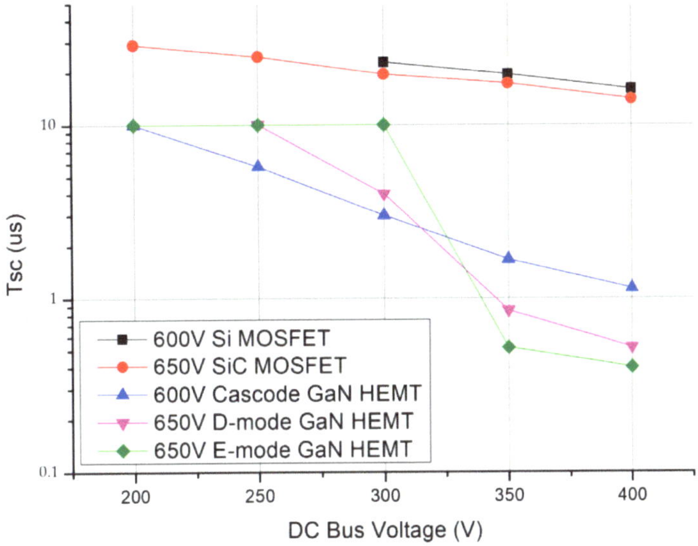

Figure 1. Short-circuit withstand times with voltage at 600 V/650 V for power semiconductors [3–6].

Additionally, at DC bus voltages below 350 V, E-mode GaN HEMTs exhibit relatively superior short-circuit endurance times (T_{SC} > 10 µs). Cascode GaN HEMTs demonstrate shorter short-circuit withstand times compared to E-mode GaN HEMTs at voltages below 300 V. However, at voltages of 350 V and above, Cascode GaN HEMTs exhibit short-circuit withstand times in the range of 1~2 µs [3–6].

Therefore, for GaN-based power semiconductors, which require very fast short-circuit protection compared to Si-based power semiconductors, it is generally challenging to apply commercially available gate drivers with saturation protection functions that require response times of 1 µs or more. Therefore, several short-circuit detection and protection techniques applicable to GaN-based power semiconductors have been explored. Incorporating a Shunt Resistor into the power loop in series to detect short-circuit currents presents a simple and universally adaptable solution for all systems. This current sensing method necessitates high-precision resistors and swift ADCs to ensure signal accuracy and detection speed. Refs [7,8] demonstrate the application of Shunt Resistors for short-circuit detection and protection in systems employing SiC and GaN devices, with detection and protection times recorded at 150 ns and 60 ns, respectively. However, the primary disadvantage of using Shunt Resistors for current sensing lies in power loss; high currents in high-power systems can lead to significant power loss across the Shunt Resistors, and low-power systems might require larger resistors to maintain signal accuracy, thereby reducing efficiency in low-power applications. Additionally, incorporating Shunt Resistors can increase parasitic inductance within the circuit, potentially affecting switching performance. Another method is to detect the high di/dt that occurs during a short circuit using the stray inductance inside the power semiconductor package. In [9–11], short-circuit protection is based on the stray inductance inside the SiC MOSFET package, and the times required for short-circuit detection and final protection are 60 ns, 140 ns and 1 µs, respectively. Although this approach can be used to achieve fast short-circuit protection, it requires a device with a Kelvin source to use the internal stray inductance and requires the measurement of the internal stray inductance. In addition, unlike the SiC used in the reference, GaN devices have very small packages and have an internal stray inductance close to zero, making it difficult to apply the method even if a Kelvin source is present. Another short-circuit protection method is the use of Rogowski coils. In [12,13], a PCB-based auxiliary Rogowski coil is used to implement short circuit and overcurrent protection by di/dt measurements. The time required for the final protection is about 700 ns [12]. The method using a Rogowski coil allows fast operation by directly measuring the current, but the Rogowski coil current sensor requires an integrator circuit, making the circuit and implementation complex, and an output offset voltage is also generated due to the operational amplifier characteristics of the integrator circuit. In addition, the Rogowski coil has to be shielded as switching noise affects the current measurement. To detect overcurrent situations, refs. [14,15] utilize a portion of the power loop inductance. This is achieved by optimizing the PCB layout and reducing detection loss. The time required for short-circuit protection using this method is approximately 370 ns and 250 ns, respectively. The power loop inductance is implemented using PCB traces to minimize volume impact. When a high rate of change in the current is applied to the circuit due to a short circuit, the voltage induced in the power loop inductance exceeds a preset threshold, resulting in protective action. However, the practicality of this method is limited due to the inability to directly detect the current amplitude and voltage variability induced in the coil under different short-circuit conditions [16]. Another commonly used method is desaturation protection, which senses the drain-to-source voltage of a device. Desaturation protection can be easily implemented using commercially available gate driver devices. Refs. [17,18] utilized desaturation protection to achieve protection within 360 ns and 125 ns. However, this desaturation protection method was delayed due to the blanking time to avoid failure [19], and the additional diodes added capacitive load in parallel with the output capacitance on the power semiconductor, increasing switching losses [20]. Furthermore, the temperature dependence of $R_{ds,on}$ makes it challenging to define reference levels for desaturation protection [21]. Recently, research has been conducted on short circuit detection methods based on the DC bus voltage drop. In [6,22,23], short-circuit protection is performed based on the DC bus voltage, which is the voltage across the high-side device drain and low-side device source. The DC bus voltage exhibits an instantaneous voltage drop when a short circuit occurs because a high di/dt current flows through parasitic inductances present in the power loop. This voltage drop is used to detect whether a short circuit has occurred, and the response time of the detection circuit can be within hundreds of nanoseconds. In [22,23], the circuit is protected from a short circuit within 280 ns and 370 ns, respectively. However, when using DC bus voltage sensing, the voltage dip caused by the normal switching of the device can result in false short-circuit protection behavior.

In this paper, short circuits are detected by monitoring instantaneous drops in the DC bus voltage. In addition, a method is proposed that uses a bandpass filter to distinguish between voltage drops that occur during normal switching and short circuits. Section 2 of this paper explains the instantaneous voltage drop that occurs during a short circuit due to parasitic inductance in the circuit. Additionally, a bandpass filter is introduced to distinguish this from similar voltage drops that occur during normal switching. All analyses are performed based on PSPICE simulations.

Section 3 presents a short-circuit detection and protection circuit by applying the content covered in Section 2, and the operation of the circuit is confirmed through the PSPICE simulation. Experimental results with the proposed circuit are presented in Section 4. Section 5 concludes the paper.

2. Short Circuit Detection Method Based on DC Bus Voltage Drop

A half-bridge structure circuit is commonly used in power converters and has the structure shown in Figure 2. Among the components constituting the circuit, the DC-Link Capacitor serves to balance the instantaneous power difference between the input power and output load and minimize the voltage change in the DC-Link, and electrolytic capacitors with relatively large capacitance are mainly used. Additionally, filter capacitors are placed near power semiconductors and DC-Link Capacitors to minimize the stray inductance components in the power loop. Ceramic capacitors with excellent high-frequency characteristics are preferred for this purpose.

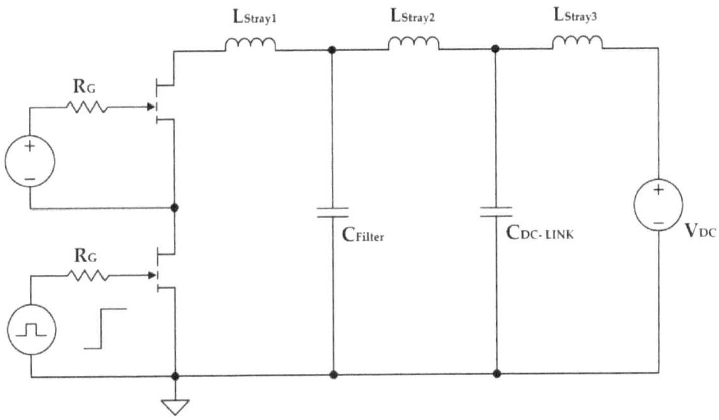

Figure 2. Short-circuit test circuit based on GaN HEMTs.

Stray inductance components denoted as L_{Stray}, are mostly formed by the PCB layout and can interfere with the stable operation of high-speed-switching GaN devices. Consequently, various research efforts are underway to minimize these components to ensure reliable operation. Despite these stray inductance components inevitably disrupting normal system operation, they can be effectively utilized to detect short circuits within the circuit.

2.1. Instantaneous Voltage Drop Due to Short Circuit

To measure the drop in the DC bus voltage during a short circuit, it is crucial to minimize the influence of stray inductance. This can be achieved by measuring the high-side device drain and low-side device source sides, as depicted in Figure 3a. Under normal switching conditions, as shown in Figure 3b, the DC bus voltage remains equal to the input voltage V_{DC}. However, during the occurrence of a short circuit, an instantaneous large di/dt current flows through the half-bridge, resulting in an instantaneous voltage drop in the DC bus voltage, as illustrated in Figure 3c, due to the circuit's stray inductance. The amplitude of this voltage drop is proportional to the voltage across L_{Stray1}, which is equal to Equation (1), and the stray inductance in the circuit is determined during PCB fabrication, so the amplitude of the voltage drop is determined by di/dt.

$$V_{L\_stray1} = L_{Stray1}\frac{di}{dt} \quad (1)$$

The slope at the time of a short circuit is determined by the switch's turn-on time, which depends on the gate resistance value. Therefore, when detecting a short circuit through a drop in the DC bus voltage, it is essential to consider the gate resistance value used in the circuit. Although some power semiconductor datasheets may provide a correlation between the gate resistance value and switching time, this information is not available for the GS-065-011-1-L from GaNsystem used in this paper. Therefore, the SPICE model provided by the manufacturer can be used to determine the main phenomena that occur during a short circuit. Therefore, in this paper, an LTSPICE-based simulation circuit was constructed using the SPICE model of the device, as shown in Figure 4. Then, an analysis was performed on the instantaneous voltage drop that appears in the DC bus voltage [24].

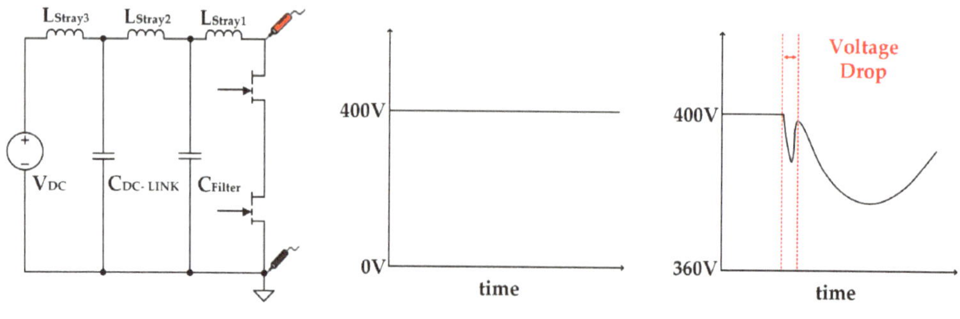

(a) (b) (c)

Figure 3. DC bus voltage-based short-circuit detection method (**a**) detection point. (**b**) DC bus voltage waveform during normal operation. (**c**) DC bus voltage waveform when a short circuit occurs.

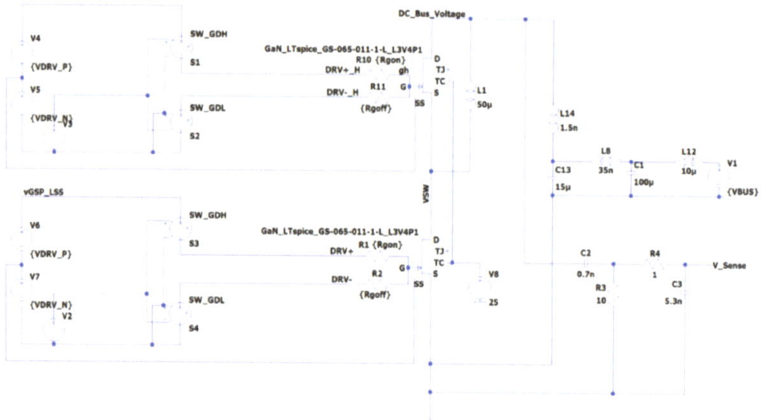

Figure 4. Simulation circuit diagram based on LT spice.

First, a simulation was performed to determine the instantaneous drop in DC bus voltage that occurs in a short circuit situation. Figure 4 shows the circuit used in the simulation, where the high-side GaN HEMT is always on, and the low-side GaN HEMT turns on 20 ns after the start of the simulation. There may be differences in simulation results depending on the presence or absence of parasitic resistance components present in the circuit. However, because the parasitic resistance components present in the PCB pattern and DC power connection line are sufficiently small, they do not have a significant impact on the simulation results. Therefore, the resistance component was not considered in the simulation. Additionally, because the size of the capacitor in the circuit was sufficiently large, changes in stray inductance components except L_{Stray1} did not affect the DC bus voltage drop. Therefore, in the simulation, the voltage drop level of the DC bus voltage was checked by changing the values of L_{Stray1} and C_{Filter}, which are key components that directly affect the instantaneous drop in the DC bus voltage.

Table 1 displays the main parameters used in the simulation, and Figure 5 shows the results. Figure 5a illustrates the effect of changing the value of L_{Stray1}, the stray inductance between the filter capacitor and the half-bridge in the circuit, on the voltage drop, and it can be seen that the value of L_{Stray1} directly affects the magnitude of the voltage drop that occurs at the moment of the short circuit, and the larger the value of L_{Stray1}, the larger the amplitude of the voltage drop. Figure 5b shows the voltage drop waveform at the short circuit according to the value of the filter capacitor, and it can be seen that the value of the filter capacitor does not affect the magnitude of the voltage drop at the short circuit, but it does affect the magnitude of the voltage in the saturation region after the instantaneous voltage drop.

Table 1. Simulation parameter.

	L_{Stray1}	L_{Stray2}	C_{Filter}	$C_{DC\text{-}Link}$
Value	0.2 nH 0.5 nH 1 nH 1.5 nH	35 nH	0.2 μF 0.5 μF 1 μF 1.5 μF	100 μF

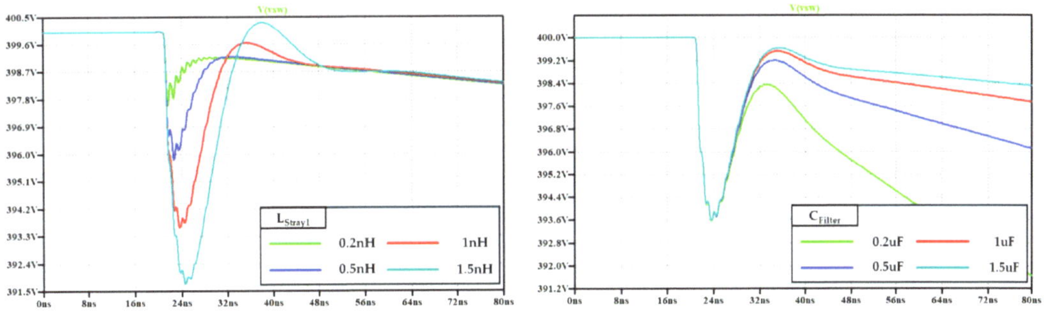

(**a**) (**b**)

Figure 5. Simulation results of DC bus voltage drop by parameter changes in (**a**) stray inductance (L_{Stray1}). (**b**) Filter capacitor (C_{Filter}).

The simulation results show that the value of L_{Stray1} and the gate resistors present in the circuit have a significant impact on the instantaneous voltage drop of the DC bus voltage used in the short-circuit protection circuit. In PCB design, the value of L_{Stray1} typically ranges from a few to tens of nanoHenrys. Additionally, the gate resistor is selected based on factors such as the dv/dt, di/dt, and EMI of the circuit. Depending on the values of these factors, an instantaneous drop in the DC bus voltage is expected to provide meaningful results for short-circuit detection. Figure 6 illustrates the amplitude of the DC bus voltage drop as a function of gate resistance and the L_{Stray1} value when the input voltage is 400 V.

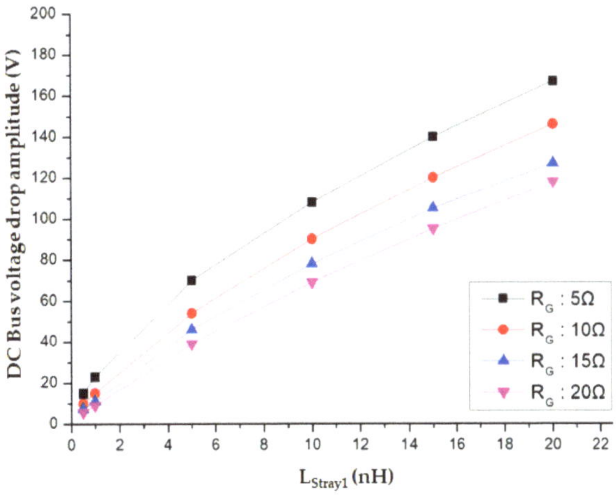

Figure 6. DC bus voltage drops depending on gate resistance and L_{Stray1}.

2.2. Instantaneous Voltage Drop Due to Normal Switching

However, the instantaneous voltage drop in the DC bus voltage does not only occur during short-circuit conditions. Figure 7a illustrates the situation when the power semiconductor device under test (DUT) is turned on during a typical double pulse test. At this point, the current flows through the load inductor to form the test current waveform. Figure 7b illustrates that when the DUT is turned on while the upper device is already on, the short circuit current flows through both the upper and lower devices, causing an instantaneous voltage drop similar to that observed earlier. Figure 8 shows the simulation results of implementing a typical double pulse test, observing the DC bus voltage at the moment the low-side device turns on. Similar to the voltage drop observed during a short circuit, a voltage drop is also identified during normal switching operations. It is important to differentiate between the voltage drop in the DC bus voltage that occurs during normal switching and that which occurs during a short circuit.

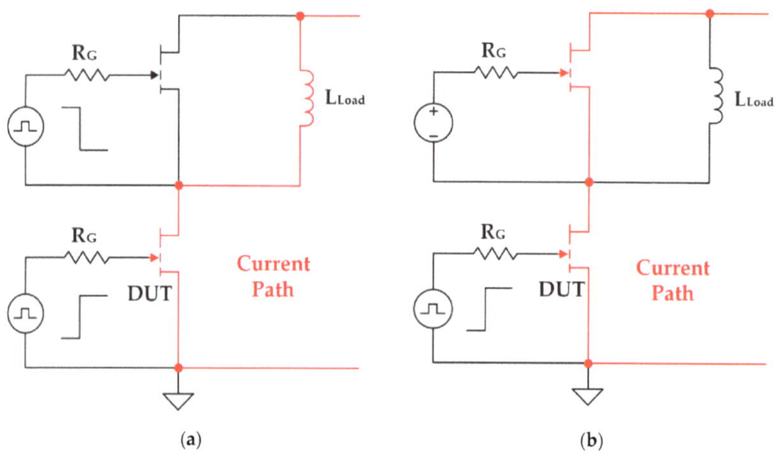

Figure 7. Current path when DC bus voltage drops: (**a**) double pulse test; (**b**) short circuit situation.

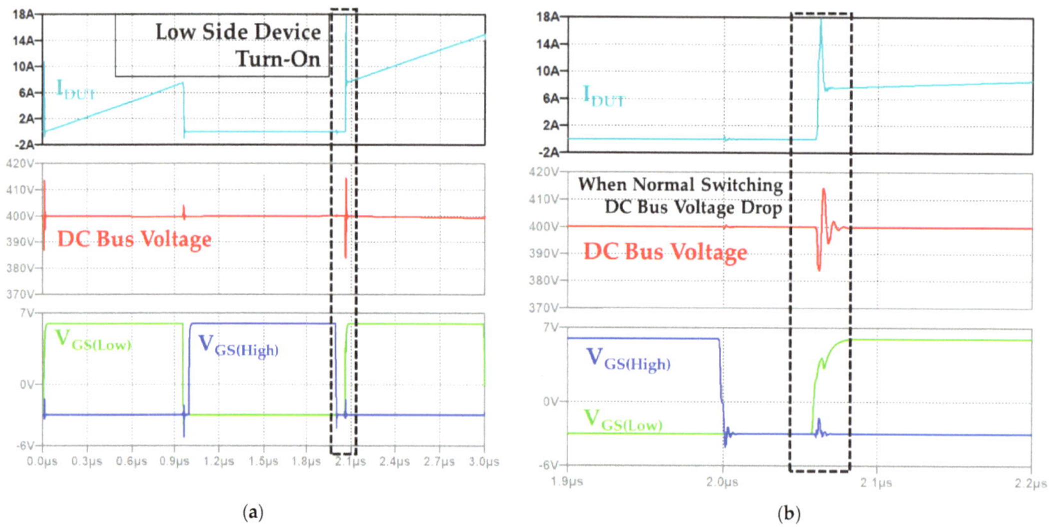

Figure 8. Double pulse tests waveforms: (**a**) full waveform; (**b**) zoomed waveform.

Figure 9a shows the voltage drops observed by SPICE simulations during switching and short-circuiting events. It can be seen that the voltage drop during a short-circuit condition is of a higher magnitude compared to the voltage drop during a switching condition. Bandpass filtering can be used to distinguish between these two voltage drops. Figure 9b presents the FFT analysis of the DC bus voltage during switching and short circuit events. Since the difference between the two voltages is greatest in the 20~30 MHz frequency range, the voltage drops can be distinguished by selecting the appropriate frequency band as the bandwidth using a bandpass filter. Therefore, the bandpass filter is designed to pass this frequency range. Figure 10 shows the schematic of the proposed bandpass filter, which consists of components C_1, R_1, C_2, and R_2.

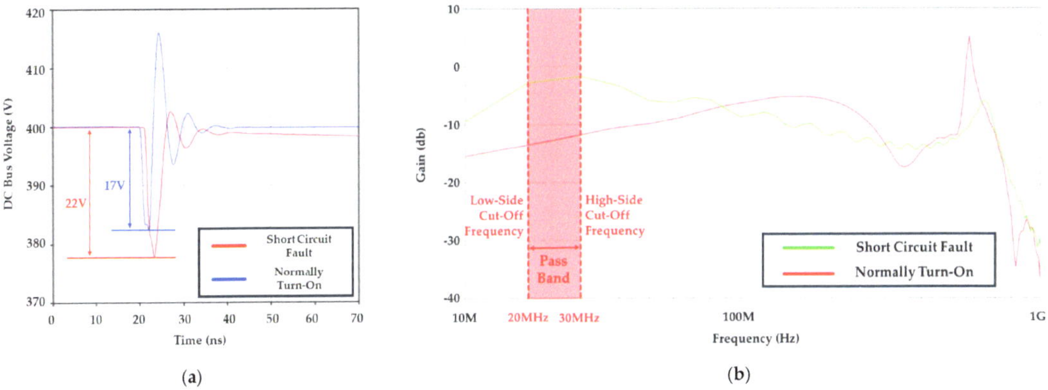

Figure 9. Comparison of switching and short circuit events (**a**) DC bus voltage drop; (**b**) fast Fourier transform (FFT).

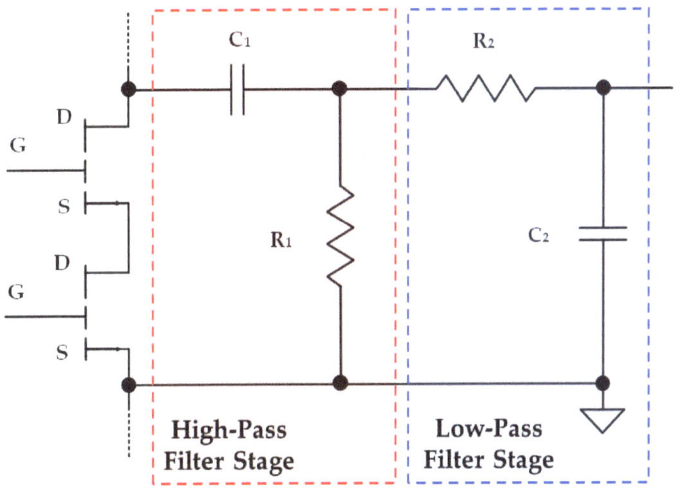

Figure 10. Bandpass filter for short circuit signal detection.

The bandpass filter is composed of a combination of low-pass and high-pass filters, which are utilized to detect voltage drop signals while blocking the DC voltage and high-frequency switching noise. As shown in Figure 10, the first stage of the circuit consists of a passive RC high-pass filter. This filter allows signals with frequencies above the low cutoff frequency ($f_{c\text{-}low}$) to pass while attenuating signals with frequencies below $f_{c\text{-}low}$. The low cutoff frequency is given by Equation (2).

$$f_{c-low} = \frac{1}{2\pi R_1 C_1} \quad (2)$$

The second stage of the circuit is a passive RC low-pass filter. This filter allows signals with frequencies below the high cutoff frequency ($f_{c\text{-}high}$) to pass while attenuating signals with frequencies above $f_{c\text{-}high}$. The high cutoff frequency is calculated using Equation (3).

$$f_{c-high} = \frac{1}{2\pi R_2 C_2} \quad (3)$$

The specific frequency range through which a bandpass filter allows signals to pass is called the bandwidth. The bandwidth is calculated as the difference between the high and low cutoff frequencies, as described by Equation (4).

$$\text{Bandwidth} = f_{c-high} - f_{c-low} \quad (4)$$

In Figure 9b, the high-pass frequency is set to 20 MHz, and the low-pass frequency is set to 30 MHz, resulting in a bandwidth of 10 MHz. As can be seen from Equations (2) and (3), there are infinite combinations of resistors and capacitors that can satisfy the relevant frequency. Therefore, among the combinations that can satisfy the corresponding frequency, C_1: 790 pF, R_1: 10 Ω, R_2: 1 Ω and C_2: 5.3 nF were promoted. However, to apply this value, since there were no exact matching values among commercial products, resistors and capacitors with the closest values among commercial products were used. The filter element values of the final bandpass filter used were C_1: 750 pF, R_1: 10 Ω, R_2: 1 Ω and C_2: 5.1 nF. Therefore, the actual applied high-pass frequency was 21 MHz, and the low-pass frequency was 31 MHz.

3. Proposed Short-Circuit Protection Circuit

This section presents a proposed fast protection method based on DC bus voltage detection using the proposed detection signal and detection circuit, as shown in Figure 11. The voltage detected in the detection circuit is transferred to the input of the comparator. When a short circuit occurs, a voltage drop in the DC bus voltage is detected, which triggers the detection signal in the comparator. Since the voltage drop in the DC bus voltage that occurs during a short circuit is transient, a logic control circuit such as an SR latch can be used, as shown in Figure 11. Finally, the generated error signal is applied to the disable pin of the gate driver to disable it.

To verify the proper operation of the proposed circuit, SPICE-based simulations were performed to simulate normal switching and fault conditions and confirm the fault signal detection and protection actions. Figure 12 shows the simulation results using the double-pulse test circuit, where the switching instance of the DC bus voltage and the occurrence of the fault signal are observed. At approximately 2.05 μs, the low-side device was turned on, resulting in an instantaneous voltage drop in the DC bus voltage. However, the fault signal detected by the sensing circuit, shown as V_{Sense}, only dropped to −1.8 V and did not fall below the −2 V threshold. Consequently, the signal required to disable the gate driver was not generated.

Figure 13 shows the waveform that confirms the DC bus voltage and short-circuit signal generation during a short circuit using the same circuit. Since the high-side device is always turned on, a short circuit occurs at 20 ns when the low-side device turns on. At this time, an instantaneous voltage drop in the DC bus voltage and the short-circuit signal V_{Sense} detected by the short-circuit detection circuit drops below the −2 V threshold. The short-circuit signal finally generates a protective operation signal through the comparator, and the signal is formed 93 ns after the short-circuit occurs and is applied to the gate driver's disable pin to protect the circuit from short-circuit.

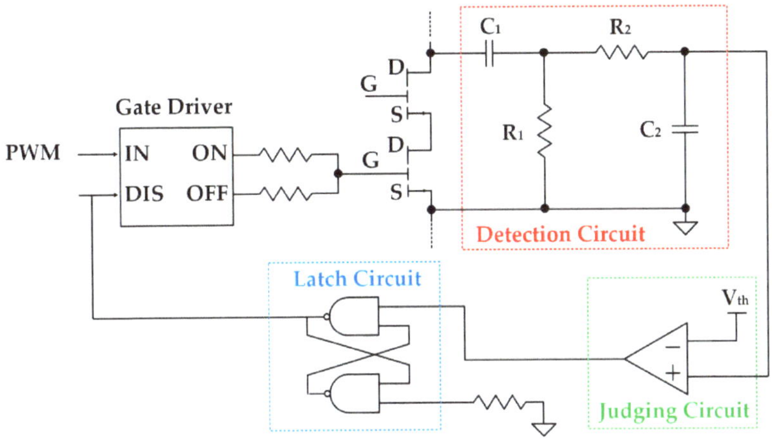

Figure 11. Proposed short-circuit protection circuit.

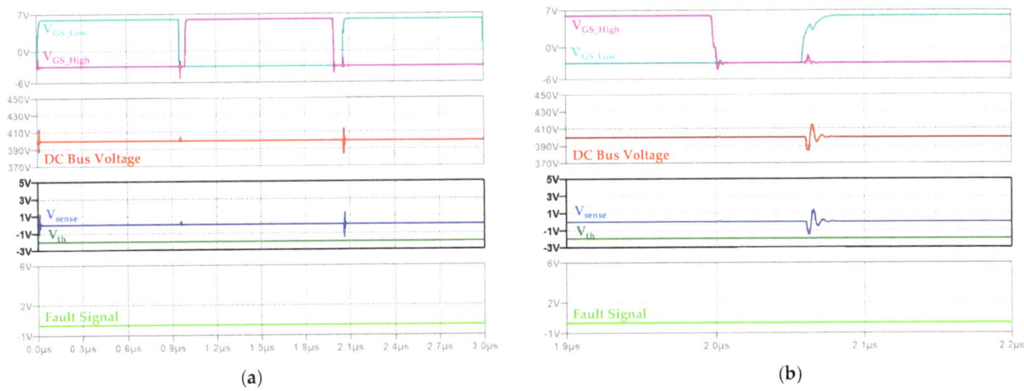

Figure 12. Simulation waveforms during switching (**a**) full waveforms. (**b**) Zoomed waveforms.

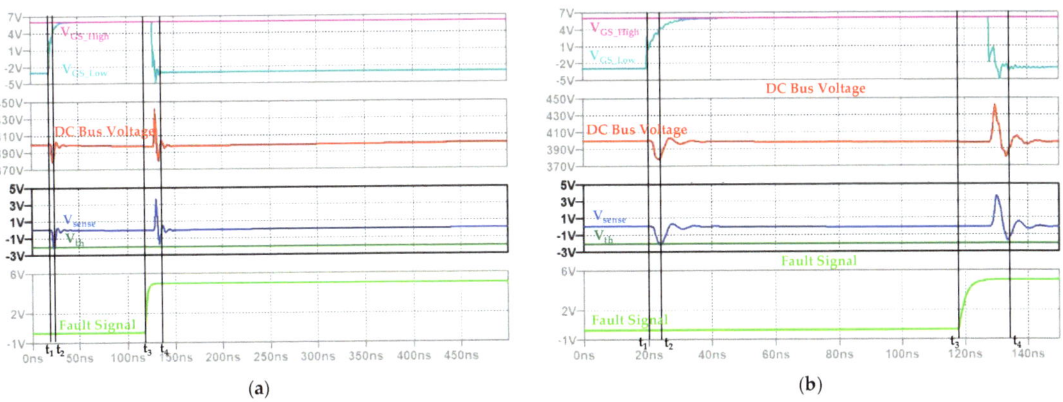

Figure 13. Short circuit simulation waveform (**a**) full waveform. (**b**) Zoomed waveform.

The simulation results show that when a short-circuit occurs in a circuit using GaN HEMT, the protection based on the short-circuit detection circuit proposed in this paper is performed correctly. The simulation waveform shows that the final signal for short-circuit protection is formed 93 ns after the short-circuit occurs, and it takes only 105 ns to turn off the GaN HEMT. The proposed circuit is more efficient in protecting against short circuits than the existing desaturation method, as it requires a much shorter time for short-circuit protection.

4. Experimental Results

To verify the feasibility of our proposed circuit, a test board was fabricated like the circuit in Figure 14 using the same structure as in the simulation. The test board used in the experiment is shown in Figure 15a, and the complete test bed setup is shown in Figure 15b. The parameters used in the experiment are detailed in Table 2. The test bed enables the execution of double pulse and short circuit tests, and the GaN system's GS-065-011-1-L device is used for the experiment. The experiment was carried out at an ambient temperature of 25 °C.

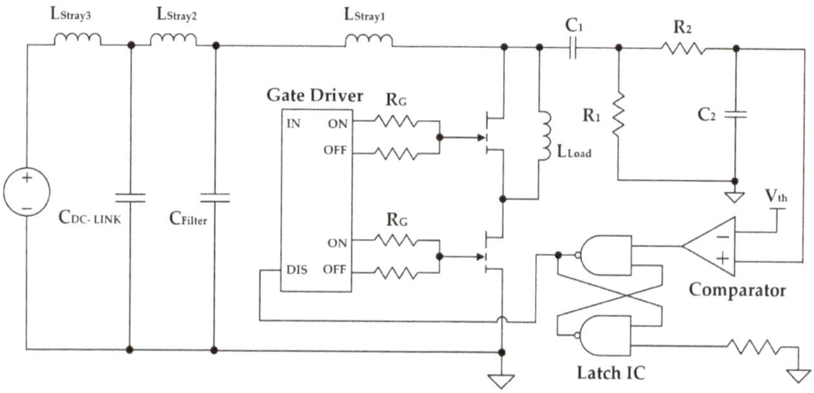

Figure 14. Circuit diagram of the board used in the experiment.

(**a**) (**b**)

Figure 15. Experimental setup: (**a**) test board. (**b**) Test setup.

Table 2. Experimental parameter.

Parameter	$C_{DC\text{-}Link}$	C_{Filter}	L_{Stray1}	L_{Stray1}
Value	100 μF	4 μF	1.5 nH	35 nH
Parameter	V_{DC}	L_{Load}	R_{G_ON}	V_{On}/V_{Off}
Value	400 V	100 μH	15 Ω	6 V/−3 V
Parameter	R_1	C_1	R_2	C_2
Value	10 Ω	750 pF	1 Ω	5.1 nF

Figure 16 shows the protective operation of the proposed circuit. Channel 1 (CH 1) represents the gate–source voltage of the low-side GaN HEMT, while Channel 2 (CH 2) represents the DC bus voltage. Channel 3 (CH 3) is the output signal of the bandpass filter, which detects voltage drops, and Channel 4 (CH 4) represents the fault signal applied to the disable pin of the gate driver.

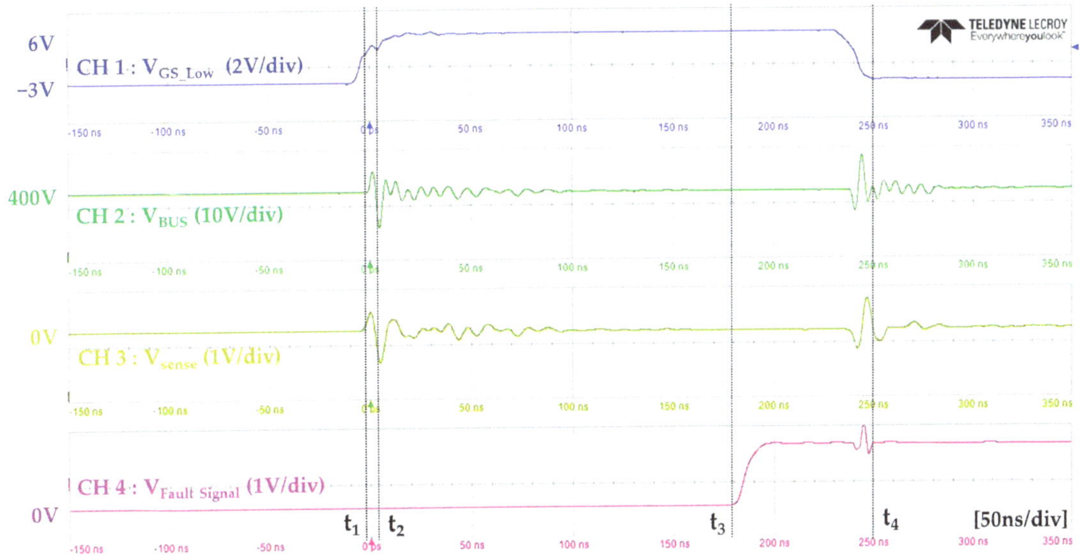

Figure 16. Experimental waveforms when a short circuit occurs.

As the gate voltage of the low-side GaN HEMT, corresponding to CH 1 in Figure 16 rises from −3 V to 6 V, the low-side device turns on. At point t_1, when the low-side gate–source voltage exceeds the device's threshold voltage, the short-circuit current begins to flow, and distortion occurs in the DC bus voltage. At time t_2, the instantaneous voltage drop in the DC bus voltage corresponding to CH 2 appears the largest. At this time, there is a voltage drop of 25 V from 400 V to 375 V. This voltage drop passes through a bandpass filter and appears on CH 3. The output of the bandpass filter maintains V_{Sense} at 0 V in a normal condition. However, when a short circuit occurs, the voltage shows an instantaneous drop similar to the DC bus voltage, decreasing to −2.2 V at time t_2. This is 10 ns after the device is turned on and the short circuit occurs. Since the magnitude of the voltage V_{th} applied to the negative pin of the comparator is −2 V, a fault signal is formed through the output of the comparator when V_{Sense} becomes less than V_{th}. The fault signal is initially generated from the comparator output, proceeds through the latch circuit to maintain the high signal and is then applied to the gate driver's disable pin. It appears in fault signal CH 4 and rises from 0 V to 5 V after 177 ns from t_2, and the delay that appears between t_2 and t_3 is caused by the propagation delay of the comparator and latch IC. After the fault signal is applied to the disable pin of the gate driver, about 50 ns later, the low-side gate source voltage of CH 1 begins to decrease. It takes about 20 ns, and the device is completely turned off at time t_4. Therefore, it can be confirmed that the device is completely turned off a total of 257 ns after the short circuit occurs, thereby protecting the circuit.

Since it has been confirmed that the proposed short-circuit protection circuit operates normally during a short circuit, it has been confirmed that the short-circuit protection circuit does not operate during normal switching operations. The experiment can be seen in Figure 17. Figure 17 is an enlarged waveform of the moment when the second pulse is applied during the double pulse test operation. CH 1 represents the gate–source voltage of the low-side GaN HEMT, CH 2 represents the DC bus voltage, CH 3 represents the voltage drop detection signal, and CH 4 represents the fault signal. At time t_1, the gate–source voltage of the low-side GaN HEMT exceeds the threshold, causing distortion of the DC bus voltage. At t_2, the DC bus voltage drop reaches its maximum, and the voltage, which is maintained at 400 V in the normal state, decreases by 22 V to 378 V. The V_{Sense} voltage of CH 3 shows a similar form to the DC bus voltage and has the lowest value at time t_2, but it can be seen that it is greatly attenuated after passing through the bandpass filter. In the normal state, the voltage is maintained at 0 V but decreases to -1 V at time t_2. The V_{Sense} voltage is applied to the positive pin of the comparator and does not become less than -2 V, which is the voltage being applied to the negative pin. Therefore, the output of the comparator continues to remain at 0 V and no fault signal is formed.

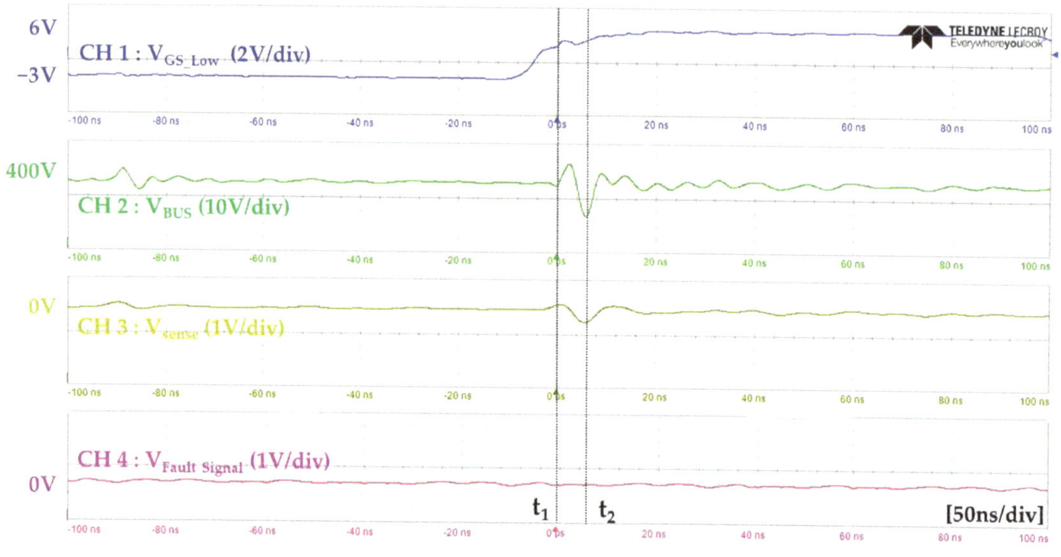

Figure 17. Experimental waveforms during normal switching.

During a short circuit, the DC bus voltage drops by 25 V, while during normal switching (turn-on), the voltage drop is 22 V, which is larger during a short circuit. However, after passing through a bandpass filter, the voltage drop decreases to 2.2 V during a short circuit and 1 V during normal switching. As such, it can be seen that there is a significant difference in the attenuation level of the DC bus voltage that appears as it passes through the bandpass. It is confirmed that the proposed circuit can distinguish between short-circuit and normal switching and effectively provide short-circuit protection.

Table 3 compares the proposed short-circuit protection method with an existing study related to short-circuit protection of wide-bandgap power semiconductors. In the existing study, various methods were applied to protect SiC and GaN-based power semiconductors, and the time required for short-circuit protection was on the order of hundreds of ns. The proposed method did not have the fastest protection operation speed compared to the existing study. However, it had an above-average protection operation speed. And it can also be implemented with fewer elements compared to the average number of elements used. In addition, the proposed short-circuit protection method effectively prevents malfunctions during normal switching through a bandpass filter, achieving the high reliability of short-circuit protection operations.

Table 3. Comparison with existing studies.

Reference	Device	Protection Method	Component (Passive/Active)	Protection Time
[7]	SiC	Shunt Resistor	4/3	150 ns
[8]	GaN	Shunt Resistor	-	60 ns
[9]	SiC	Package Internal Inductance	5/4	60 ns
[10]	SiC	Package Internal Inductance	6/7	100 ns
[12]	SiC	Rogowski coil	4/5	700 ns
[13]	SiC	Rogowski coil	8/5	-
[14]	GaN	Stray inductance	-	250 ns
[15]	GaN	Stray inductance	3/5	250 ns
[17]	GaN	Desaturation	4/4	360 ns
[18]	GaN	Desaturation	8/5	125 ns
[22]	GaN	DC bus voltage	4/3	280 ns
[23]	GaN	DC bus voltage	8/5	370 ns
Proposed	GaN	DC bus voltage	5/2	257 ns

5. Conclusions

In this paper, a short-circuit detection and protection circuit for GaN HEMT is proposed and verified. GaN HEMT has a shorter short-circuit withstand time compared to other types of power semiconductors, so it requires a short-circuit protection circuit that has high reliability and operates at high speeds. The proposed short-circuit protection method uses a method of detecting the DC bus voltage for fast short-circuit protection. The high di/dt current that appears during a short circuit causes an instantaneous voltage drop in the parasitic inductance present in the circuit. Therefore, when a short circuit occurs, an instantaneous voltage drop occurs in the inductance component present in the power loop, and an instantaneous voltage drop also occurs in the DC bus voltage, which remains the same as the DC link voltage in the normal state. This DC bus voltage drop is a quick and clear indicator of the occurrence of a short circuit and can be used to effectively protect the circuit from short circuits. However, similar voltage drops can occur in normal switching situations as well. In the proposed short-circuit protection method, the frequency analysis of the DC bus voltage drop was performed based on the SPICE simulation. Based on that analysis, a bandpass filter was used to effectively distinguish between the two types of voltage drops. When using the proposed short-circuit protection method, short-circuit protection was finally completed 257 ns after the short-circuit occurred. In practice, after a short circuit occurs, it takes 10 ns for the output signal of the bandpass filter to be compared to the comparator threshold. Therefore, it is expected that short-circuit protection will be possible at a faster rate if the propagation delay of the comparator IC and latch IC used in the experiment is reduced. Future research should be conducted to improve the time required for short-circuit protection in the proposed method.

Author Contributions: Conceptualization, C.-M.K.; Methodology, C.-M.K.; Software, C.-M.K.; Validation, C.-M.K.; Formal analysis, C.-M.K. and N.-J.K.; Investigation, H.-S.Y.; Resources, N.-J.K.; Data curation, C.-M.K.; Writing—original draft, C.-M.K.; Writing—review and editing, J.-S.K. and N.-J.K.; Visualization, C.-M.K. and H.-S.Y.; Supervision, N.-J.K.; Project administration, N.-J.K. All authors have read and agreed to the published version of the manuscript.

Funding: This work was supported by the Ministry of Education of the Republic of Korea and the National Research Foundation of Korea (NRF-2021R1F1A10506811).

Data Availability Statement: The data are available in a publicly accessible repository.

Conflicts of Interest: The authors declare no conflicts of interest.

References

1. Lidow, A.; Strydom, J. *Gallium Nitride (GaN) Technology Overview*; White Paper; Efficient Power Conversion Corporation (EPC-co): El Segundo, CA, USA, 2012.
2. Li, H.; Li, X.; Wang, X.; Wang, J.; Alsmadi, Y.; Liu, L.; Bala, S. E-mode GaN HEMT short circuit robustness and degradation. In Proceedings of the 2017 IEEE Energy Conversion Congress and Exposition (ECCE), Cincinnati, OH, USA, 1–5 October 2017; pp. 1995–2002. [CrossRef]
3. Huang, X.; Lee, D.Y.; Bondarenko, V.; Baker, A.; Sheridan, D.C.; Huang, A.Q.; Baliga, B.J. Experimental study of 650V AlGaN/GaN HEMT short-circuit safe operating area (SCSOA). In Proceedings of the IEEE International Symposium on Power Semiconductor Devices & IC's (ISPSD), Waikoloa, HI, USA, 15–19 June 2014; pp. 273–276.
4. Badawi, N.; Awwad, A.E.; Dieckerhoff, S. Robustness in short-circuit mode: Benchmarking of 600V GaN HEMTs with power Si and SiC MOSFETs. In Proceedings of the IEEE Energy Conversion Congress and Exposition (ECCE), Milwaukee, WI, USA, 18–22 September 2016; pp. 1–7.
5. Fernández, M.; Perpiñà, X.; Roig, J.; Vellvehi, M.; Bauwens, F.; Jordà, X.; Tack, M. P-GaN HEMTs Drain and Gate Current Analysis Under Short-Circuit. *IEEE Trans. Electron Device Lett.* **2017**, *38*, 505–508. [CrossRef]
6. Kim, C.-M.; Kim, J.-S.; Kim, N.-J. Study on GaN FET Short Circuit Characteristics and Development of Effective Short Circuit Protection Method. In Proceedings of the 2022 25th International Conference on Electrical Machines and Systems (ICEMS), Chiang Mai, Thailand, 29 November–2 December 2022; pp. 1–5.
7. Awwad, A.E.; Dieckerhoff, S. Short-circuit evaluation and overcurrent protection for SiC power MOSFETs. In Proceedings of the 2015 17th European Conference on Power Electronics and Applications (EPE'15 ECCE-Europe), Geneva, Switzerland, 8–10 September 2015; pp. 1–9.
8. Zhang, W.; Wang, F.; Zhang, Z.; Holzinger, B. Fast Wide-bandgap Device Overcurrent Protection with Direct Current Measurement. In Proceedings of the 2019 10th International Conference on Power Electronics and ECCE Asia (ICPE 2019–ECCE Asia), Busan, Republic of Korea, 27–30 May 2019; pp. 1–6. [CrossRef]
9. Xue, J.; Xin, Z.; Wang, H.; Loh, P.C.; Blaabjerg, F. An improved di/dt-RCD detection for short-circuit protection of SiC MOSFET. *IEEE Trans. Power Electron.* **2021**, *36*, 12–17. [CrossRef]
10. Wang, Z.; Shi, X.; Xue, Y.; Tolbert, L.M.; Wang, F.; Blalock, B.J. Design and performance evaluation of overcurrent protection schemes for Silicon Carbide (SiC) power MOSFETs. *IEEE Trans. Ind. Electron.* **2014**, *61*, 5570–5581. [CrossRef]
11. Sun, K.; Wang, J.; Burgos, R.; Boroyevich, D. Design, Analysis, and Discussion of Short Circuit and Overload Gate-Driver Dual-Protection Scheme for 1.2-kV, 400-A SiC MOSFET Modules. *IEEE Trans. Power Electron.* **2020**, *35*, 3054–3068. [CrossRef]
12. Stecca, M.; Tiftikidis, P.; Soeiro, T.B.; Bauer, P. Gate Driver Design for 1.2 kV SiC Module with PCB Integrated Rogowski Coil Protection Circuit. In Proceedings of the 2021 IEEE Energy Conversion Congress and Exposition (ECCE), Vancouver, BC, Canada, 10–14 October 2021; pp. 5723–5728. [CrossRef]
13. Rafiq, A.; Pramanick, S. Ultrafast Protection of Discrete SiC MOSFETs With PCB Coil-Based Current Sensors. *IEEE Trans. Power Electron.* **2023**, *38*, 1860–1870. [CrossRef]
14. Alemdar, O.S.; Karakaya, F.; Keysan, O. PCB Layout Based Short-Circuit Protection Scheme for GaN HEMTs. In Proceedings of the 2019 IEEE Energy Conversion Congress and Exposition (ECCE), Baltimore, MD, USA, 29 September–3 October 2019; pp. 2212–2218. [CrossRef]
15. Karakaya, F.; Alemdar, O.S.; Keysan, O. Layout-Based Ultrafast Short-Circuit Protection Technique for Parallel-Connected GaN HEMTs. *IEEE J. Emerg. Sel. Top. Power Electron.* **2021**, *9*, 6385–6395. [CrossRef]
16. Jiang, W.L.; Murray, S.K.; Zaman, M.S.; De Vleeschouwer, H.; Moens, P.; Roig, J.; Trescases, O. An Integrated GaN Overcurrent Protection Circuit for Power HEMTs Using SenseHEMT. *IEEE Trans. Power Electron.* **2022**, *37*, 9314–9324. [CrossRef]
17. Sadik, D.-P.; Colmenares, J.; Tolstoy, G.; Peftitsis, D.; Bakowski, M.; Rabkowski, J.; Nee, H.-P. Short-Circuit Protection Circuits for Silicon-Carbide Power Transistors. *IEEE Trans. Ind. Electron.* **2016**, *63*, 1995–2004. [CrossRef]
18. Hou, R.; Lu, J.; Chen, D. An Ultrafast Discrete Short-Circuit Protection Circuit for GaN HEMTs. In Proceedings of the 2018 IEEE Energy Conversion Congress and Exposition (ECCE), Portland, OR, USA, 23–27 September 2018; pp. 1920–1925. [CrossRef]
19. Acuna, J.; Walter, J.; Kallfass, I.; Gan, G.N. Very fast short circuit protection for gallium-nitride power transistors based on printed circuit board integrated current sensor. In Proceedings of the 2018 20th European Conference on Power Electronics and Applications (EPE'18 ECCE Europe), Riga, Latvia, 17–21 September 2018; pp. 1–10. Available online: https://ieeexplore.ieee.org/document/8515547 (accessed on 13 January 2024).
20. Jones, E.A.; Williford, P.; Wang, F. A fast overcurrent protection scheme for GaN GITs. In Proceedings of the 2017 IEEE 5th Workshop on Wide-Bandgap Power Devices and Applications, Albuquerque, NM, USA, 30 October–1 November 2017; pp. 277–284.

21. Sun, K.; Wang, J.; Burgos, R.; Boroyevich, D.; Kang, Y.; Choi, E. Analysis and design of an overcurrent protection scheme based on parasitic inductance of SiC MOSFET power module. In Proceedings of the 2018 IEEE Applied Power Electronics Conference and Exposition (APEC), San Antonio, TX, USA, 4–8 March 2018; pp. 2806–2812.
22. Li, H.; Lyu, X.; Wang, K.; Abdullah, Y.; Hu, B.; Yang, Z.; Wang, J.; Liu, L.; Bala, S. An Ultra-Fast Short Circuit Protection Solution for E-mode GaN HEMTs. In Proceedings of the 2018 1st Workshop on Wide Bandgap Power Devices and Applications in Asia (WiPDA Asia), Xi'an, China, 17–19 May 2018; pp. 187–192. [CrossRef]
23. Zhou, Y.; Bai, Z.; Gong, X. Study of a Fast Over-Current Protection Based on Bus Voltage Drop for Gallium Nitride Power Device. *IOP Conf. Ser. Mater. Sci. Eng.* **2020**, *782*, 022085. [CrossRef]
24. Datasheet GS-065-011, GaN Systems. Available online: https://gansystems.com/gan-transistors/gs-065-011-1-l/ (accessed on 13 January 2024).

electronics MDPI

Article

A Magnetic Integration Mismatch Suppression Strategy for Parallel SiC Power Devices Applications

Shikai Sun [1,2], Jialin Liu [2], Lei Chen [2,*], Zhenlin Lu [2,*], Yuan Wang [1], Wenhao Yang [3], Yuyin Sun [3] and Hui Guo [3]

1 School of Integrated Circuits, Peking University, Beijing 100871, China; sunshikai@mxtronics.com (S.S.); wangyuan@pku.edu.cn (Y.W.)
2 Beijing Microelectronics Technology Institute, Beijing 100076, China; liujialin@mxtronics.com
3 School of Microelectronics, Xidian University, Xi'an 710071, China; yang_wenhao@stu.xidian.edu.cn (W.Y.); yuyinsun@stu.xidian.edu.cn (Y.S.); guohui@mail.xidian.edu.cn (H.G.)
* Correspondence: chenleinpu@vip.126.com (L.C.); luzhenlin@mxtronics.com (Z.L.)

Abstract: A new magnetic integrated parallel current sharing control method for parallel silicon carbide (SiC) power devices is presented in this article. The problem of the application of parallel connected SiC power devices is analyzed. The coupled inductance method is adopted to solve the problem. Based on the active-back converter, we establish the theoretical model of the coupled inductance, and figure out its working mechanism. The integrated magnetic device is designed based on the working mechanism, and the effectiveness is determined through simulation. A 12 V/10 A output magnetic integrated active-flyback converter prototype is fabricated and tested to verify the strategy. Measurement results show that, with the proposed magnetic integrated method, the mismatch voltage is suppressed to 0.1 V under all load conditions, and the efficiency increases by at most 6.52% under full load conditions.

Keywords: SiC power devices; magnetic integrated; parallel current sharing; working mechanism

Citation: Sun, S.; Liu, J.; Chen, L.; Lu, Z.; Wang, Y.; Yang, W.; Sun, Y.; Guo, H. A Magnetic Integration Mismatch Suppression Strategy for Parallel SiC Power Devices Applications. *Electronics* **2024**, *13*, 954. https://doi.org/10.3390/electronics13050954

Academic Editor: Nakkeeran Kaliyaperumal

Received: 29 December 2023
Revised: 1 February 2024
Accepted: 11 February 2024
Published: 1 March 2024

1. Introduction

Silicon Carbide (SiC) material can push the power density and efficiency of semiconductor devices and power systems to higher limits due to its wide band gap, high critical field, and high thermal conductivity [1–3]. With the development of SiC technology, the application of SiC power devices is becoming more and more popular [4,5]. The low switching-loss characteristic facilitates a reduction in power loss and an improvement in working frequency, which leads to the use of smaller passive components and improving power density [6,7].

The development of power electronics demands higher and higher current ratings, which promotes the parallel connection of power devices. When the parallel connection is used, the current imbalance among the paralleled power devices becomes a major concern [8]. The current imbalance is caused by the mismatch in device parameters among the paralleled semiconductors or the mismatch in the parasitic parameters of their corresponding circuits when the circuit layouts are asymmetrical. The condition may result in conduction and switching losses, which may further cause thermal distribution problems.

For the SiC power device applications, the value of ON-resistance is smaller than that of the counterpart Si devices. A little mismatch may lead to a large percentage change. Thus, the SiC power devices are more sensitive to the variation of device parameters in paralleled applications. The current mismatch phenomenon has already appeared in the paralleled applications of SiC devices [9,10]. Ref. [11] analyzed the influence of the variability of device parameters on the current sharing of parallel-connected SiC MOSFETs. Experimental investigations of static and transient current sharing were carried out in ref. [12]. The parallel-connected application of packaged SiC power devices was evaluated in Ref. [13]. The above articles provide a detailed analysis of the mismatch mechanism of the devices. Many studies have been performed on the imbalance of current suppression. From the view of the study object, the imbalance current suppression method can be classified into three categories: device classification; device operating condition monitoring; and circuit topology [14–19]. The chip screening method is proposed to solve the mismatch introduced by the asymmetric layout [20].

The typical representative of the device classification view is the transfer curve distance coefficient classification criterion proposed in ref. [14]. This paper evaluates the factors of the device characteristics and finds that the transfer characteristic contains the main influences. The strategy realizes the mismatch suppression by weighting the distance coefficients of the device transfer curves. The strategy needs to test every device, which limits its massive applications and universality. The typical representation of the device operating condition monitoring is the SiC MOSFET gate driving scheme with a dynamic current equalization mechanism for over-current protection proposed in Ref. [21]. The scheme realizes the simultaneous turn-on of SiC MOSFETs with different threshold voltages by monitoring the device current cycle by cycle to achieve the mismatch current suppression of the parallel device. However, the strategy needs to add extra devices to suppress the mismatch. The circuit structure route to suppress the mismatch current is typified by a parallel current feedback equal-current resonant converter [22]. The strategy adopts a two-stage structure including an interleaved parallel boost converter and a double magnetically coupled half-bridge LLC resonant conversion. This scheme adds two inductors to realize parallel current equalization, which increases the number of magnetic devices in the converter, as well as the iron and copper losses.

In this paper, a novel magnetic integration strategy is proposed to achieve parallel equalization control without increasing the number and size of the converter cores. To verify the proposed strategy, a prototype converter is designed, fabricated, and tested. The measured results show that power efficiency is enhanced by at most 6.52% in the whole load range.

2. Operating Principle of the Proposed Strategy

2.1. The Topology Evolution

To solve the problem of the mismatch current distribution of SiC power devices in parallel applications, this paper proposes a control method to solve the problem at the topology level. Its topology evolution is shown in Figure 1. Filter inductance is generally used in parallel to reduce copper losses in high-current applications, as shown in Figure 1a. The parallel SiC power devices and filter inductance decoupled to form different branches, as shown in Figure 1b. Since the impedance of the secondary filter inductance is much larger than the on-resistance of the SiC power device, the influence of the device characteristics on the current distribution is converted into the influence of the filter inductance. The topology introduces magnetic coupling by sharing the common magnetic core to suppress the mismatch current, which is shown in Figure 1c. This topology evolution transforms the SiC power devices mismatch into the inconsistency of the filter inductance and further reduces the influence using coupled inductance. And, the final topology does not increase the number of devices.

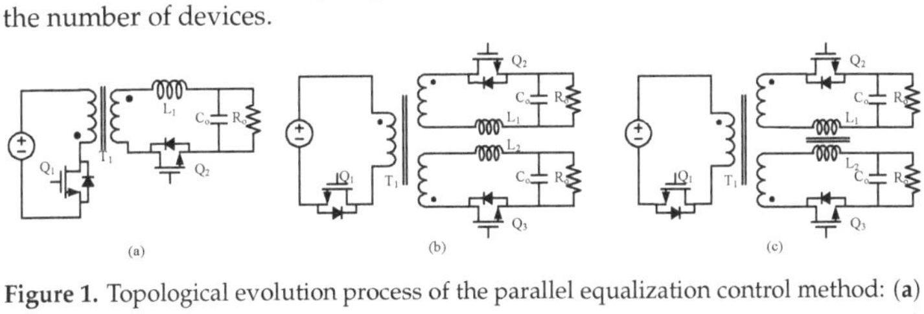

Figure 1. Topological evolution process of the parallel equalization control method: (**a**) conventional circuit output flyback converter; (**b**) parallel two-output flyback converter; and (**c**) coupled inductance flyback converter.

2.2. The Operational Principle of the Coupled Inductance

There are two main ways of performing coupled inductance, namely flux mutual and flux cancellation. As shown in Figure 2, coil1 corresponds to inductance L_1 and has N_1 turns whilst coil2 corresponds to inductance L_2 and has N_2 turns.

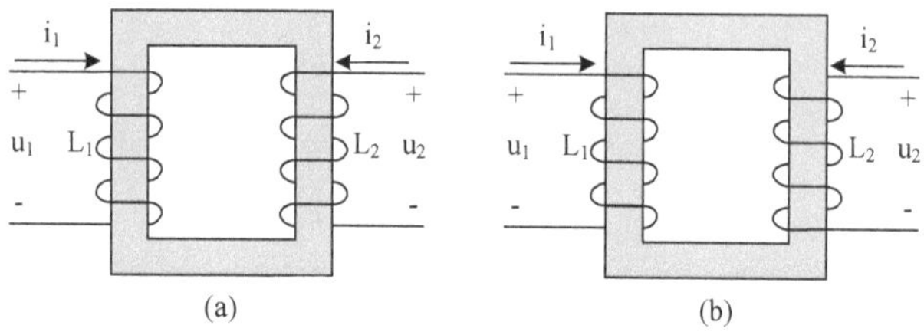

Figure 2. The two main means of coupled inductance: (**a**) flux mutual; and (**b**) flux cancellation.

$$\Psi_{11} = N_1\Phi_{11} = L_1 i_1 \tag{1}$$

$$\Psi_{21} = N_2\Phi_{21} = M i_1 \tag{2}$$

When a current i_1 flows through coil1 in Figure 2a, the total self-induced magnetic flux linkage can be expressed as (1), and the mutual magnetic flux linkage can be expressed as (2), where Ψ_{11} is the self-induced magnetic flux linkage, Φ_{11} is the mutual magnetic flux generated by cycle of coil1, L_1 is the self-induction of coil1, and Ψ_{21} is the mutual magnetic flux linkage generated by coil1 and affecting coil2, Φ_{21} is the mutual magnetic flux by cycle of coil2, while M is the mutual inductance of coil1 and coil2.

$$\Psi_{22} = N_2\Phi_{22} = L_2 i_2 \tag{3}$$

$$\Psi_{12} = N_1\Phi_{12} = M i_2 \tag{4}$$

Similarly, coil2 also generates a self-induced magnetic flux linkage and mutual magnetic flux linkage. Its self-induced magnetic flux linkage is denoted by Ψ_{22} and its mutual-induced magnetic flux linkage is denoted by Ψ_{12}, as Ψ_{22} is the self-induced magnetic flux linkage, L_2 is the self-induction of coil2, Ψ_{12} is the mutual magnetic flux linkage generated by coil2 and affecting coil1, Φ_{12} is the mutual magnetic flux generated by coil2 and affecting coil1, and M is the mutual inductance of coil1 and coil2.

Under linear conditions, $M_{12} = M_{21} = M$, and hereafter M is used to denote mutual inductance. According to the right-handed helix rule, the self- and mutual-inductive flux of the two coils shown in Figure 2a go in the same direction, which is defined as flux mutual, and the total magnetic flux linkage of coil1 and coil2 is denoted by (5) and the port voltage is denoted by (6).

$$\begin{cases} \Psi_1 = \Psi_{11} + \Psi_{12} = L_1 i_1 + M i_2 \\ \Psi_2 = \Psi_{22} + \Psi_{21} = L_2 i_2 + M i_1 \end{cases} \tag{5}$$

$$\begin{cases} u_1 = \dfrac{d\Psi_1}{dt} = L_1\dfrac{di_1}{dt} + M\dfrac{di_2}{dt} \\ u_2 = \dfrac{d\Psi_2}{dt} = L_2\dfrac{di_2}{dt} + M\dfrac{di_1}{dt} \end{cases} \tag{6}$$

The two coils shown in the corresponding Figure 2b have their self-inductive and mutual-inductive fluxes in opposite directions, which is defined as flux cancellation, and the total magnetic flux linkage of coil1 and coil2 is denoted by (7), and the port voltage can be denoted by (8).

$$\begin{cases} \Psi_1 = \Psi_{11} - \Psi_{12} = L_1 i_1 - M i_2 \\ \Psi_2 = \Psi_{22} - \Psi_{21} = L_2 i_2 - M i_1 \end{cases} \tag{7}$$

$$\begin{cases} u_1 = \dfrac{d\Psi_1}{dt} = L_1 \dfrac{di_1}{dt} - M\dfrac{di_2}{dt} \\ u_2 = \dfrac{d\Psi_2}{dt} = L_2 \dfrac{di_2}{dt} - M\dfrac{di_1}{dt} \end{cases} \tag{8}$$

To simplify the description of the port voltage, the coupling coefficient k is introduced. The coupling coefficient represents the geometric mean of the ratio of mutual inductance to the self-induced inductance chain of the two coils and is expressed by Equation (9).

$$k = \sqrt{\frac{\Phi_{12}\Phi_{21}}{\Phi_{11}\Phi_{22}}} \tag{9}$$

Substituting the magnetic flux linkages separately gives the coupling coefficient expression (10).

$$k = \sqrt{\frac{\Phi_{12}\Phi_{21}}{\Phi_{11}\Phi_{22}}} = \frac{M}{\sqrt{L_1 L_2}} \tag{10}$$

Quantitatively describing coupled coils in terms of coupling coefficients and leakage inductance allows the modelling of coupled coils to be directly embedded in the port voltages of coil1 and coil2, which can be expressed as (11) and (12), respectively.

$$\begin{cases} u_1 = L_{k1}\dfrac{di_1}{dt} + k\sqrt{\dfrac{L_1}{L_2}}u_2 \\ u_2 = L_{k2}\dfrac{di_2}{dt} + k\sqrt{\dfrac{L_2}{L_1}}u_1 \end{cases} \tag{11}$$

$$\begin{cases} u_1 = L_{k1}\dfrac{di_1}{dt} - k\sqrt{\dfrac{L_1}{L_2}}u_2 \\ u_2 = L_{k2}\dfrac{di_2}{dt} - k\sqrt{\dfrac{L_2}{L_1}}u_1 \end{cases} \tag{12}$$

Based on the above analysis, it can be seen that the coupling coefficient is less than or equal to 1, i.e., $k \leq 1$, and the leakage inductance of the two coils can be expressed as $L_{k1} = (1-k^2)L_1$ and $L_{k2} = (1-k^2)L_2$, respectively. The coupled inductance voltage-current relationship is reconstructed to create a symmetrical coupled inductance model. Assuming that the coupled inductance has equal values in terms of excitation inductance, Equations (11) and (12) can be expressed as (13) and (14).

$$\begin{cases} L_{k1}\dfrac{di_1}{dt} = u_1 - ku_2 \\ L_{k2}\dfrac{di_2}{dt} = u_2 - ku_1 \end{cases} \tag{13}$$

$$\begin{cases} L_{k1}\dfrac{di_1}{dt} = u_1 + ku_2 \\ L_{k2}\dfrac{di_2}{dt} = u_2 + ku_1 \end{cases} \tag{14}$$

The equivalent circuit of the two coils is shown in Figure 3, where the controlled voltage source represents the coupling effect between the two coils, and the inductance is the respective leakage inductance of the coupled coils. The voltage of the coupled coils in the converter secondary side can be expressed by Equation (15).

$$\begin{cases} u_1 = u_{sec1} - V_{out} \\ u_2 = u_{sec2} - V_{out} \end{cases} \quad (15)$$

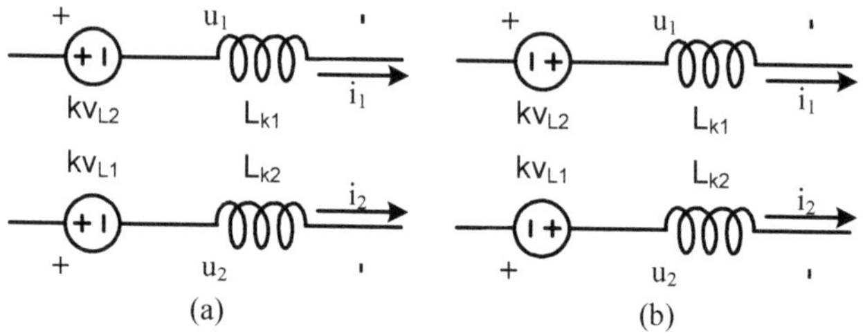

Figure 3. The equivalent circuit of the two coils: (**a**) flux mutual; and (**b**) flux cancellation.

Since the mismatch is only affected by the secondary side, the following analysis focuses on it. Embedding coupled inductance models into the topological secondary side, the equivalent circuit is shown in Figure 4. The reference points x_1 and x_2 for the voltages of flux mutual and flux cancellation can be expressed by (16) and (17), respectively.

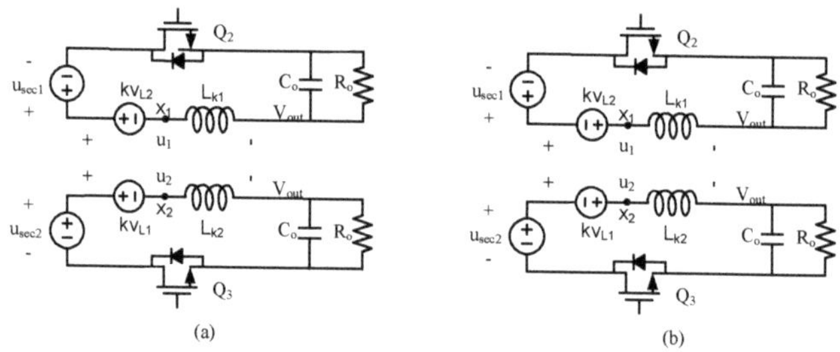

Figure 4. The equivalent circuit of the secondary side of the topology: (**a**) flux mutual; and (**b**) flux cancellation.

$$\begin{cases} u_{x1} = u_{sec1} - kv_{L2} = (u_{sec1} - kv_{sec2}) + kV_{out} \\ u_{x2} = u_{sec2} - kv_{L1} = (u_{sec2} - kv_{sec1}) + kV_{out} \end{cases} \quad (16)$$

$$\begin{cases} u_{x1} = u_{sec1} + kv_{L2} = (u_{sec1} - kv_{sec2}) + V_{out} \\ u_{x2} = u_{sec2} + kv_{L1} = (u_{sec2} - kv_{sec1}) + V_{out} \end{cases} \quad (17)$$

Unifying the equivalent voltage source generated by the coupled inductance into the voltage source of the secondary excitation inductance, the complex model of coupled inductance is simplified into the equivalent model of the voltage source and the leakage inductance.

Based on the coupled inductance equivalent model established above, the output current change rate of the converter flux mutual aid and flux cancellation coupled inductance is expressed by Equations (18) and (19).

$$S_F = -\frac{(1-k)V_{out}}{L_k} = -\frac{V_{out}}{(1+k)L} \quad (18)$$

$$S_F = -\frac{(1+k)V_{out}}{L_k} = -\frac{V_{out}}{(1-k)L} \quad (19)$$

The peak value of the current during the steady-state operation of the converter can be obtained according to the converter operating principle, and the peak value of the current for flux mutual and flux cancellation can be expressed by Equations (20) and (21), respectively.

$$\Delta I_{pp} = S_F(1-D)T = \frac{V_{out}}{(1+k)L}(1-D)T \tag{20}$$

$$\Delta I_{pp} = S_F(1-D)T = \frac{V_{out}}{(1-k)L}(1-D)T \tag{21}$$

From the above analysis, it can be seen that the output current ripple suppression effect is positively correlated with the coupling coefficient in the flux mutual; and the output current ripple suppression effect is negatively correlated with the coupling coefficient in the flux cancellation.

The main reasons for the mismatch in current distribution include the mismatch of on-resistance and parasitic inductance at the device level, the passive components at the circuit level, and the parasitic mismatch of the layout. The above mismatches can be expressed by correcting the device model, where R_{ds} denotes the different on-resistance of the two branches and L_{ds} denotes the different parasitic inductance of the two branches. Embedding the modified device model into the output model, the secondary side equivalent circuit of the conventional flyback topology, the flux mutual coupled inductance, and the flux cancellation are shown in Figure 5.

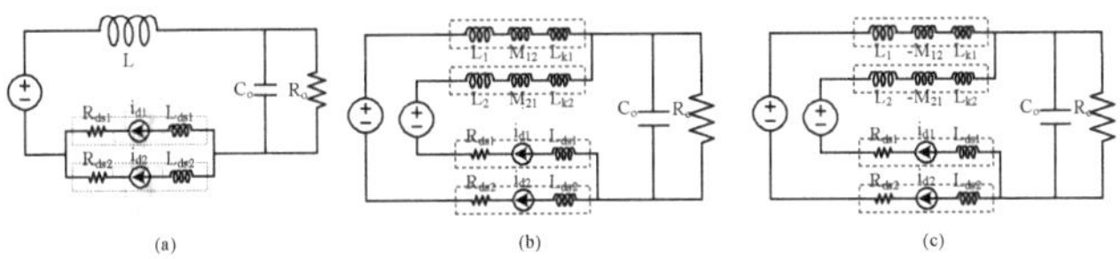

Figure 5. The equivalent circuit of the secondary side of the topology: (**a**) without coupled inductance; (**b**) flux mutual; and (**c**) flux cancellation.

Referring to Figure 5a, the mismatch resistance can be expressed as $\Delta R_{ds} = R_{ds1} - R_{ds2}$ and the mismatch inductance can be expressed as $\Delta L_{ds} = L_{ds1} - L_{ds2}$. Since the on-resistance mismatch of the MOSFET is at the mΩ level, its parasitic inductance and that of the circuit layout are at the level of a few nH, while the filtering inductance is at the level of a few tens of μH, and the non-ideal effect can be ignored when performing loop current calculations. Under these conditions, the converter's secondary side current is consistent with the typical current of the converter. The total current at the secondary side in this case can be used in (22).

$$\begin{aligned} I_{sec} &= \frac{V_{out}}{sL + (R_{ds1} + sL_{ds1})//(R_{ds2} + SL_{ds2})} \\ &= \frac{V_{out}[R_{ds1} + R_{ds2} + s(L_{ds1} + L_{ds2})]}{sL[R_{ds1} + R_{ds2} + s(L_{ds1} + L_{ds2}))] + (R_{ds1} + SLds1)(R_{ds2} + SLds2)} \end{aligned} \tag{22}$$

This current is split between the two branches, and according to Kirchhoff's current law (KCL) and Kirchhoff's voltage law (KVL), the current distribution between the two branches is inversely proportional to the total impedance of the branches, which can be expressed as (23).

$$\begin{cases} i_{d1} = I_{sec}\dfrac{R_{ds2} + sL_{ds2}}{R_{ds1} + R_{ds2} + s(L_{ds1} + L_{ds2})} \\ i_{d2} = I_{sec}\dfrac{R_{ds1} + sL_{ds1}}{R_{ds1} + R_{ds2} + s(L_{ds1} + L_{ds2})} \end{cases} \tag{23}$$

The mismatch current in the absence of coupling inductance can be expressed by (23), where the parasitic inductance size is roughly at the level of a few nH, while the filtering inductance is in the order of tens to tens of μH and $L \gg L_{ds}$. Therefore, a further simplified representation of the mismatch current can be made.

$$
\begin{aligned}
\Delta i &= |i_{d1} - i_{d2}| \\
&= |i_{sec}\frac{\Delta R_{ds} + s\Delta L_{ds}}{R_{ds1} + R_{ds2} + s(L_{ds1} + L_{ds2})}| \\
&= |\frac{V_{out}(\Delta R_{ds} + s\Delta L_{ds})}{sL[R_{ds1} + R_{ds2} + s(L_{ds1} + L_{ds2})] + (R_{ds1} + sL_{ds1})(R_{ds2} + sL_{ds2})}| \\
&\approx |\frac{V_{out}(\Delta R_{ds} + s\Delta L_{ds})}{sL[R_{ds1} + R_{ds2}]}|
\end{aligned}
\tag{24}
$$

From the above analysis, it can be seen that the current distribution between the two MOSFETs is independent of filter inductance, and only of the resistance and parasitic inductance of the MOSFETs and the layout. The current mismatch is directly determined by the MOSFETs resistance and the parasitic inductance mismatch. Referring to the flux mutual coupling inductance model shown in Figure 5a, the voltage equations for the two branches can be listed as (25) and (26), respectively.

$$
V_{Z1} + V_{sec} + V_{L1} + V_{M12} = V_{out} \tag{25}
$$

$$
V_{Z2} + V_{sec} + V_{L2} + V_{M21} = V_{out} \tag{26}
$$

where V_{Z1} and V_{Z2} represent the voltage drop across the MOSFET, which can be expressed by Equation (27).

$$
\begin{cases} V_{Z1} = i_{d1} \cdot (R_{ds1} + sL_{ds1}) \\ V_{Z2} = i_{d2} \cdot (R_{ds2} + sL_{ds2}) \end{cases} \tag{27}
$$

V_{sec} denotes the equivalent voltage source generated by the transformer coupling to the secondary side, which is determined by the converter parameters and is a constant in steady-state operation. V_{L1} and V_{L2} denote the voltage drops generated by the coupled excitation inductance and leakage inductance, respectively, which can be expressed by the Equation (28).

$$
\begin{cases} V_{L1} = L_1\dfrac{di_{d1}}{dt} \\ V_{L2} = L_2\dfrac{di_{d2}}{dt} \end{cases} \tag{28}
$$

V_{M12} denotes the voltage drop corresponding to the mutual inductance generated by inductance L_2 over inductance L_1, and V_{M21} denotes the voltage drop corresponding to the mutual inductance generated by inductance L_1 over inductance L_2, which can be expressed by (29).

$$
\begin{cases} V_{M12} = M_{12}\dfrac{di_{d2}}{dt} \\ V_{M21} = M_{21}\dfrac{di_{d1}}{dt} \end{cases} \tag{29}
$$

Since the coupled inductance is wound by the PCB, its consistency and symmetry are extremely high, and the difference generated by the leakage inductance is negligible compared with the excitation inductance and mutual inductance, so it can be assumed that $L_1 = L_2 = L$ and $M_{12} = M_{21} = M$.

Substituting (28) and (29) into Equations (25) and (26) yields (30) and (31).

$$
V_{Z1} + V_{sec} + L\frac{di_{d1}}{dt} + M\frac{di_{d2}}{dt} = V_{out} \tag{30}
$$

$$
V_{Z2} + V_{sec} + L\frac{di_{d2}}{dt} + M\frac{di_{d1}}{dt} = V_{out} \tag{31}
$$

Subtract (30) from (31) using the formula (32).

$$V_{Z1} - V_{Z2} + (L - M)(\frac{di_{d1}}{dt} - \frac{di_{d2}}{dt}) = 0 \tag{32}$$

Then, Equation (32) can be reduced to (33).

$$\frac{di_{d1}}{dt} - \frac{di_{d2}}{dt} = \frac{V_{Z1} - V_{Z2}}{M - L} \tag{33}$$

By defining $\Delta i = i_{D1} - i_{D2}$ according to the difference subtraction relation, we can denote (33) by (34).

$$\frac{d\Delta i}{dt} = \frac{d(i_{d1} - i_{d2})}{dt} = \frac{V_{Z1} - V_{Z2}}{M - L} \tag{34}$$

Due to the intrinsic properties of the coupled inductance, $M - L < 0$. When $i_{d1} > i_{d2}, V_{Z1} > V_{Z2}$, there are $\Delta i > 0, \frac{d\Delta i}{dt} < 0$, and the coupling inductance suppresses the mismatch current with a suppression rate of $|\frac{V_{Z1}-V_{Z2}}{M-L}|$. When $i_{d1} < i_{d2}, V_{Z1} < V_{Z2}$ with $\Delta i < 0, \frac{d\Delta i}{dt} > 0$, the coupled inductance will suppress the mismatch current, and the suppression rate is still $|\frac{V_{Z1}-V_{Z2}}{M-L}|$. When $i_{d1} > i_{d2}, V_{Z1} < V_{Z2}$ with $\Delta i > 0, \frac{d\Delta i}{dt} > 0$, the coupling inductance will increase the mismatch current to equalize the voltage drop of the two branches at a rate of $|\frac{V_{Z1}-V_{Z2}}{M-L}|$. When $i_{d1} < i_{d2}, V_{Z1} > V_{Z2}$ with $\Delta i < 0, \frac{d\Delta i}{dt} < 0$, the coupling inductance increases the mismatch current to equalize the voltage drops of the two branches, and the rate of increase remains $|\frac{V_{Z1}-V_{Z2}}{M-L}|$.

The difference between a flux cancellation coupled inductance and a flux mutual coupled inductance is in the polarity of the mutual inductance. The two-branch voltage relationships of flux cancellation coupled inductance are shown in (25) and (26), the voltage drop and self-inductance voltage relationships are shown in (27) and (28), and the mutual inductance voltage drop is different from that of flux mutual coupling, which can be expressed as (35).

$$\begin{cases} V_{M12} = M_{12}\dfrac{di_{d2}}{dt} \\ V_{M21} = M_{21}\dfrac{di_{d1}}{dt} \end{cases} \tag{35}$$

The solution process is the same as the flux mutual approach, which will not be repeated in this paper, and the obtained mismatch current transformation rate can be expressed as (36)

$$\frac{d\Delta i}{dt} = \frac{d(i_{d1} - i_{d2})}{dt} = \frac{V_{Z1} - V_{Z2}}{-(M + L)} \tag{36}$$

In the flux cancellation, when $i_{d1} > i_{d2}, V_{Z1} > V_{Z2}$, there are $\Delta i > 0, \frac{d\Delta i}{dt} < 0$, and the coupled inductance suppresses the mismatch currents with a suppression rate of $|\frac{V_{Z1}-V_{Z2}}{M+L}|$. When $i_{d1} < i_{d2}, V_{Z1} < V_{Z2}$ with $\Delta i < 0, \frac{d\Delta i}{dt} > 0$, the coupled inductance will suppress the mismatch current, and the suppression rate is still $|\frac{V_{Z1}-V_{Z2}}{M+L}|$. When $i_{d1} > i_{d2}, V_{Z1} < V_{Z2}$ with $\Delta i > 0, \frac{d\Delta i}{dt} > 0$, the coupling inductance will increase the mismatch current to equalize the voltage drop of the two branches at a rate of $|\frac{V_{Z1}-V_{Z2}}{M+L}|$. When $i_{d1} < i_{d2}, V_{Z1} > V_{Z2}$, there are $\Delta i < 0, \frac{d\Delta i}{dt} < 0$, the coupling inductance will increase the mismatch current to realize the voltage drop of the two branches are equal, and the rate of increase is still $|\frac{V_{Z1}-V_{Z2}}{M+L}|$.

To summarize the above, the coupled inductance scheme mismatch current to the two-branch MOSFET on the voltage drop is equally as critical when the two-way MOSFET current size and the voltage drop size trend are the same, which suppresses the mismatch current; when the two-way MOSFET current size and the voltage drop size of the opposite, the mismatch current is increased. The unbalanced voltage drop is suppressed, centered around the two MOSFET voltage drops being equal, and the suppression speed is $|\frac{V_{Z1}-V_{Z2}}{M-L}|$ in the flux mutual and $|\frac{V_{Z1}-V_{Z2}}{M+L}|$ in the flux cancellation.

3. Coupled Inductance Magnetic Device Design

The parameters of the converter used are shown in Table 1. Since the conventional transformer design is familiar to the electrical engineer, this paper only explains the coupled inductance design process.

Table 1. The parameters of the converter.

Characters	Value
Input voltage V_{in}	100 V
Output voltage V_{out}	12 V
Output current I_o	10A
Single branch current i_{single}	5 A
Output voltage ripple V_{ripple}	0.5 V
Output capacitance C_o	470 μF
Single branch filter inductance L_{single}	2.2 μH

3.1. Magnetically Integrated Flux Analysis

The core's magnetic flux is calculated in flux mutual and flux cancellation, and the effect of the two cases on the magnetic flux is analyzed. According to the introduction of magnetic device requirements, the parameters related to the converter core selected in this chapter are shown in Figure 6. The lengths are shown in millimeters (mm).

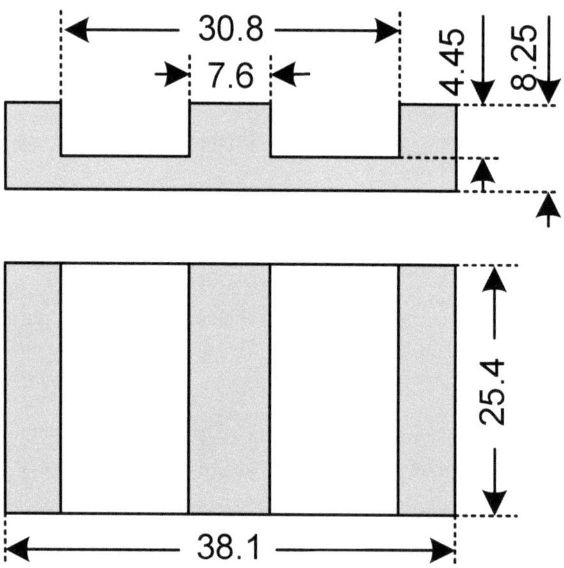

Figure 6. The core of the coupled inductance.

Based on the core, the coupled inductance is designed. The current direction and equivalent flux in the core are shown in Figure 7. The flux mutual forms the same direction in the core, superimposed upon each other; the flux cancellation forms the opposite direction of flux, and cancel each other. There is almost no energy stored in the flux cancellation.

According to the above flux analysis, it can be seen that the flux mutual needs the core size to meet the energy storage, while the flux mutual and flux cancellation offset one another, which results in the core size being smaller. The flux cancellation impact on the flux distribution of the core is very small, and there is an opportunity to realize integration with the main transformer.

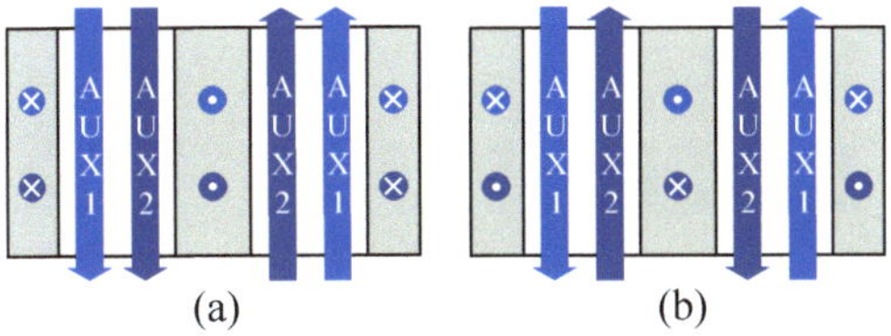

Figure 7. The analysis of equivalent flux: (**a**) flux mutual; and (**b**) flux cancellation.

The flyback converter working process is the first half cycle core energy storage, and the second half cycle of the core stored energy is released to the output. The coupled inductance is only related to the second half cycle. The flux distribution state of the magnetic core during the second half cycle is analyzed. The output current and flux distributions of the two branches are shown in Figure 8. The flux-coupled filter inductance of the converter is integrated into the transformer core to realize the double utilization of the core.

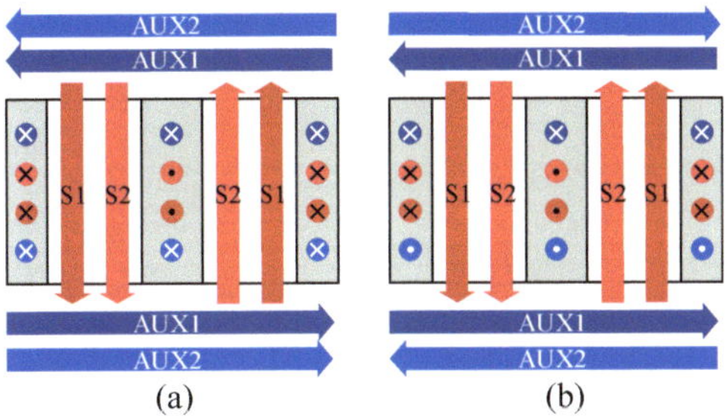

Figure 8. The magnetic integration analysis: (**a**) flux mutual; and (**b**) flux cancellation.

As shown in Figure 8, the two secondary currents are donated by S1 and S2, so the two filter currents are denoted by AUX1 and AUX2, respectively. The coils of the coupled inductance surround the core, and the entire core and the peripheral air form a closed flux loop with a low coupling coefficient. The two auxiliary coils generate the magnetic flux on the core in flux mutual in the same direction while generating flux cancellation in the opposite direction.

Since the two filter currents are equal in magnitude and arranged in the same way in mutual cancellation, the flux generated by the two auxiliary coils can cancel each other out. Based on this, the coupled coils do not affect the flux distribution, and the auxiliary coils can be integrated into the main core. The integrated strategy reduces the filter core and improves the power density of the converter.

3.2. The Simulation of the Magnetic Integration

To verify the analysis results, the integrated magnetic device is modeled in the MAXWELL module of ANSYS software, and the coupling coefficients of the primary winding, secondary winding and auxiliary winding are simulated and analyzed. The model in MAXWELL is shown in Figure 9.

The winding and core relationship is schematically shown in the front view section in Figure 10. To improve the coupling coefficient between the primary and secondary sides as well as the consistency between the two windings of the secondary side, the secondary1 and secondary2 windings are arranged in symmetrical positions above and below the primary winding. The two auxiliary windings are on top of the secondary winding. The primary winding, the secondary1 and secondary2 windings as well as the auxiliary1 and auxiliary2 windings and the isolation medium FR4 are represented. The auxiliary and main windings are discrete monolithic structures that realize the disassembly and assembly of the magnetic integration.

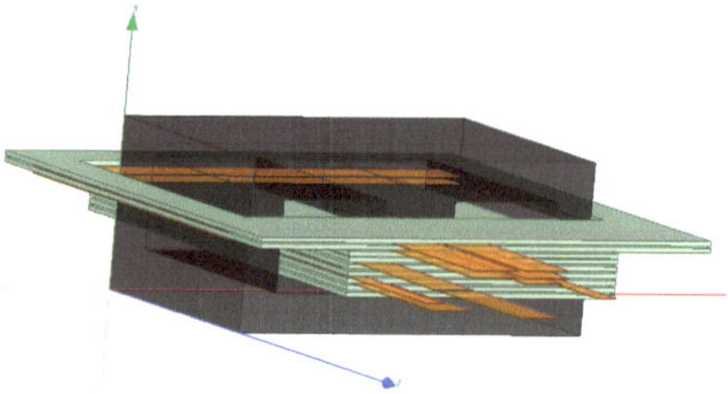

Figure 9. The integrated magnetic device model.

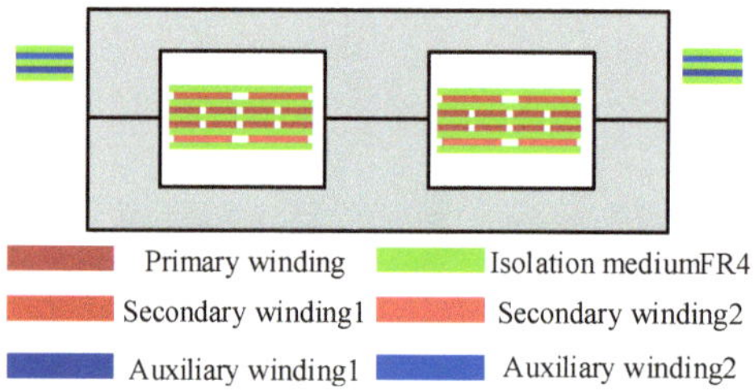

Figure 10. The front cutaway view of the integrated magnetic device model.

The winding width, thickness, and other related parameters are shown in Table 2. Since the secondary and auxiliary winding currents are equal, the winding width is set as equal. Based on the manufacturing cost and on-resistance relationship, a 2 ounce copper thickness was selected, and its thickness is 70 μm.

Table 2. The parameters of the transformer's windings.

Winding	Width	Copper Thickness
Primary winding	2 mm	70 μm
Secondary winding1	4.2 mm	70 μm
Secondary winding2	4.2 mm	70 μm
Auxiliary winding1	4.2 mm	70 μm
Auxiliary winding1	4.2 mm	70 μm

Based on the arrangement of the windings, the transformers without/with auxiliary windings were simulated separately to form a comparison. The simulation flux distributions of the main transformer without/with auxiliary windings are shown in Figure 11. Comparing the flux distributions between the transformer with and without auxiliary windings, the magnetic flux is essentially the same. The coupling coefficients are derived from flux-related data for quantitative analysis. The coupling coefficients of the transformers without/with auxiliary windings are listed in Tables 3 and 4, respectively.

Comparing the coupling coefficients without/with auxiliary windings, the coupling coefficients between the primary and secondary windings increase from 0.991 to 0.992. Since the changes in the coupling coefficients are small enough, the effect can be neglected. The coupling coefficient between the two auxiliary windings is 0.150. From (36), the 0.150 coupling coefficient can work effectively in the flux cancellation application.

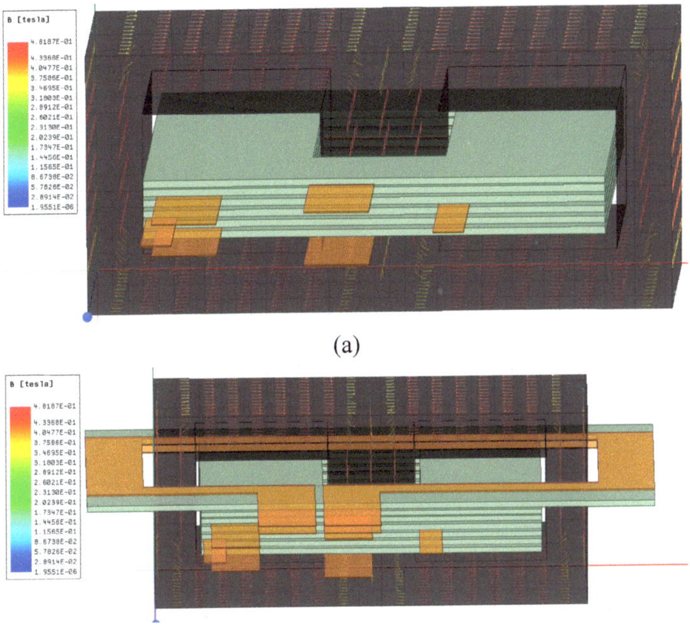

(a)

(b)

Figure 11. The simulation flux distributions of the main transformer without/with magnetic integration: (**a**) without magnetic integration; and (**b**) with magnetic integration.

Table 3. The simulation results of the coefficients of the transformer without the auxiliary windings.

	Primary	**Secondary1**	**Secondary2**
Primary	1	0.991	0.991
Secondary1	0.991	1	0.984
Secondary2	0.991	0.984	1

Table 4. The simulation results of the coefficients of the transformer with the auxiliary windings.

	Primary	**Secondary1**	**Secondary2**	**Auxiliary1**	**Auxiliary2**
Primary	1	0.992	0.992	0.030	0.027
Secondary1	0.992	1	0.984	0.048	0.027
Secondary2	0.992	0.984	1	0.031	0.024
Auxiliary1	0.030	0.048	0.031	1	0.150
Auxiliary2	0.027	0.027	0.024	0.150	1

4. Experimental Results

4.1. Verification of the Magnetic Integrated Transformer

The magnetically integrated layout was designed, fabricated and tested according to the magnetically integrated design scheme, in which the auxiliary and main windings are discrete monolithic types. The physical photos of the main and auxiliary windings and the integrated transformer are shown in Figure 12. The primary winding ports are on its left side, whilst the two secondary side windings are distributed in the two ports above and below the winding, respectively. The auxiliary windings have a total of six ports, comprising the upper three ports which correspond to the secondary1 winding and the lower three ports which correspond to the secondary2 winding.

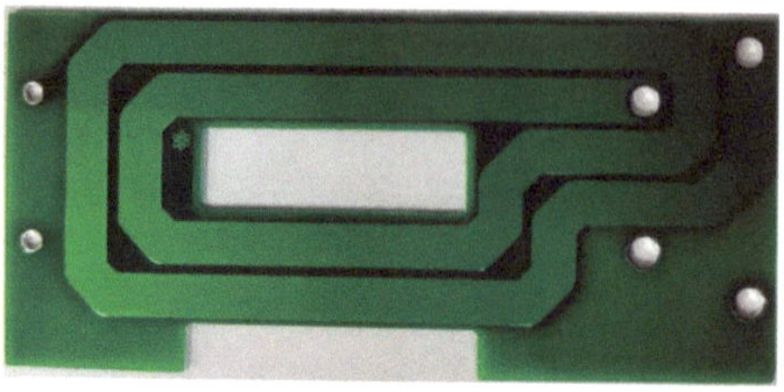

(a)

(b)

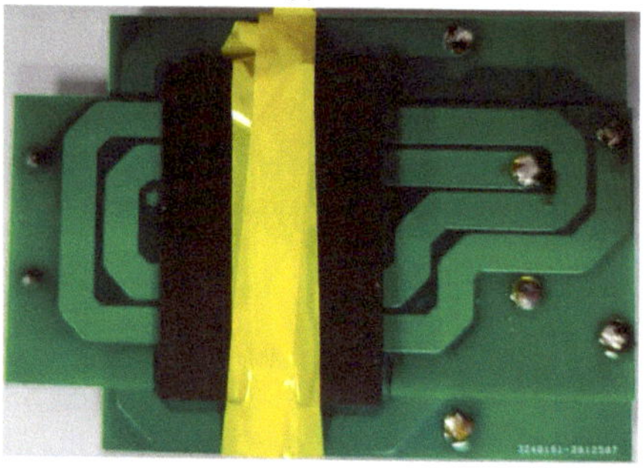

(c)

Figure 12. The experimental prototype of the integrated transformer: (**a**) the main windings; (**b**) the auxiliary windings; and (**c**) the integrated transformer.

The coupling coefficients of the integrated transformer are measured to verify the simulation results, and the test results are shown in Table 5.

Comparing the simulation and test coupling coefficients, the coupling coefficients of the primary and secondary windings in the test results are higher than the simulation results by about 0.006. According to the definition of the coupling coefficients, their influence on the converter's performance is very small and can be ignored. The test coupling coefficient between the two auxiliary windings is higher than the simulation result. This is because ANSYS uses the finite element simulation method, its simulation area is the air box that is immediately adjacent to the auxiliary winding, and the magnetic circuit formed by the air around the auxiliary winding is calculated. The simulation results of the transformer and the test results match well with the key parameters.

Table 5. The tested results of the coefficients of the transformer with the auxiliary windings.

	Primary	Secondary1	Secondary2	Auxiliary1	Auxiliary2
Primary	1	0.998	0.998	0.027	0.025
Secondary1	0.998	1	0.994	0.021	0.026
Secondary2	0.998	0.994	1	0.025	0.022
Auxiliary1	0.027	0.021	0.025	1	0.210
Auxiliary2	0.025	0.026	0.022	0.210	1

4.2. *Verification of Suppression Strategy*

The proposed suppression is verified using SiC diodes as rectifier devices. The primary switch of the converter adopts the CPM309000065B SiC MOSFET from CREE. In general, the diode mismatch is relatively small, the mismatch characteristics are difficult to test and demonstrate. The more severe test conditions were constructed to verify the effectiveness of this strategy. The rectifier diodes on the secondary side are selected as the ASD10120C SiC diode from AnBon, and MUR1520 Si diode from Onsemi. The specific parameters are shown in Table 6.

Table 6. The parameters of the diodes used.

Characteristics	ASD10120C	MUR1520
Repetitive Peak Reverse Voltage	1200 V	200 V
Forward Voltage @ 10 A	1.6 V	0.9 V
Average Rectified Forward Current	29 A	15 A

A comparison group without auxiliary windings was constructed to verify the effectiveness of the magnetic integration transformer, and the schematic of the experimental prototypes and the prototypes themselves are shown in Figures 13 and 14, respectively.

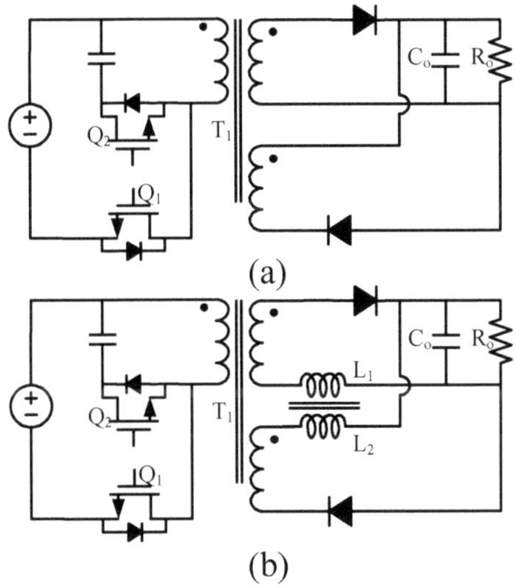

Figure 13. The schematic of the experimental prototypes: (**a**) without magnetic integration; and (**b**) with magnetic integration.

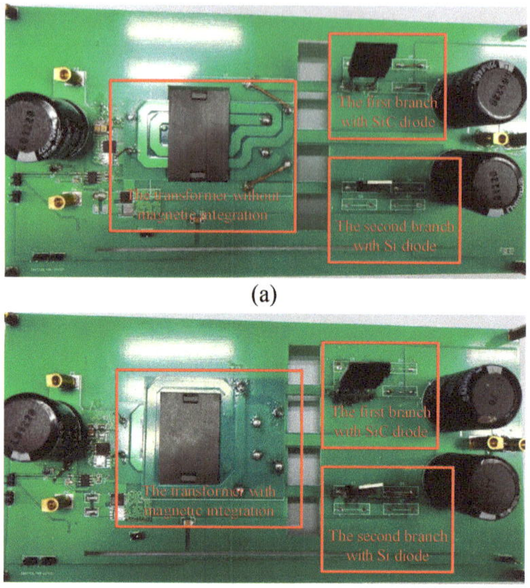

(a)

(b)

Figure 14. The experimental prototypes: (**a**) without magnetic integration; and (**b**) with magnetic integration.

The experimental principle prototypes use ASD10120C in the first branch and MUR1520 in the second branch to achieve the mismatch of the two branches. The test results in different loads without/with magnetic integration are shown in Figure 15.

The three test lines indicate the diode voltage drop of the first branch, the diode voltage drop of the second branch, and the difference in the voltage drop between the two branches, respectively. Comparing the test results without/with magnetic integration in the same load, the voltage difference between the two branches of the experimental prototype with the magnetic integration is significantly smaller. Among them, the second branch test waveform of the prototype without magnetic integration shows drastic voltage fluctuations during the diode conduction process. The test results reflect that the two branches have a large mismatch without magnetic integration and the second branch is unstable. There is a mismatch in the voltages of the two branches of about 0.1 V on the experimental prototype with the magnetic integration under all load conditions, which is the result of the mismatch between the coupled inductance of the two branches. Due to the actual manufacturing process, the main winding PCB has a notch in the lower part of the main winding, which has a certain effect on the self-inductance of the secondary main winding, and in the mismatch suppression process, there is a small voltage difference of 0.1 V on the diode to match the mismatch of the secondary winding of the two branches. The consistency of the voltage drop difference under different loads verifies the analysis. The stable operation under different load cases verifies the stability of the magnetic integration control method proposed in this paper. The converter has a stable mismatch rejection capability under different load cases. This is consistent with the previous analysis that the coupled inductance suppresses the voltage drop mismatch reduction direction of the two branches.

According to the operation of the converter, the effect of the mismatch of the two branches on the core bias is tested. After 5 min of steady-state operation with a 4 A load, the thermal distribution of the core and its windings was stable and photographed. The thermal distribution is shown in Figure 16.

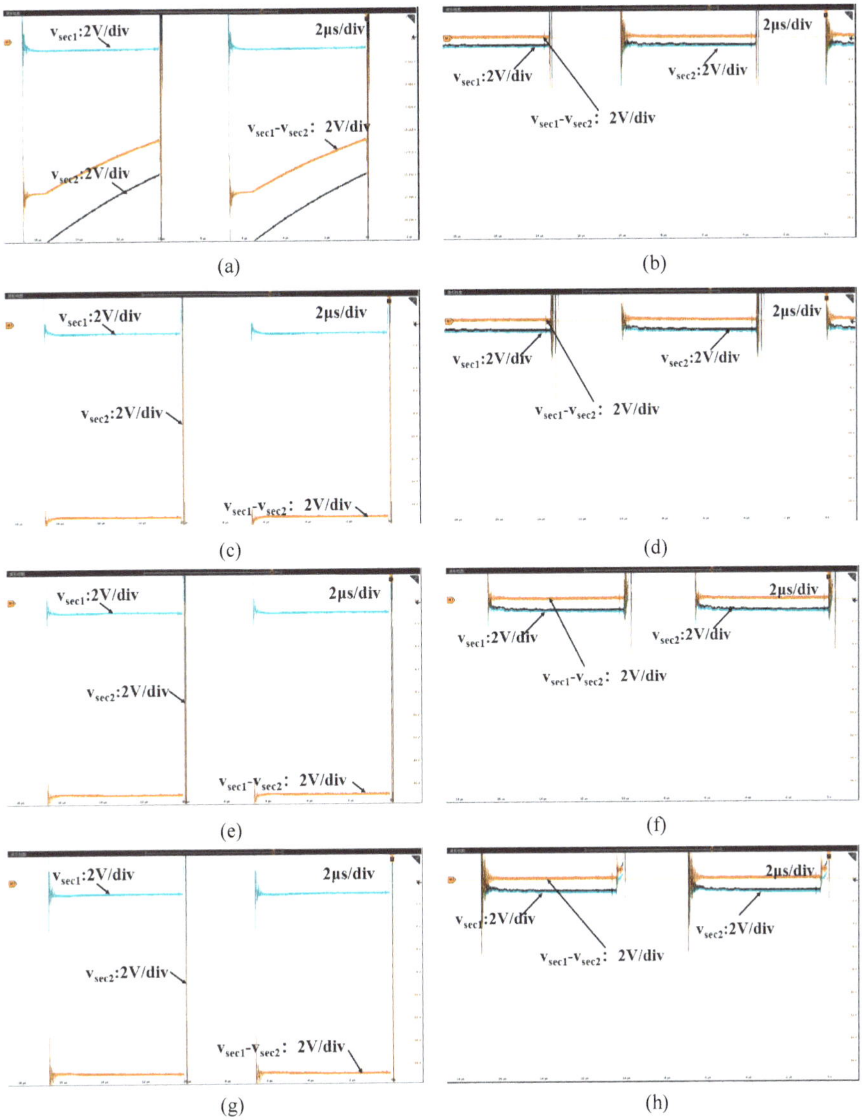

Figure 15. The voltage drop characteristics of the two branches: (**a**) 1 A load without magnetic integration; (**b**) 1 A load with magnetic integration; (**c**) 4 A load without magnetic integration; (**d**) 4 A load with magnetic integration; (**e**) 7 A load without magnetic integration; (**f**) 7 A load with magnetic integration; (**g**) 10 A load without magnetic integration; (**h**) 10 A load with magnetic integration.

In the case of the 4 A load, the maximum temperature of the transformer without core integration is 39.3 °C, the average temperature is 37.6 °C, the maximum temperature of the transformer with magnetic integration is 35.8 °C, and the average temperature is 34.1 °C. The overall operating temperature of the transformer with magnetic integration is significantly lower than that of the transformer without magnetic integration. The temperature distribution diagram shows that the magnetic integration scheme has little effect on the magnetic energy distribution of the core, and its suppression of current loss can effectively reduce the working temperature of the transformer, which has a good effect on the thermal distribution and reliability of the converter.

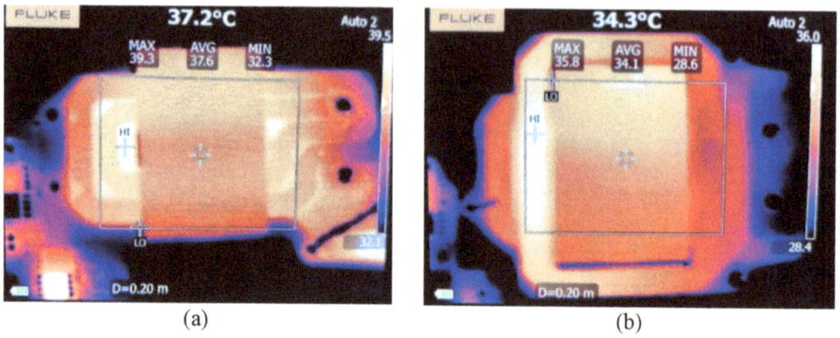

Figure 16. The temperature distribution of the transformer: (**a**) without magnetic integration; and (**b**) with magnetic integration.

According to the test waveform and the calculation results of the mismatch current, it can be seen that the prototype of the magnetic integration principle can effectively suppress the current mismatch caused by the mismatch. The efficiency test with/without magnetic integration is carried out under different load conditions, and the efficiency test results are shown in Figure 17.

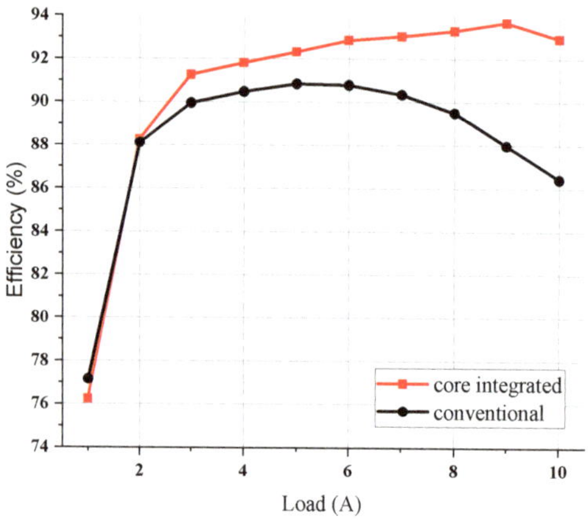

Figure 17. The efficiency comparison.

When the load is 1 A, the efficiency of the converter with magnetic integration is lower than that of the traditional converter, which is due to the new auxiliary winding increasing the copper loss of the transformer. This part of the conduction loss accounts for a large proportion in light load, which reduces the efficiency of the principle prototype with magnetic integration. With the load increasing, the efficiency of the converter with magnetic integration monotonically increases in the load range of 1–9 A, while the efficiency of the traditional converter only increases in the range of 1–5 A. With the load increasing, the current is distributed in the two branches, and the diode conduction and reverse recovery loss is small, while the diode conduction loss of the traditional converter rapidly increases due to the uneven current distribution. The efficiency of the converter without magnetic integration decreases after the 5 A load because the diode conduction and reverse recovery loss account for the loss dominance. As the load continues to increase, the efficiency of the magnetic integration prototype reaches its maximum at 9 A, which is 93.68%. Through the above efficiency analysis, it can be seen that the magnetic integrated control method can effectively improve the efficiency of the current offset converter, especially under the condition of heavy load, as the efficiency of the converter is more obvious.

The efficiency improvement of this strategy was tested at different temperatures, and the test results are shown in Figure 18. This strategy worked effectively in the range from −25 °C to 75 °C. The enhancement effect is more obvious under heavy-load and high-temperature conditions. This is because the on-state voltage drops of Si and SiC diodes become opposites as the operating temperature increases. The on-state voltage of the Si diode drops while the SiC diode rises as the operating temperature increases, which leads to a more serious mismatch in the conventional method. The mismatch reduces the efficiency of the conventional converter. The proposed strategy suppresses the mismatch, and the efficiency improvement is more obvious.

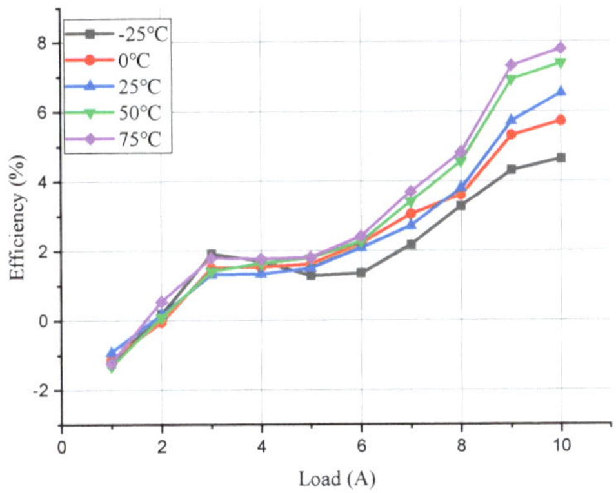

Figure 18. The efficiency improvements at different temperatures.

For long-term stability and effectiveness, the life test was performed. The principle prototype operated continuously for 10 h with 10 A load at room temperature. The efficiency variations are shown in Figure 19. The efficiency degradation is within 3%, which is acceptable in commercial applications.

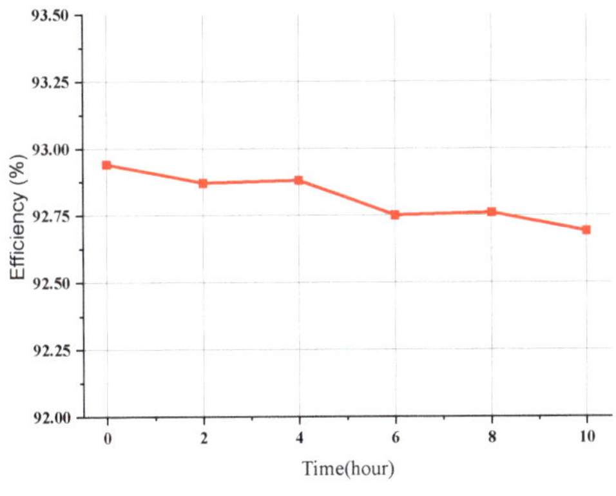

Figure 19. The efficiency variations in long-term operating.

5. Conclusions

In high-frequency applications, the mismatch between devices and circuits can have extremely harsh effects. To take full advantage of the high-frequency characteristics of SiC power devices, suppressing the mismatch has become an urgent problem. Starting from the current mismatch problem of parallel devices, this paper analyzes the working mechanism of the coupling inductance to suppress the mismatch and reveals that the mismatch suppression of the coupling inductance on different branches is essentially the essence of the suppression of the voltage mismatch of different branches. Furthermore, a novel magnetic integration strategy is proposed to suppress the mismatch. The strategy integrates the coupling inductance into the main core and balances the mismatch of the two branches at the output end by suppressing the voltage difference. To verify the control method, the design and manufacture of the experimental prototype were carried out, and the effectiveness of the control method was verified in the whole load by constructing severe mismatch conditions. The mismatch voltage of the two branches is controlled within 0.1 V. Compared with the comparison group, the proposed strategy suppresses the loss caused by mismatch and improves the efficiency of the converter. The efficiency of the magnetic integrated converter at full load is 92.94%, which is 6.52% higher than that of the traditional converter, and the prototype of the magnetic integration principle has a maximum efficiency of 93.68% at a load of 9 A. The technical advantages are analyzed above, whilst the scalability and mass production are analyzed as follows. The principle prototype produced in this paper is based on a commercial printed-circuit-board (PCB) preparation process and mature commercial devices, so it is a perfect match for existing technological fabrication processes. The magnetic integration strategy reduces an auxiliary winding core, and uses separated windings to avoid increasing the number of layers of PCB, which effectively controls the cost of the converter. Therefore, mass production is not a problem in terms of manufacturing and cost.

Author Contributions: Conceptualization, L.C. and Z.L.; methodology, S.S.; software, J.L.; validation, J.L., W.Y. and Y.S.; formal analysis, H.G.; investigation, H.G.; resources, Y.W.; data curation, S.S.; writing—original draft preparation, H.G.; writing—review and editing, Y.W.; visualization, Y.W.; supervision, L.C.; project administration, Z.L.; funding acquisition, L.C. All authors have read and agreed to the published version of the manuscript.

Funding: The APC was funded by National High-Level Personnel of Special Support Program (2023) and Independent Research and Development Program of CASC (YJ2345CX).

Data Availability Statement: The authors confirm that the data supporting the findings of this study are available within the article.

Conflicts of Interest: The authors declare no conflicts of interest.

References

1. Millán, J.; Godignon, P.; Perpiñà, X.; Pérez-Tomás, A.; Rebollo, J. A Survey of Wide Bandgap Power Semiconductor Devices. *IEEE Trans. Power Electron.* **2014**, *29*, 2155–2163. [CrossRef]
2. Palmour, J.W. Silicon carbide power device development for industrial markets. In Proceedings of the 2014 IEEE International Electron Devices Meeting, San Francisco, CA, USA, 15–17 December 2014; pp. 1.1.1–1.1.8. [CrossRef]
3. Song, Q.; Yang, S.; Tang, G.; Han, C.; Zhang, Y.; Tang, X.; Zhang, Y.; Zhang, Y. 4H-SiC Trench MOSFET With L-Shaped Gate. *IEEE Electron Device Lett.* **2016**, *37*, 463–466. [CrossRef]
4. She, X.; Huang, A.Q.; Lucía, O.; Ozpineci, B. Review of Silicon Carbide Power Devices and Their Applications. *IEEE Trans. Ind. Electron.* **2017**, *64*, 8193–8205. [CrossRef]
5. Song, Q.; Yuan, H.; Sun, Q.; Han, C.; Tang, X.; Zhang, Y.; Yuan, L.; Yang, S.; Zhang, Y. Reverse-Bias Stress-Induced Electrical Parameters Instability in 4H-SiC JBS Diodes Terminated Nonequidistance FLRs. *IEEE Trans. Electron Devices* **2019**, *66*, 3935–3939. [CrossRef]
6. Jiang, X.; Zhang, Y.; Zhang, Y.; Song, Q.; Tang, X. A 1MHz Gate Driver for Parallel Connected SiC MOSFETs with Protection Circuit. In Proceedings of the 2018 1st Workshop on Wide Bandgap Power Devices and Applications in Asia (WiPDA Asia), Xi'an, China, 17–19 May 2018; pp. 6–11. [CrossRef]
7. Zhang, Y.; Rong, G.; Qu, S.; Song, Q.; Tang, X.; Zhang, Y. A High-Power LED Driver Based on Single Inductor-Multiple Output DC–DC Converter With High Dimming Frequency and Wide Dimming Range. *IEEE Trans. Power Electron.* **2020**, *35*, 8501–8511. [CrossRef]
8. Nakamura, Y.; Shintani, M.; Sato. T. Dominant Model Parameter Extraction for Analyzing Current Imbalance in Parallel Connected SiC MOSFETs. In Proceedings of the 2021 IEEE Energy Conversion Congress and Exposition (ECCE), Vancouver, BC, Canada 10–14 October 2021; pp. 5622–5628. [CrossRef]

9. Colmenares, J.; Peftitsis, D.; Nee, H.P.; Rabkowski, J. Switching performance of parallel-connected power modules with SiC MOSFETs. In Proceedings of the 2014 International Power Electronics Conference (IPEC-Hiroshima 2014—ECCE ASIA), Hiroshima, Japan, 18–21 May 2014; pp. 3712–3717.
10. Fabre, J.; Ladoux, P. Parallel connection of SiC MOSFET modules for future use in traction converters. In Proceedings of the 2015 International Conference on Electrical Systems for Aircraft, Railway, Ship Propulsion and Road Vehicles (ESARS), Aachen, Germany, 3–5 March 2015; pp. 1–6.
11. Nakamura, Y.; Kuroda, N.; Yamaguchi, A.; Nakahara, K.; Shintani, M.; Sato, T. Influence of Device Parameter Variability on Current Sharing of Parallel-Connected SiC MOSFETs. In Proceedings of the 2020 IEEE 29th Asian Test Symposium (ATS), Penang, Malaysia, 23–26 November 2020; pp. 1–6.
12. Sadik, D.P.; Colmenares, J.; Peftitsis, D.; Lim, J.K.; Rabkowski, J.; Nee, H.P. Experimental investigations of static and transient current sharing of parallel-connected silicon carbide MOSFETs. In Proceedings of the 2013 15th European Conference on Power Electronics and Applications (EPE), Lille, France, 2–6 September 2013; pp. 1–10.
13. Wang, G.; Mookken, J.; Rice, J.; Schupbach, M. Dynamic and static behavior of packaged silicon carbide MOSFETs in paralleled applications. In Proceedings of the 2014 IEEE Applied Power Electronics Conference and Exposition—APEC 2014, Fort Worth, TX, USA, 16–20 March 2014; pp. 1478–1483.
14. Ke, J.; Zhao, Z.; Sun, P.; Huang, H.; Abuogo, J.; Cui, X. Chips Classification for Suppressing Transient Current Imbalance of Parallel-Connected Silicon Carbide MOSFETs. *IEEE Trans. Power Electron.* **2020**, *35*, 3963–3972. [CrossRef]
15. Hua, M.; Li, R.; Chen, H.; Zhou, X.; Jiang, H. Avalanche capability degradation of the parallel-connected SiC MOSFETs, Microelectronics Reliability. *Microelectron. Reliab.* **2023**, *142*, 114926.
16. Ding, S.; Wang, P.; Wang, W.; Xu, D.; Blaabjerg, F. Current Sharing Behavior of Parallel Connected Silicon Carbide MOSFETs Influenced by Parasitic Inductance. In Proceedings of the 2019 10th International Conference on Power Electronics and ECCE Asia, Busan, Republic of Korea, 27–30 May 2019; pp. 2846–2852.
17. Du, Z.; Sang, L.; Tian, L.; Zhang, W.; Wu, P.; Zhang, Y.; Zhao, Z.; Yang, F. Dispersion analysis of dynamic and static characteristic parameters of 1200V SiC MOSFET. In Proceedings of the 2021 IEEE 1st International Power Electronics and Application Symposium, Shanghai, China, 13–15 November 2021; pp. 1–4.
18. Haihong, Q.; Ying, Z.; Ziyue, Z.; Dan, W.; Dafeng, F.; Shishan, W.; Chaohui, Z. Influences of circuit mismatch on paralleling silicon carbide MOSFETs. In Proceedings of the 2017 12th IEEE Conference on Industrial Electronics and Applications (ICIEA), Siem Reap, Cambodia, 18–20 June 2017; pp. 556–561.
19. Bertelshofer, T.; März, A.; Bakran, M.M. Modelling parallel SiC MOSFETs: Thermal selfstabilisation vs. switching imbalances. *IET Power Electron.* **2019**, *12*, 1071–1078. [CrossRef]
20. Liu, Y.; Dai, X.; Jiang, X.; Zeng, Z.; Qi, F.; Liu, Y.; Ke, P.; Wang, Y.; Wang, J. A New Screening Method for Alleviating Transient Current Imbalance of Paralleled SiC MOSFETs. In Proceedings of the 2020 IEEE 1st China International Youth Conference on Electrical Engineering (CIYCEE), Wuhan, China, 1–4 November 2020; pp. 1–6. [CrossRef]
21. Zhang, Y.; Song, Q.; Tang, X.; Zhang, Y. Gate driver for parallel connection SiC MOSFETs with over-current protection and dynamic current balancing scheme. *J. Power Electron.* **2020**, *20*, 319–328. [CrossRef]
22. Lin, B.R.; Lin, Y. Parallel current-fed resonant converter with balance current sharing and no input ripple current. *IET Power Electron.* **2019**, *12*, 212–219. [CrossRef]

electronics

MDPI

Communication

Design of a 1.2 kV SiC MOSFET with Buried Oxide for Improving Switching Characteristics

Hyowon Yoon [1] and Ogyun Seok [2,*]

[1] Department of Semiconductor System Engineering, Kumoh National Institute of Technology, Gumi 39177, Republic of Korea; hwyoon@kumoh.ac.kr

[2] School of Electronic Engineering, Kumoh National Institute of Technology, Gumi 39177, Republic of Korea

* Correspondence: ogseok@kumoh.ac.kr

Abstract: The 1.2 kV SiC MOSFET with a buried oxide was verified to be effective in improving switching characteristics. It is crucial to reduce the gate–drain charge (Q_{GD}) of devices to minimize switching loss (E_{total}). The SiC MOSFET with a split gate and device with a buffered oxide have been proposed by previous studies to reduce the Q_{GD} of the devices. However, both devices have a common issue of the concentration of the electric field at the gate oxide. In this paper, we propose the 1.2 kV SiC MOSFET with a buried oxide to reduce the Q_{GD} and suppress the electric field crowding effect at the gate oxide. We analyzed the specific on-resistance ($R_{on,sp}$), Q_{GD} and the maximum electric field at the gate oxide in the off state ($E_{ox,max}$) according to the width (W_{BO}) and thickness of the buried oxides (T_{BO}). The device with the buried oxide, under optimal conditions, showed lower $E_{ox,max}$ and E_{total} without significant increase in $R_{on,sp}$ in comparison to the device with a conventional structure. These results indicate that the buried oxide can improve the switching characteristics of 1.2 kV SiC MOSFETs.

Keywords: SiC; MOSFET; gate charge; electric field crowding effect; switching characteristics

Citation: Yoon, H.; Seok, O. Design of a 1.2 kV SiC MOSFET with Buried Oxide for Improving Switching Characteristics. *Electronics* **2024**, *13*, 962. https://doi.org/10.3390/electronics13050962

Academic Editor: Yahya M. Meziani

Received: 14 February 2024
Revised: 25 February 2024
Accepted: 28 February 2024
Published: 1 March 2024

1. Introduction

Silicon carbide (SiC) power semiconductors exhibit superior performance compared to silicon (Si) power semiconductors owing to their wide bandgap energy, high critical electric field and thermal conductivity [1–5]. SiC devices with wide bandgap energy demonstrate reliable operation even at elevated temperatures due to their lower intrinsic carrier density [1,2]. The high critical electric field of SiC enables the design of a thinner N-drift layer with higher doping concentration compared to Si devices operating at the same rated voltage. The resistance of the N-drift layer accounts for the majority of the total resistance, so SiC devices with thin and high-doped N-drift layer exhibit lower on-resistance and conduction loss. Additionally, SiC devices have high thermal conductivity, ensuring stable performance even at high temperatures [3,4].

SiC devices such as SiC Schottky barrier diodes (SBDs) and SiC metal oxide semiconductor field effect transistors (MOSFETs) are widely used in the rated voltage range from 600 V to over 3.3 kV due to their superior breakdown characteristics and low resistance. Particularly, 1.2 kV SiC MOSFETs are extensively utilized as switching components in electric vehicle on-board chargers, inverters and DC–DC converters [4,5]. However, the gate charge of 1.2 kV SiC MOSFETs are required to be minimized because a large portion of the energy loss occurs during the switching operation of the devices. The gate–drain charge (Q_{GD}) is a main parameter that influences the switching characteristics of SiC MOSFETs. High voltage and current are applied simultaneously to the devices when the gate–drain capacitance (C_{GD}) is being charged [5,6]. Several studies to improve the switching characteristics of SiC MOSFETs by reducing the Q_{GD} have been reported [7–17].

Baliga's research group reported the use of a 1.2 kV SiC MOSFET with a split gate to reduce Q_{GD} of the device. The split gate has a split structure of a poly-Si gate over the

JFET region. The device with the split gate is effective for reducing Q_{GD} because the area overlapped between the gate and the drain is reduced [8–12]. However, the device with the split gate presents the issue that the electric field is concentrated at the gate oxide [8,9]. In subsequent studies, they added a p-type layer of high concentration to the bottom of the p-base region to suppress the electric field crowding effect at the device. This p-type layer of high concentration serves to protect the gate oxide by widening the depletion layer in the p-base junction [10–12].

Buffered oxide is another structure that can reduce the Q_{GD} of SiC MOSFETs. The buffer oxide has the structure where a thick oxide is deposited over the JFET region. The 1.2 kV SiC MOSFET with buffered oxide has lower Q_{GD} compared to the device with a conventional structure due to a decrease in C_{GD} [13,14]. However, the protruding shape of the poly-Si gate leads to the concentration of the electric field at the corners of the buffered oxide. The 1.2 kV SiC MOSFET with a tapered buffer oxide has been introduced to address the issue of electric field crowding effect in the device with the buffered oxide [14]. The tapered buffer oxide is characterized by the addition of an angle to the buffered oxide. The device with tapered buffer oxide has shown the ability to disperse the concentration of the electric field at the corners of the gate oxide [7,14].

The device with the split gate and the p-type layer with high concentration needs an additional step for ion implantation [10–12,15]. In the case of the device with the tapered buffer oxide, the maximum electric field at the gate oxide in the off state ($E_{ox,max}$) of the device is highly dependent on the angle of the tapered buffer oxide [14]. Several studies that have investigating the gate reliability of SiC MOSFETs using an analysis of time dependent dielectric breakdown have reported that SiC MOSFETs have reliable electrical characteristics when the $E_{ox,max}$ has been designed below 3 MV/cm [16–18]. Therefore, it is necessary to develop the devices for low $E_{ox,max}$ as well as for low Q_{GD}.

In this paper, the 1.2 kV SiC MOSFET with buried oxide which has improved switching and electric field characteristics is proposed. We used TCAD simulations to verify the performance of the 1.2 kV SiC MOSFET with the buried oxide. The device with the buried oxide, shown in Figure 1b, has a structure in which the JFET region is etched and filled with oxide. We analyzed the specific on-resistance ($R_{on,sp}$), Q_{GD} and $E_{ox,max}$ of the devices with the buried oxides according to the width (W_{BO}) and thickness of the buried oxide (T_{BO}) to obtain optimal conditions for the buried oxide. In addition, the switching characteristics of the device with the conventional structure and the device with the buried oxide under optimal conditions were compared. The results showed that buried oxide is effective in improving the switching characteristics of 1.2 kV SiC MOSFETs without deteriorating other electrical characteristics.

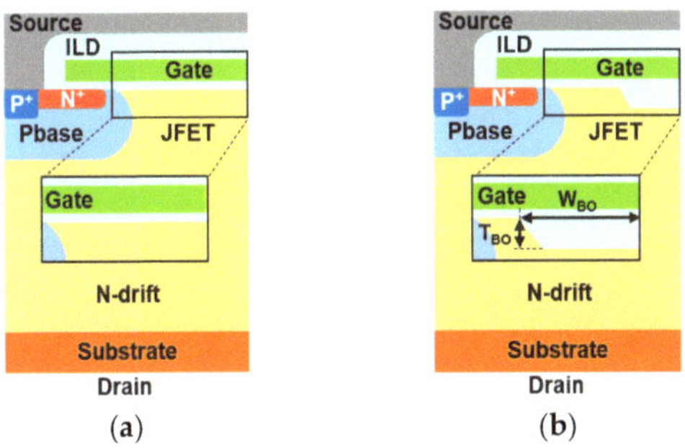

Figure 1. Cross-sectional views of 1.2 kV SiC MOSFETs. (**a**) The device with the conventional structure and (**b**) the device with the buried oxide.

2. Methods of Simulations

Sentaurus TCAD simulations were used to evaluate the performance of the 1.2 kV SiC MOSFET with buried oxide. For comparison, the 1.2 kV SiC MOSFET with the conventional structure and the 1.2 kV SiC MOSFETs with buried oxides were designed. The simulations were conducted based on the size of a half-cell. The target values for breakdown voltage (BV) and $R_{on,sp}$ were set at 1560 V and 5 mΩ·cm^2, respectively. The cell pitch, channel length, width of the JFET, concentration and thickness of the N-drift layer were designed considering the target BV and $R_{on,sp}$ [19–21]. The width of the JFET from the surface of the N-drift layer, based on the half-cell, is 0.7 μm. The aluminum (Al) concentration of the p-base is highest at a distance of 0.4 μm from the surface of the N-drift layer. At this location of the highest Al concentration in the p-base, the width of the JFET, based on the half-cell, is 0.5 μm. To ensure that the buried oxide does not completely block the JFET region, the W_{BO} was designed to be between 0.1 and 0.5 μm and T_{BO} was designed to be between 0.1 and 0.3 μm. Design parameters of the device with the conventional structure and the device with the buried oxide are summarized in Table 1. To optimize the design conditions for buried oxide, we compared the $R_{on,sp}$, Q_{GD} and $E_{ox,max}$ of the devices with buried oxides according to the W_{BO} and T_{BO}. The switching loss (E_{total}) of the device with the conventional structure and that with the buried oxide under optimal conditions were then compared, revealing that buried oxide is effective in improving the switching characteristics of 1.2 kV SiC MOSFETs.

Table 1. Design parameters of the 1.2 kV SiC MOSFET with the conventional structure and devices with the buried oxides.

Design Parameters [Unit]	Value
Width of the half-cell [μm]	3
Length of the channel [μm]	0.5
Concentration of the N-drift layer [cm^{-3}]	1×10^{16}
Thickness of the N-drift layer [μm]	10
Width of the JFET in half-cell [μm]	0.7
Width of the buried oxide (W_{BO}) [μm]	0.1, 0.2, 0.3, 0.4, 0.5
Thickness of the buried oxide (T_{BO}) [μm]	0.1, 0.2, 0.3

3. Results and Discussion

Figure 2 is the $R_{on,sp}$ of the device with the conventional structure and devices with the buried oxides according to the W_{BO} and T_{BO}.

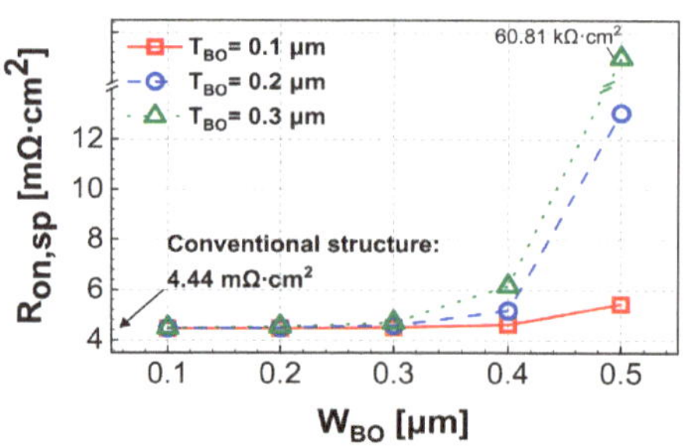

Figure 2. $R_{on,sp}$ of the 1.2 kV SiC MOSFETs with the conventional structure and devices with buried oxides according to the W_{BO} and T_{BO} (V_{GS} = 18 V).

The 1.2 kV SiC MOSFETs with the buried oxides have higher $R_{on,sp}$ than the device with the conventional structure. The $R_{on,sp}$ of the devices with buried oxides increases sharply with the increasing T_{BO} when the W_{BO} is 0.4 μm or higher. To analyze the sharp increase in the $R_{on,sp}$ of the devices with buried oxides when the W_{BO} is 0.4 μm or greater, the total current density of the devices with buried oxides was analyzed.

Figure 3 illustrates the total current density of 1.2 kV SiC MOSFETs with the buried oxides. The total current density was derived when the devices with the buried oxides had

V_{DS} and V_{GS} of 10 and 18 V, respectively. The buried oxide approaches the p-base region as the W_{BO} increases from 0.3 to 0.5 μm, leading to a reduction in the current density flow through the JFET region. For the devices with the T_{BO} of 0.3 μm, it is evident that the buried oxide significantly impedes current flow through the JFET region. This obstruction of the current flow in the JFET region results in an increase in accumulation resistance [19–21]. In the case of the device with the buried oxide, where the W_{BO} is 0.5 and T_{BO} is 0.3 μm, the $R_{on,sp}$ of the device is 60.81 kΩ·cm^2 because the current is rarely flowing through the JFET region. These results indicate that the W_{BO} needs to be 0.4 μm or less to reduce the $R_{on,sp}$.

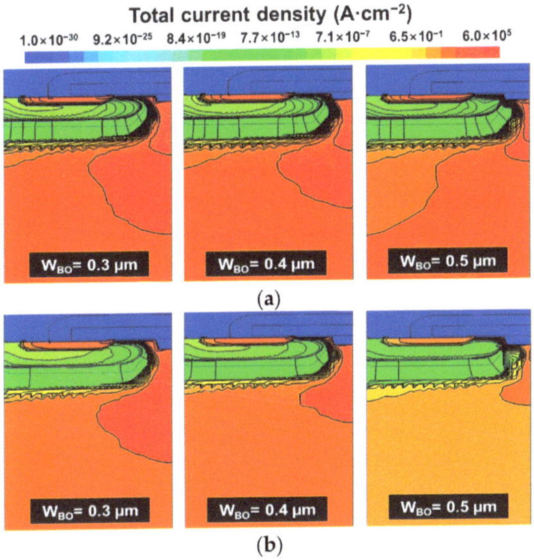

Figure 3. Total current density of the 1.2 kV SiC MOSFETs with the buried oxides (V_{DS}= 10 and V_{GS}= 18 V). The devices with the buried oxide with (**a**) a T_{BO} of 0.2 μm and a W_{BO} from 0.3 to 0.5 μm and (**b**) a T_{BO} of 0.3 μm and a W_{BO} from 0.3 to 0.5 μm.

The Q_{GD} of the 1.2 kV SiC MOSFET with the conventional structure and devices with the buried oxides according to the W_{BO} and T_{BO} are shown in Figure 4.

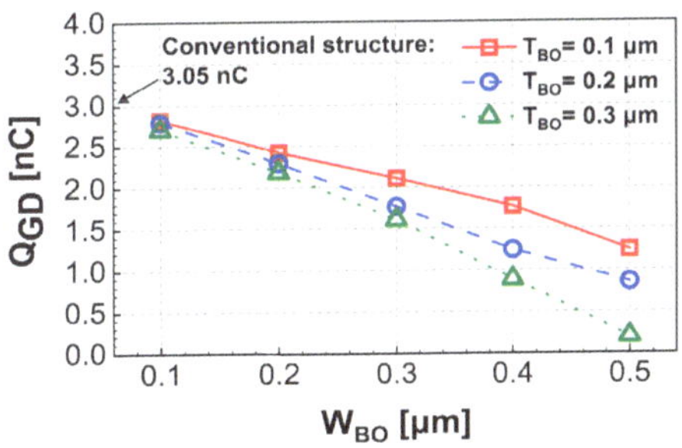

Figure 4. Q_{GD} of the 1.2 kV SiC MOSFETs with the conventional structure and devices with the buried oxides according to the W_{BO} and T_{BO}.

The Q_{GD} was determined by calculating the shift in gate charge between the two points where the slope of the $V_{gs}-Q_G$ curve exhibits the maximum value [13,14]. In contrast to Figure 2, the Q_{GD} of the devices with the buried oxides is lower than that of the device with the conventional structure. This tendency becomes more noticeable as the W_{BO} and T_{BO}

increase. When the devices have buried oxides with larger W_{BO} and T_{BO}, the area where the gate and the drain overlap is reduced, leading to a decrease in C_{GD} between the gate and the drain [5,13,14].

Figure 5 shows the $E_{ox,max}$ of the 1.2 kV SiC MOSFET with the conventional structure and devices with the buried oxides according to the W_{BO} and T_{BO}. The $E_{ox,max}$ of the devices with the buried oxides is lower than that of the device with the conventional structure because the buried oxide is effective in dispersing the electric field at the gate oxide [13,14]. The $E_{ox,max}$ of the devices with buried oxide decreases as the W_{BO} increases from 0.1 to 0.3 μm. However, the $E_{ox,max}$ of the devices with buried oxide increases as the W_{BO} increases from 0.4 to 0.5 μm. The electric field distribution of 1.2 kV SiC MOSFETs with buried oxides according to the W_{BO} and T_{BO} was analyzed to figure out the cause of the increase in the $E_{ox,max}$ of the devices with the buried oxides when the W_{BO} is 0.4 μm or greater.

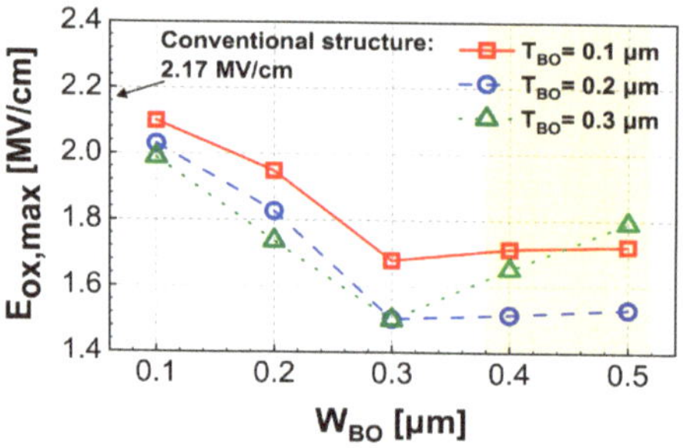

Figure 5. $E_{ox,max}$ of the 1.2 kV SiC MOSFETs with the conventional structure and devices with the buried oxides according to the W_{BO} and T_{BO}.

Figure 6a,b are the electric field distribution of the devices with the T_{BO} of 0.4 and 0.5 μm, respectively. The respective $E_{ox,max}$ and locations where the electric field concentrates at the buried oxide are noted within each figure. The p-base has the highest Al concentration at a distance of 0.4 μm from the surface of the N-drift layer. At this point, the width of the JFET is 0.5 μm based on the half-cell. Therefore, the devices with the buried oxides and with a T_{BO} of 0.3 μm are positioned closer to the junction between the p-base and the JFET than those with a T_{BO} of 0.1 and 0.2 μm. Since the electric field is concentrated at the junction between the p-type and n-type semiconductors, the devices with the buried oxides with a T_{BO} of 0.3 μm have high $E_{ox,max}$ due to the influence of the concentrated electric field at the junction between the p-base and the JFET [5,19,22,23]. When the W_{BO} is 0.3 μm or less, the buried oxide is far enough away from the junction between the p-base and the JFET to prevent this phenomenon. However, when the W_{BO} is 0.4 μm or greater, the buried oxide approaches the junction between the p-base and the JFET, causing an increase in $E_{ox,max}$. This increase in $E_{ox,max}$ is particularly pronounced when the T_{BO} is 0.3 μm.

The devices with buried oxides, where the W_{BO} is greater than 0.4 μm, exhibit a significant increase in $R_{on,sp}$ compared to the other devices with the buried oxides. Among the devices with the buried oxides where the W_{BO} is 0.3 μm or less, the device with the buried oxide with a W_{BO} of 0.3 and a T_{BO} of 0.3 μm has the smallest Q_{GD} and $E_{ox,max}$. Therefore, the optimal conditions for the W_{BO} and T_{BO} were chosen as 0.3 and 0.3 μm, respectively.

We compared the switching characteristics of the device with the conventional structure and the device with the buried oxide under optimal conditions (W_{BO} = 0.3 and T_{BO} = 0.3 μm). The used circuit for switching characteristics is shown in Figure 7. The circuit of Figure 7 was designed based on the settings of the Keysight PD1550A Double Pulse Analyzer. The gate voltage for switching the devices was switched from −5 to 18 V and the drain supply voltage was set to 800 V. Resistance of the gate and the load of the inductor were set to 10 Ω and 120 μH, respectively. The E_{total} was calculated as the integral of the product of I_d and V_{ds} during the time which the device was switched [24].

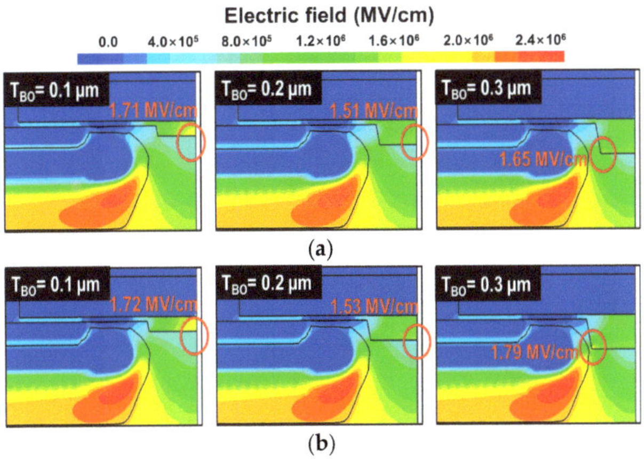

Figure 6. Electric field distribution of the 1.2 kV SiC MOSFETs with the buried oxides in the off state (V_{DS} = 1200 V) with (**a**) a W_{BO} of 0.4 and (**b**) a W_{BO} of 0.5 μm.

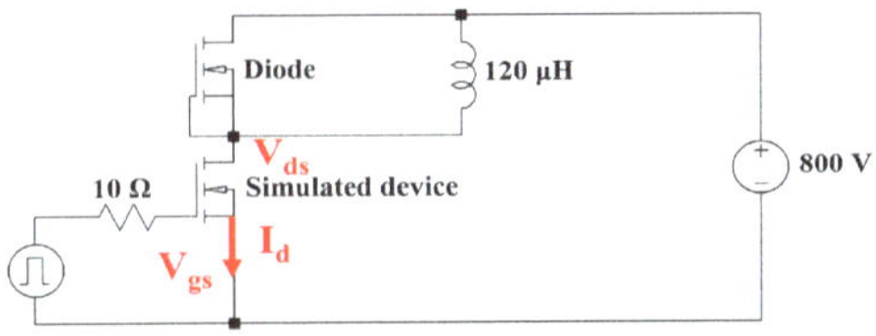

Figure 7. Circuit for switching simulations to compare the E_{total} of the device with the conventional structure and the device with the buried oxide under optimal conditions (W_{BO} = 0.3 and T_{BO} = 0.3 μm).

Figures 8 and 9 are the turn-on and turn-off switching characteristics of the 1.2 kV SiC MOSFETs with the conventional structure and the device with the buried oxide under optimal conditions (W_{BO} = 0.3 and T_{BO} = 0.3 μm), respectively. The V_{gs}, V_{ds} and I_d of the device with the buried oxide (W_{BO} = 0.3 and T_{BO} = 0.3 μm) vary more sharply during the switching process than the device with the conventional structure. Because the device with the buried oxide (W_{BO} = 0.3 and T_{BO} = 0.3 μm) has lower Q_{GD} than the device with the conventional structure, the device with the buried oxide (W_{BO} = 0.3 and T_{BO} = 0.3 μm) has fast rates of charge and discharge [5–7]. The device with the buried oxide (W_{BO} = 0.3 and T_{BO} = 0.3 μm) has lower power dissipated during the switching process than the device with the conventional structure.

The electrical characteristics of the 1.2 kV SiC MOSFETs with the conventional structure and the device with the buried oxide (W_{BO} = 0.3 and T_{BO} = 0.3 μm) are summarized in Table 2. The $R_{on,sp}$, Q_{GD} and $E_{ox,max}$ of the device with the buried oxide (W_{BO} = 0.3 and T_{BO} = 0.3 μm) are 4.71 mΩ·cm^2, 1.61 nC and 1.50 MV/cm, respectively. The device with the buried oxide (W_{BO} = 0.3 and T_{BO} = 0.3 μm) showed a 46.11% decrease in Q_{GD}, a 30.88% decrease in $E_{ox,max}$ and only a 6.10% increase in $R_{on,sp}$ compared to the device with the conventional structure. In addition, the E_{total} of the 1.2 kV SiC MOSFET with the buried oxide (W_{BO} = 0.3 and T_{BO} = 0.3 μm) is lower than the device with the conventional structure. The E_{total} of the device with the buried oxide (W_{BO} = 0.3 and T_{BO} = 0.3 μm) is 39.49 μJ, which is 20.84% lower than that of the device with the conventional structure. These results demonstrate that buried oxide can improve the switching characteristics of 1.2 kV SiC MOSFETs.

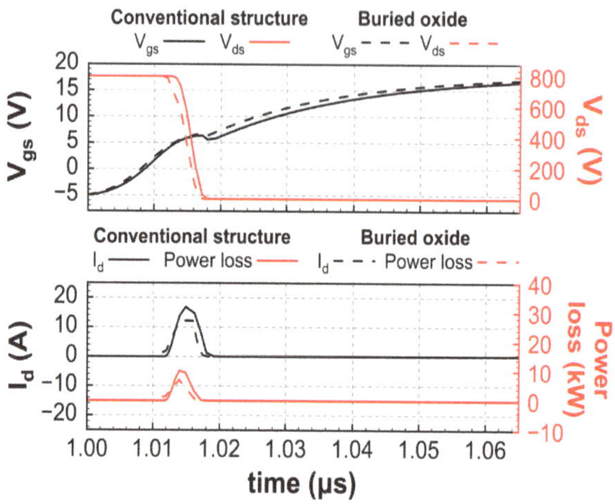

Figure 8. The turn-on switching characteristics of the 1.2 kV SiC MOSFETs with the conventional structure and the device with the buried oxide under optimal conditions (W_{BO} = 0.3 and T_{BO} = 0.3 μm).

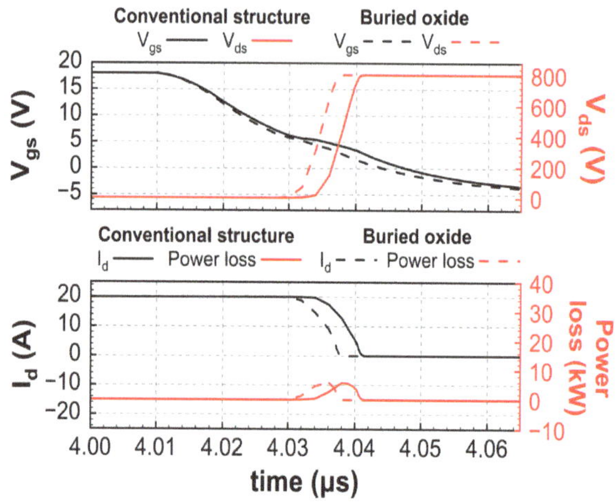

Figure 9. The turn-off switching characteristics of the 1.2 kV SiC MOSFETs with the conventional structure and device with the buried oxide under optimal conditions (W_{BO} = 0.3 and T_{BO} = 0.3 μm).

Table 2. Comparisons of the electrical characteristics of the 1.2 kV SiC MOSFETs with the conventional structure and device with the buried oxide under optimal conditions (W_{BO} = 0.3 and T_{BO} = 0.3 μm).

Electrical Characteristics	The Device with the Conventional Structure	The Device with the Buried Oxide (W_{BO} = 0.3 and T_{BO} = 0.3 μm)
$R_{on,sp}$ [mΩ·cm^2]	4.44	4.71
BV [V]	1677	1688
Q_{GD} [nC]	2.99	1.61
$E_{ox,max}$ [MV/cm]	2.17	1.50
E_{total} [μJ]	49.89	39.49

4. Conclusions

The 1.2 kV SiC MOSFET with the buried oxide was introduced to improve the switching characteristics of the device by reducing the Q_{GD}. By analyzing the $R_{on,sp}$, Q_{GD} and

$E_{ox,max}$ of the devices with the buried oxides according to the W_{BO} and T_{BO}, it was shown that the device with the buried oxide has the best performance when the W_{BO} and T_{BO} are 0.3 and 0.3 μm, respectively. The $E_{ox,max}$ and E_{total} of the device with the buried oxide (W_{BO} = 0.3 and T_{BO} = 0.3 μm) were 1.50 MV/cm and 39.49 μJ, resulting in 30.88% and 20.84% reductions compared to those with the conventional structure. The $R_{on,sp}$ of the device with the buried oxide (W_{BO} = 0.3 and T_{BO} = 0.3 μm) increased by only 6.10% compared to that of the device with the conventional structure.

Author Contributions: Conceptualization, H.Y. and O.S.; data curation, H.Y.; formal analysis, H.Y. and O.S.; investigation, H.Y.; project administration, O.S.; writing—original draft preparation, H.Y.; writing—review and editing, H.Y. and O.S. All authors have read and agreed to the published version of the manuscript.

Funding: This research received no external funding.

Data Availability Statement: All data are presented in this paper in the form of figures.

Acknowledgments: This work has been supported through the "Regional Industry-based Talent Development and Developing Innovative Technology" Program and the EDA tool was supported by the IC Design Education Center (IDEC), Republic of Korea.

Conflicts of Interest: The authors declare no conflicts of interest.

References

1. Elasser, A.; Chow, T.P. Silicon carbide benefits and advantages for power electronics circuits and systems. *Proc. IEEE* **2002**, *90*, 969–986. [CrossRef]
2. Palmour, J.W. Silicon carbide power device development for industrial markets. In Proceedings of the 2014 IEEE International Electron Devices Meeting (IEDM), San Francisco, CA, USA, 15–17 December 2014.
3. She, X.; Huang, A.Q.; Lucía, Ó.; Ozpineci, B. Review of Silicon Carbide Power Devices and Their Applications. *IEEE Trans. Ind. Electron.* **2017**, *64*, 8193–8205. [CrossRef]
4. Millán, J.; Godignon, P.; Perpiñà, X.; Pérez-Tomás, A.; Rebollo, J. A Survey of Wide Bandgap Power Semiconductor Devices. *IEEE Trans. Power Electron.* **2014**, *29*, 2155–2163. [CrossRef]
5. Baliga, B.J. *Fundamentals of Power Semiconductor Devices*, 2nd ed.; Springer: Cham, Switzerland, 2019; pp. 1–520.
6. Baliga, B.J. Power semiconductor device figure of merit for high-frequency applications. *IEEE Electron Device Lett.* **1989**, *10*, 455–457. [CrossRef]
7. Ikeda, H.; Yoshida, H.; Onikura, T. High-frequency, high-power MOS-FET. In Proceedings of the 1982 International Electron Devices Meeting (IEDM), San Francisco, CA, USA, 13–15 December 1982.
8. Han, K.; Baliga, B.J.; Sung, W. Split-Gate 1.2-kV 4H-SiC MOSFET: Analysis and Experimental Validation. *IEEE Electron Device Lett.* **2017**, *38*, 1437–1440. [CrossRef]
9. Liao, C.-N.; Chien, F.-T.; Chen, C.-W.; Cheng, C.-H.; Tsai, Y.-T. High performance power VDMOSFETs with a split-gate floating np-well design. *Semicond. Sci. Technol.* **2008**, *23*, 122001. [CrossRef]
10. Han, K.; Baliga, B.J.; Sung, W. A Novel 1.2 kV 4H-SiC Buffered-Gate (BG) MOSFET: Analysis and Experimental Results. *IEEE Electron Device Lett.* **2018**, *39*, 248–251. [CrossRef]
11. Agarwal, A.; Han, K.; Baliga, B.J. 2.3 kV 4H-SiC Accumulation-Channel Split-Gate Planar Power MOSFETs with Reduced Gate Charge. *IEEE J. Electron Devices Soc.* **2020**, *8*, 499–504. [CrossRef]
12. Xu, H.-Y.; Wang, Y.; Bao, M.-T.; Cao, F. Low Switching Loss Split-Gate 4H-SiC MOSFET With Integrated Heterojunction Diode. *IEEE J. Electron Devices Soc.* **2022**, *10*, 554–561. [CrossRef]
13. Ni, W.; Wang, X.; Xu, M.; Li, M.; Feng, C.; Xiao, H.; Jiang, L.; Li, W.; Wang, Q. Comparative Study of SiC Planar MOSFETs With Different p-Body Designs. *IEEE Trans. Electron Devices* **2020**, *67*, 1071–1076. [CrossRef]
14. Yoon, H.; Kim, J.; Kim, S.; Kim, C.; Park, Y.; Kwon, S.; Song, J.; Lee, J.; Ha, M.-W.; Seok, O. 1.2 kV SiC MOSFETs with tapered buffer oxide for the suppression of the electric field crowding effect. *Jpn. J. Appl. Phys.* **2023**, *62*, 114001. [CrossRef]
15. Zhang, Q.J.; Wang, G.; Doan, H.; Ryu, S.-H.; Hull, B.; Young, J.; Allen, S.; Palmour, J. Latest results on 1200 V 4H-SiC CIMOSFETs with Rsp, on of 3.9 mΩ·cm2 at 150 °C. In Proceedings of the 2015 27th International Symposium on Power Semiconductor Devices & IC's (ISPSD), Hong Kong, China, 10–14 May 2015.
16. Losee, P.A.; Bolotnikov, A.; Yu, L.C.; Dunne, G.; Esler, D.; Erlbaum, J.; Rowden, B.; Gowda, A.; Halverson, A.; Ghandi, R.; et al. SiC MOSFET design considerations for reliable high voltage operation. In Proceedings of the 2017 IEEE International Reliability Physics Symposium (IRPS), Monterey, CA, USA, 2–6 April 2017.
17. Yu, L.C.; Dunne, G.T.; Matocha, K.S.; Cheung, K.P.; Suehle, J.S.; Sheng, K. Reliability Issues of SiC MOSFETs: A Technology for High-Temperature Environments. *IEEE Trans. Device Mater. Reliab.* **2010**, *10*, 418–426. [CrossRef]

18. Yu, L.; Cheung, K.P.; Campbell, J.; Suehle, J.S.; Sheng, K. Oxide Reliability of SiC MOS Devices. In Proceedings of the 2008 IEEE International Integrated Reliability Workshop Final Report, South Lake Tahoe, CA, USA, 12–16 October 2008.
19. Sung, W.; Han, K.; Baliga, B.J. Optimization of the JFET region of 1.2 kV SiC MOSFETs for improved high frequency figure of merit (HF-FOM). In Proceedings of the 2017 IEEE 5th Workshop on Wide Bandgap Power Devices and Applications (WiPDA), Albuquerque, NM, USA, 30 October–1 November 2017.
20. Kim, D.; Yun, N.; Jang, S.Y.; Morgan, A.J.; Sung, W. An Inclusive Structural Analysis on the Design of 1.2 kV 4H-SiC Planar MOSFETs. *IEEE J. Electron Devices Soc.* **2021**, *9*, 804–812. [CrossRef]
21. Saha, A.; Cooper, J.A. A 1-kV 4H-SiC Power DMOSFET Optimized for Low on-Resistance. *IEEE Trans. Electron Devices* **2007**, *54*, 2786–2791. [CrossRef]
22. Seok, O.; Kang, I.H.; Moon, J.H.; Kim, H.W.; Ha, M.-W.; Bahng, W. Double p-base structure for 1.2-kV SiC trench MOSFETs with the suppression of electric-field crowding at gate oxide. *Microelectron. Eng.* **2020**, *225*, 111280. [CrossRef]
23. Park, Y.; Yoon, H.; Kim, C.; Kim, G.; Kang, G.; Seok, O.; Ha, M.-W. Design of 1.2 kV SiC trench MOSFET using tilted ion implantation for suppression of electric field crowding at the bottom of the gate oxide. *Jpn. J. Appl. Phys.* **2003**, *62*, 011001. [CrossRef]
24. JEDEC. Available online: https://www.jedec.org/standards-documents/docs/jesd-24-2 (accessed on 5 February 2024).

electronics

Article

A Hierarchical Driving Control Strategy Applied to Parallel SiC MOSFETs

Yin Luo [1], Xiaoyan Tang [1,*], Shikai Sun [2,*], Jialin Liu [2], Wenhao Yang [1] and Yuyin Sun [1]

1 School of Microelectronics, Xidian University, Xi'an 710071, China; yinluo@stu.xidian.edu.cn (Y.L.); yang_wenhao@stu.xidian.edu.cn (W.Y.); yuyinsun@stu.xidian.edu.cn (Y.S.)
2 Beijing Microelectronics Technology Institute, Beijing 100076, China; liujialin@mxtronics.com
* Correspondence: xytang@mail.xidian.edu.cn (X.T.); sunshikai@mxtronics.com (S.S.)

Abstract: SiC (silicon carbide) MOSFETs have been extensively used in the power electronics industry due to their exceptional characteristics. First, it was found in this study that their driving loss is larger than their conduction loss in high-frequency applications. Based on this finding, this study proposes a hierarchical driving control strategy for improving the parallel-converter efficiency of SiC MOSFETs under light loads. Efficiency under light loads is of great importance for battery-based energy storage systems. To minimize the sum of the conduction loss and driving loss in parallel devices, this study proposes a current-monitoring hierarchical driving strategy based on an active-clamped flyback converter. By monitoring the output current of the converter, the strategy minimizes the sum of the driving and conduction losses by switching the driving state under different loads. The results of simulations indicate the effectiveness of the load-current-monitoring strategy. To verify the effectiveness of this method, a principle prototype of two SiC MOSFETs connected in parallel at 12 V/5 A was fabricated and tested, and the test results showed that there was a maximum improvement of 1.4% in the converter's efficiency when the load current was in the range of 0.5–1.5 A.

Keywords: parallel SiC MOSFETs; conduction loss; driving loss; hierarchical driving

check for updates

Citation: Luo, Y.; Tang, X.; Sun, S.; Liu, J.; Yang, W.; Sun, Y. A Hierarchical Driving Control Strategy Applied to Parallel SiC MOSFETs. *Electronics* **2024**, *13*, 70. https://doi.org/10.3390/electronics13010070

Academic Editor: Alexander Barkalov

Received: 20 November 2023
Revised: 18 December 2023
Accepted: 20 December 2023
Published: 22 December 2023

1. Introduction

Excessive emissions of greenhouse gases have led to global warming and a resultant increase in the frequency of extreme weather, heat waves, floods, and other natural disasters, which have jeopardized the survival of all of humankind. With the deterioration of the natural environment, the suppression of climate change has become the center of focus of research in the mainstream scientific community [1,2]. Clean energy is one of the main solutions. However, clean energy sources, such as wind and solar power, are subject to external conditions, and their operating conditions fluctuate greatly. An effective solution is the application of battery energy storage systems. Battery energy storage systems must maintain high efficiency in all load ranges, especially under medium and light loads [3–5]. SiC MOSFETs offer dynamic characteristics superior to those of Si (silicon) MOSFETs and IGBTs, resulting in significant improvements in their energy conversion efficiency and power density. At the same voltage and current levels, SiC MOSFETs have the advantages of a small on-resistance, small gate charge, low input and output capacitance, and small switching loss compared with those of Si MOSFETs [6–11].

Compared with Si MOSFETs, the performance advantages of SiC MOSFETs are mainly reflected in their smaller gate charge and faster switching speed, enabling the development of high-performance converters with a higher frequency, higher efficiency, and higher power density [12–14]. Due to the low electron mobility of the SiC material itself, it requires a higher gate drive voltage to generate enough channel carriers to achieve a low on-resistance [15,16]. This means that the gain of the lower on-resistance of SiC MOSFETs comes at the cost of a higher drive voltage. Under such conditions, a higher drive voltage means that the driver circuit needs to provide a higher pulse current and greater losses. At the same time, SiC MOSFETs have significantly lower conduction and switching losses than those of Si MOSFETs, and drive losses play a more important role in the total power consumption. The change in the loss relationship is more pronounced in high-frequency applications of SiC MOSFETs, as the drive loss increases linearly with the operating frequency. In high-current applications, where multiple MOSFETs are connected in parallel, the drive loss increases exponentially with the number of devices connected in parallel. The driving loss of a device is comparable to the conduction loss in a certain load current range, so during the application of SiC MOSFETs, the driving loss becomes a necessary consideration in the converter design, especially for light loads.

SiC MOSFETs are employed in synchronous boost converters to meet the need for both high-voltage applications and soft switching under light loads [4]. Their topology utilizes a quasi-square-wave zero-voltage switching technique to improve the efficiency of the converter. When a converter is operating under light loads, the fixed-operating-frequency mode of operation causes the converter to lose its soft switching and significantly increases the reactive power. In this case, the use of SiC MOSFETs to increase the switching frequency ensures that the power device achieves soft switching without increasing the reactive power. The authors of [17] proposed a three-phase dual active bridge converter using SiC MOSFETs. This scheme improved the conventional single-phase phase-shift scheme by creating a hybrid duty cycle and phase-shift control scheme from the control method, which ensured the soft switching of the switches while increasing the operating frequency of the converter and, thus, improved the converter's efficiency under light loads. Such schemes optimize the converter's operating frequency from the perspective of topology, and SiC MOSFETs are adopted only to improve the operating frequency; their design focuses on switching losses, which are related to the realization of soft switching. The relationships between the conduction loss, driving loss, and switching loss in the application of SiC MOSFETs have not been analyzed.

The authors of [18] conducted an in-depth analysis of the driving loss and driving power structure of SiC MOSFETs. A method for the estimation of the driving loss according to the device manual was provided. The nonlinear variations in the device's gate capacitance and the distribution of the driving loss were analyzed. The driving loss was correlated with the drive voltage, the total gate charge, and the switching frequency. The report quantitatively analyzed the driving loss of SiC MOSFETs and quantitatively calculated the loss distribution in different operating states, which paved the way for a subsequent loss analysis and targeted designs for specific operating conditions. This study proposed a scheme for separating the design of a device's turn-on and turn-off gate resistors based on the device's operating state. It focused on the relationship between the driving loss and the device parameters in a single switch, but the relationship between the driving loss and the conduction loss was not examined.

The authors of [19] optimized the design of SiC MOSFETs with respect to the drive voltage versus the efficiency and reliability. This study showed that SiC MOSFETs are dominated by drift resistance and channel resistance in the on-resistance of the MOSFET due to the high critical electric field. A high gate drive voltage is required to realize the complete inversion of a device and reduce the on-resistance of a MOSFET. According to this study, the relatively high drive voltage recommended in the device manual could effectively reduce the on-resistance and turn-on loss of the device, which have a significant effect on its overall loss reduction. The relationship between the switching loss and conduction loss in a specific MOSFET was analyzed, but parallel devices were not considered.

The above studies investigated SiC MOSFETs at different drive voltages, in different drive modes, with driving loss distributions, and with drive soft-switching technology, and their contents are of great significance for the application of single SiC MOSFETs. However, there is a lack of targeted research on the driving of SiC power devices under high-current and high-frequency operating conditions. When SiC MOSFETs are used in parallel in high-current and high-frequency converters, their on-resistance decreases under the same withstanding voltage, and an increase in the switching speed leads to a significant decrease in the on-state and switching losses compared with those of Si MOSFETs. Although the gate charge of SiC MOSFETs has been reduced in comparison with that of Si MOSFETs, the increase in the drive voltage makes the driving loss a larger part of the device loss, which is different from the loss characteristics of conventional Si MOSFETs.

This study analyzes the relationship of the switching and the conduction losses in the application of SiC MOSFETs connected in parallel and proposes a hierarchical driving strategy to minimize the total loss. The strategy is verified with simulations and experiments. The experimental results showed that the loss was reduced by up to 9.4% for light loads, and the efficiency was not affected by heavy loads.

2. Quantitative Analysis of Losses and the Design of the Switching Point

The strategy was validated by applying it to an active-clamped flyback topology, which is shown in Figure 1. Since the operating principle of active-clamped flyback converters is familiar to that of engineers, this study focused on the operating mechanism of a hierarchical driving control strategy for parallel SiC MOSFETs. For heavy-load applications, parallel MOSFETs are the main solution for improving efficiency by reducing the conduction loss. However, parallel MOSFETs introduce driving loss under light loads.

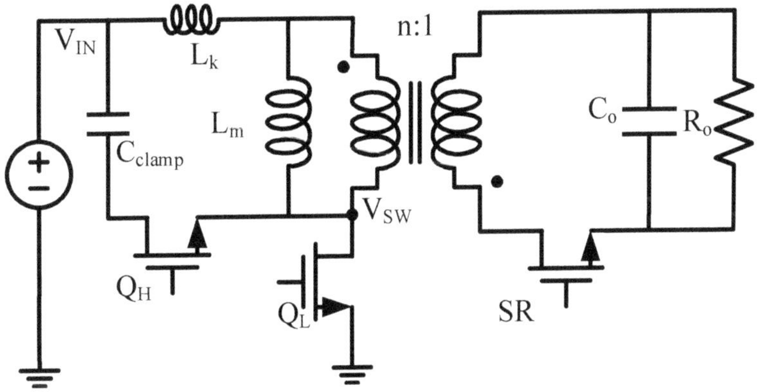

Figure 1. The topology of the active-clamped flyback converter.

A quantitative analysis of the driving and conduction losses of silicon and silicon carbide MOSFETs in parallel applications is presented. MOSFETs with a voltage rating of 1200 V and a current rating of 30 A were chosen for comparison. The CREE C3MP0075120D SiC MOSFET and Microsemi APT28M120L Si MOSFET were used to calculate the driving and conduction losses of the devices at the designed operating conditions. Since the active-clamped flyback converter has the characteristic of soft switching, the advantage of SiC MOSFETs in terms of switching characteristics can be neglected. So, this study only quantitatively analyzes the driving loss and conduction loss.

The conduction loss of the MOSFET can be expressed as (1).

$$P_{con} = \frac{1}{n}\overline{I}^2 R_{ds} \tag{1}$$

where n represents the number of MOSFETs driven and $\overline{I}$ denotes the rms current through the device. In a converter in steady-state operation, the rms current through the MOSFET can be expressed as (2).

$$\overline{I} = \sqrt{\frac{\int_0^{DT} i^2\, dt}{T}} \tag{2}$$

According to the relevant parameters of the two MOSFETs, the conduction losses of SiC and Si MOSFETs can be expressed as $P_{con-SiC} = 75\overline{I}^2$ mW and $P_{con-Si} = 450\overline{I}^2$ mW, respectively.

The driving loss can be expressed by Equation (3).

$$P_{dri} = n \cdot V_{DD} \cdot Q_G \cdot f_{SW} \tag{3}$$

where n represents the number of MOSFETs driven, V_{DD} represents the driving voltage of the MOSFETs, and Q_G represents the amount of charge stored in the gate capacitance per cycle. f_{SW} represents the switching frequency of the MOSFETs. The operating frequency is normalized to 100 KHz, allowing the frequency coefficient $\kappa = f_{SW}/100$ KHz to be defined, under which the driving losses of SiC and Si MOSFETs in a single comparison device can be expressed as $P_{dri-SiC} = 102.6 \cdot \kappa$ mW and $P_{dri-Si} = 300 \cdot \kappa$ mW.

Based on the qualitative analysis, the conduction loss of a Si MOSFET is 1.5 times the driving loss at 100 KHz/1 A. Parallel-connected Si MOSFETs can reduce the sum of the conduction and driving losses. For SiC MOSFETs, the driving loss is 1.37 times the conduction loss at 100 KHz/1 A, and parallel-connected SiC MOSFETs introduce more loss into the load. Additionally, the working frequency of the SiC MOSFETs can reach several MHz, making the driving loss higher. A loss comparison is made between APT28M120L and C3MP0075120D in dual-MOSs application; the results are shown in Table 1. The highest frequency represents the recommended highest working frequency, and the effective current represents the effective current range at the recommended highest working frequency. The hierarchical driving control strategy only can work effectively in 0–0.87 A in parallel Si MOSFET applications.

Table 1. The driving and conduction loss comparison between APT28M120L and C3MP0075120D.

Device Type	Device Part Number	Highest Frequency	Effective Current
Si MOSFET	APT28M120L	100 KHz	0–0.83 A
SiC MOSFET	C3MP0075120D	2 MHz	0–7.95 A

The CPM309000065B SiC MOSFET chip was used for the design of the prototype, and its specific electrical parameters are shown in Table 2.

Table 2. The key parameters of the CPM309000065B SiC MOSFET.

Symbol	Parameter	Typical Value	Test Conditions
R_{ds}	On-state resistance	65 mΩ	V_{GS} = 15 V
Q_G	Total gate charge	30.4 nC	V_{DS} = 400 V, V_{GS} = −4/15 V, I_D = 20 A
V_{DS}	Drain-source breakdown voltage	900 V	I_{DS} = 100 μA

Referring to the previous equations for the driving and conduction losses of power MOSFETs and the relevant parameters of the prototype, we plotted the total power consumption of the driving and conduction losses of the single-MOSFET and dual-MOS operating states at 100 KHz (Figure 2) and looked for the root-mean-square currents when the power consumption of the two modes was equal to determine the state-switching point.

The conduction and driving losses of the single MOSFET and the dual-MOSs were parabolas centered at x = 0 with only the second quadrant. The offset of the driving loss was at y = 0, and the number of parallel devices determined the total on-resistance, that is, with a constant parameter for the quadratic term, which determined the size of the parabola's "mouth"; a greater total on-resistance resulted in a smaller mouth and a quicker increase in power consumption with the continuation of the current. The intersections of the driving power consumption of the single MOSFET and dual-MOSs were equal. As shown in the figure, the power consumption of the single MOSFET's drive was smaller;

with the increase in the current, the on-resistance loss accounted for a proportion of the total loss, and the single MOSFET's driving loss increased more quickly. Under the operating condition of 200 KHz, the state-switching point of the converter was 1.89 A, and the total power consumption of driving and conduction was 347.5 mW.

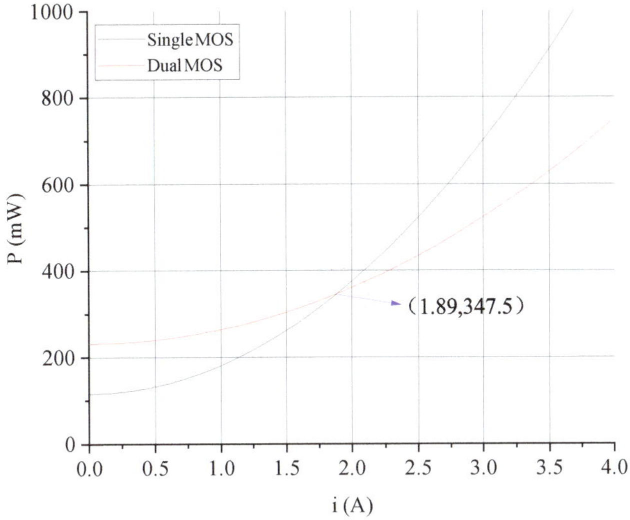

Figure 2. The sum of the driving and conduction losses in dual-MOSs at 200 KHz under different loads.

Based on the loss analysis, the state-switching point was generalized to plot the total power consumption of conduction and drive for single- and dual-MOS drive under operating frequency conditions of 100 KHz–1 MHz, and the results are depicted in Figure 3.

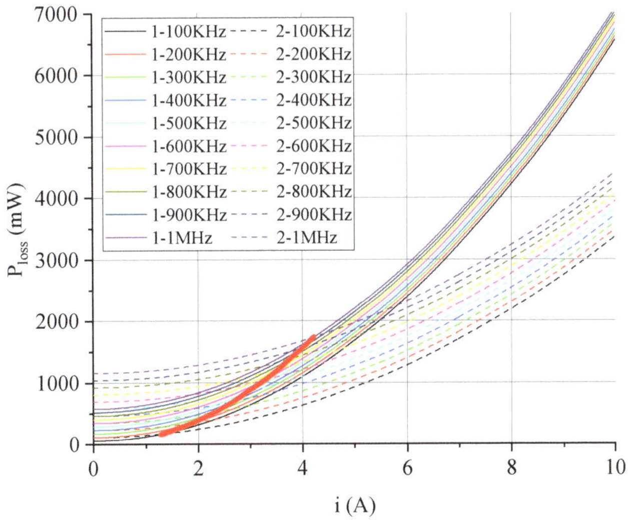

Figure 3. The sum of the driving and conduction losses at different frequencies under different loads.

The red line in the figure depicts the transformation of the state-switching point with the single- and dual-MOS drive as the operating frequency increased. The state-switching current increased as the operating frequency increased. At an operating frequency of 1 MHz, the state-switching point was 4.21 A, and the total power consumption of driving and conduction at this time was 1729.667 mW. The effective range of the hierarchical driving strategy increased with the increase in the operating frequency.

To minimize the total conduction and driving losses, the strategy divided the mode of operation into those of single MOS, dual-MOSs, and triple-MOSs in hierarchical driving applications with three parallel MOSFETs. The design of the switching points of the three parallel MOSFETs at 200 KHz is shown in Figure 4. The effective range of the hierarchical driving strategy increased with the increase in the number of parallel MOSFETs.

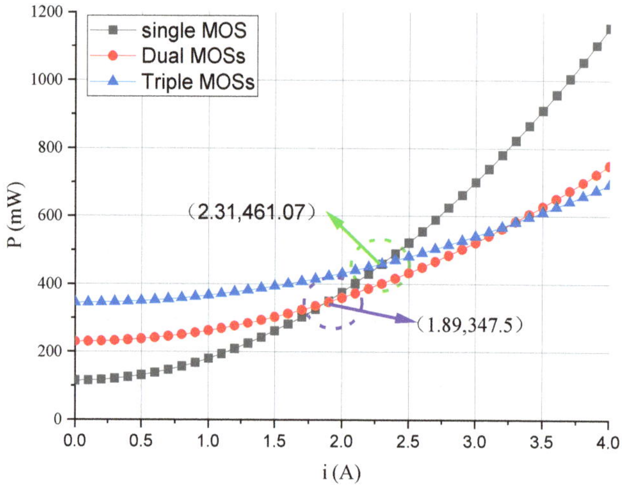

Figure 4. The sum of the driving and conduction losses in triple-MOSs at 200 KHz under different loads.

The effective range of the hierarchical driving control strategy increased with the increase in the operating frequency and number of parallel devices. The increase in the operating frequency matched the trend towards higher power densities, and the increase in the number of parallel devices matched the high current requirements.

3. Operating Principle of the Hierarchical Driving Strategy

According to the above analysis, the single-MOSFET working mode had the lowest conduction and switching losses under a load of 1.89 A, and the dual-MOS mode had better efficiency under loads above 1.89 A at 200 KHz. The hierarchical driving strategy aimed to realize switching in the single-MOSFET and dual-MOS modes under a load of 1.89 A.

Since the output current directly reflected the load state, an output-current-monitoring method could improve the sensitivity of the converter to load changes. In addition, the load had low fluctuation during steady-state operation, which contributed to the stability of the switching point.

In applications with precise input voltage requirements and real-time status detection, a system power management and protection module should be added to the output of the finished power supply module. The protection module has to monitor the output voltage and current in real time. The vice-side current detection method has compatibility with this system's power management and protection module and only needs to externalize the current detection signal of the protection module to realize the multiplexing of the functions. The current detection circuit is schematically shown in Figure 5.

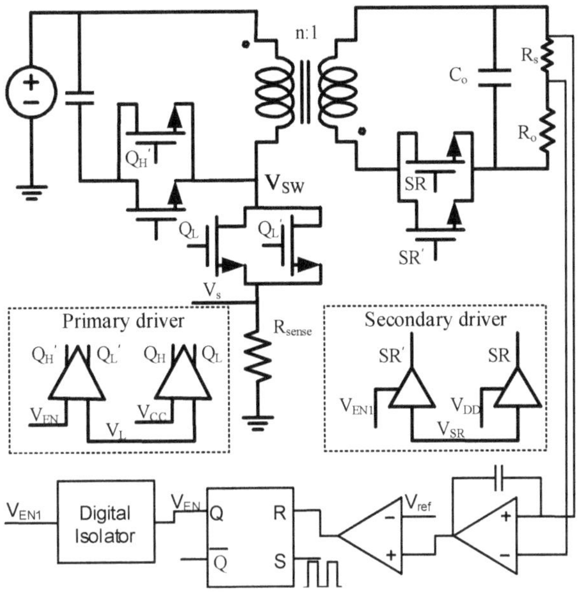

Figure 5. Schematic of the current detection circuit.

Here, Q_H and Q_H' are the parallel high-end MOSFETs on the primary side, and Q_L and Q_L' are the parallel low-end MOSFETs on the primary side. SR and SR' are the parallel synchronous rectification MOSFETs on the secondary side. *driver*1 is the driver of Q_H and Q_L, *driver*2 is the driver of Q_H' and Q_L', *driverSR*1 is the driver of SR, and *driverSR*2 is the driver of SR'. Q_H, Q_L, *driver*1, SR, and *driverSR*1 represent the first set of MOSFETs and drivers. Q_H', Q_L', *driver*2, SR', and *driverSR*2 represent the second set of MOSFETs and drivers. R_s represents the sense resistor for monitoring the output current. V_{EN} is the enable signal of the driver2, and V_{EN1} is the enable signal of driverSR2.

In the current detection scheme, a stable output current produced a stable voltage drop across R_s. The converter's output current was generally stable and only fluctuated when the load was switched. A schematic of the operating waveform in the current detection strategy is shown in Figure 6.

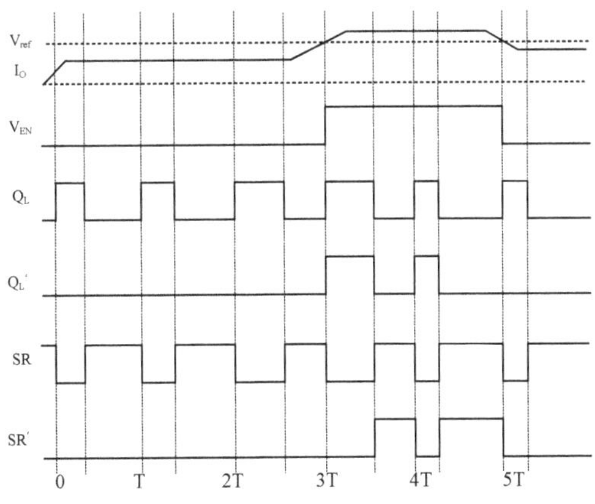

Figure 6. The schematic waveforms of the current detection circuit.

As shown in Figure 6, *driver*1 and *driverSR*1 worked in the entire load range. The working states of *driver*2 and *driverSR*2 depended on the load current. When the load current was higher than the reference value, the enabling signal V_{EN}, output V_{EN1}, *driver*2, and *driverSR*2 worked. The two sets of MOSFETs worked together to reduce the sum of the driving and conduction losses and improve the efficiency.

4. Simulations and Experimental Results

Based on the above analysis, the effectiveness of the hierarchical driving strategy was verified with simulations and experiments. The active-clamped flyback topology was used to complete the verification of the hierarchical drive control method. The relevant parameters of the prototype designed in this section are shown in Table 3.

Table 3. The key parameters of prototype.

Characteristic	Symbol	Typical Value
Input voltage	V_{IN}	100 V
Output voltage	V_{OUT}	12 V
Output current	I_O	5 A
Working frequency	f_{SW}	200 KHz
Driving voltage	V_{dri}	15 V
Primary excitation inductance	L_m	220 µH
Primary leakage inductance	L_k	10 µH
Resonant capacitance	C_c	10 µF

4.1. Simulation Results

Simulations were used to verify the effectiveness of the current-monitoring scheme. The simulation model was set based on the designed prototype, and the hierarchical driving strategy was constructed in the Pspice17.2 from the Cadence company (San Jose, CA, USA). Some devices without simulation files were replaced by devices with the same function. The simulation models used in the simulation were downloaded from the manufacturer and are listed in Table 4. The transformer was modeled from the library files in the software.

Table 4. The main models used in the prototype.

Device	Manufacturer	Part Number
Current Sense Amplifiers	INA180	TEXAS INSTRUMENT
SiC MOSFET	IMW120R060M1H	infineon
MOSFET driver	UCC20225	TEXAS INSTRUMENT
Digital isolator	ISO7721DR	TEXAS INSTRUMENT

The simulation results are shown in Figure 7. The simulation results mainly contained I_O, V_{OUT}, the representative driving signal of the second set of MOSFETs (V_{SR2}), and the representative driving signal of the first set of MOSFETs (V_{SR1}).

From the simulation results, it can be seen that at 34.01 ms, the load current started to rise from 0.1 A to 5 A with a slope of 50 A/ms; the load current was maintained at 5 A for 1 ms, and then it started to fall to 0.1 A with a slope of 50 A/ms. The whole process of the variation in the load current was maintained for 3 ms. During this process, the range of fluctuation in the output voltage ΔV_{OUT} was 0.5 V, which met the design requirements. During the whole working process, the synchronous rectifier driver of the first set, V_{SR1}, always maintained a normal output. When the load current reached 1.9 A, the synchronous rectifier driver of the second set, V_{SR2}, started to enable the output. The specific waveform of the output-enabling process is shown in the upper-left corner of the figure (V_{SR2}), and the output-disabling process is shown in the upper-right sub-figure. When the load current dropped from 5 A to 2.1 A, V_{SR2} stopped outputting, which was mainly caused by the inconsistency between the turn-on and turn-off voltages of the driver's enabler.

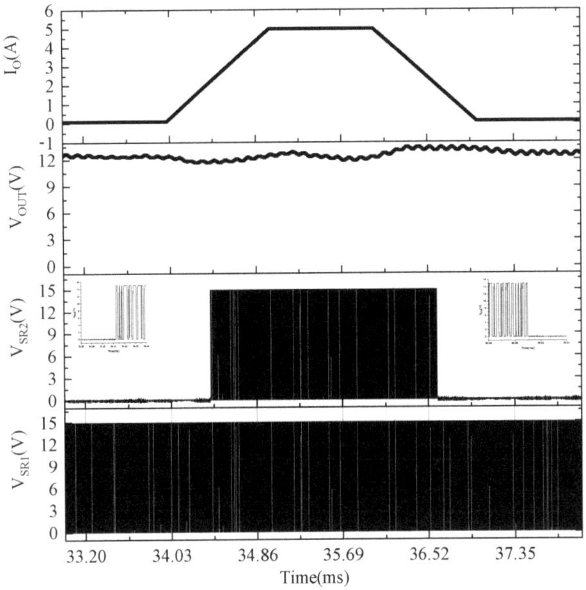

Figure 7. The simulation results of the hierarchical driving control strategy.

4.2. Experimental Results

After the experimental scheme was effectively verified with simulations, we fabricated and tested a prototype to verify the hierarchical driving strategy. The experimental prototype is shown in Figure 8. The power circuit of the power module was mainly distributed on the front side of the prototype, and the control logic was distributed on the back side of the prototype, thus realizing the separation of power and control and reducing the influence of the power circuit on the feedback and control logic. The drivers and power switches of the prototype were in the form of a single package, and each driver corresponded to one power switch, thus realizing driver control with a single switch. An output-current-monitoring module monitored the output current in real time, and its output allowed enabling the control of the second-stage driver to meet the requirements of the hierarchical drive control scheme. The test waveform and the working details are shown in Figure 9.

The test waveform included I_O, V_{OUT}, V_{SR2}, and V_{SR1}. The load current started to rise from 0.1 A to 5 A with a slope of 5 A/ms in the first 1.5 ms; the load current was maintained at 5 A for 1.5 ms and then started to fall to 0.1 A with a slope of 5 A/ ms. Throughout the process of variation in the load, the output voltage fluctuated in the range of $\Delta V_{OUT} \leq 0.5$ V, and its dynamic response to changes in the load met the design requirements. V_{SR1} always worked normally in all of the load fluctuation processes. Under a load of 2 A, V_{SR2} started to enable the output. When the load current dropped from 5 A to 1.6 A, V_{SR2} stopped outputting. The asymmetry between the enabling and disabling thresholds was due to the different turn-on and turn-off thresholds of the drive-enabling signals [20], which limited the responsiveness of the hierarchical driving strategy. In the vicinity of the enabling and disabling thresholds of the second-level drive, the second-level drive signal was discrete, which was due to the continuous output current transformation; the gain of the current detection chip was fixed, and there was a period of inaccurate judgment of the enabling port voltage threshold of the driver chip. The overall scope of the hierarchical drive strategy allowed it to realize the function of single-MOSFET operation under light loads and dual-MOSs operation under heavy loads.

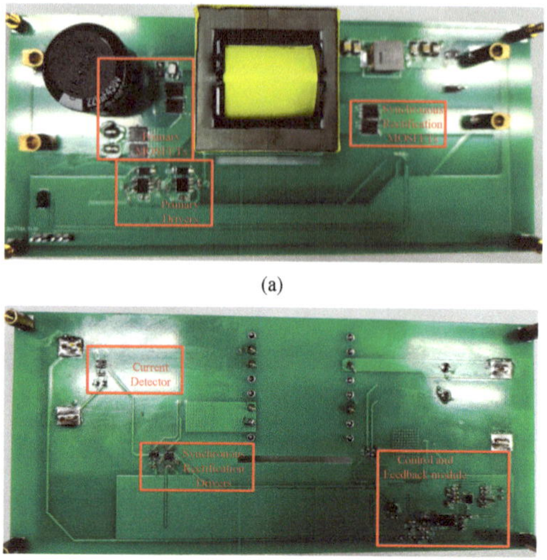

(a)

(b)

Figure 8. The experimental prototype: (**a**) the front side; (**b**) the back side.

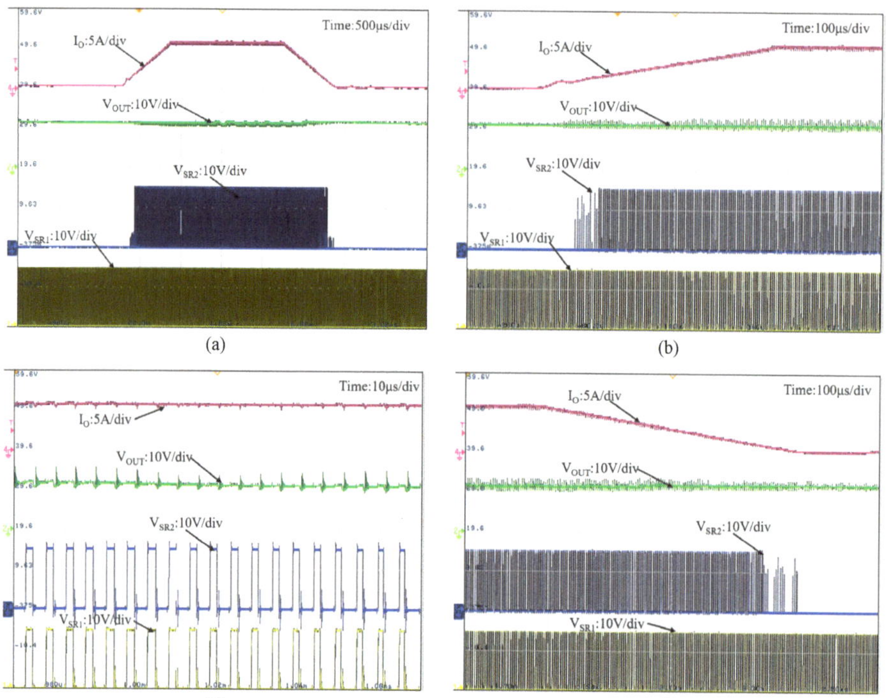

(a) (b)

(c) (d)

Figure 9. The working process of the hierarchical driving strategy: (**a**) the whole working process; (**b**) the turning-on process; (**c**) the working details; (**d**) the turning-off process.

After the functional validation of the proposed strategy, the efficiency of this study's converter and the conventional converter was tested under different load conditions to evaluate the improvement in efficiency of the proposed scheme for the converter. The efficiency of the converter was defined according to (4).

$$\eta = \frac{P_{OUT}}{P_{IN}} = \frac{V_{OUT} \cdot I_O}{V_{IN} \cdot I_{IN}} \tag{4}$$

where P_{OUT} is the output power and P_{IN} is the input power; the input voltage V_{IN}, input current I_{IN}, output voltage V_{OUT}, and output current I_O were tested with an MSOX4154A oscilloscope from Keysight.

Graphs of the efficiency tests on the two converters are shown in Figure 10. According to these graphs, it can be seen that in the light-load state—especially with a load current in the range of 0.5–1.5 A—the control scheme proposed in this study effectively improved the efficiency of the converter; the most obvious effect was at 0.5 A, where the efficiency improvement reached 1.4%. With the increase in the load current, the loss brought about by the drive accounted for a smaller and smaller proportion of the total loss, so the efficiency improvement became smaller and smaller. When the load current was 2 A, the control scheme proposed in this study lost its effect on the improvement in efficiency, and the efficiency of the traditional parallel-drive converter was slightly higher than that of this experimental scheme, which was mainly caused by the additional control losses brought about by it. With the further increase in the load, the proportion of the extra loss introduced in the total loss became smaller and smaller, and the efficiency of the two schemes was almost equal.

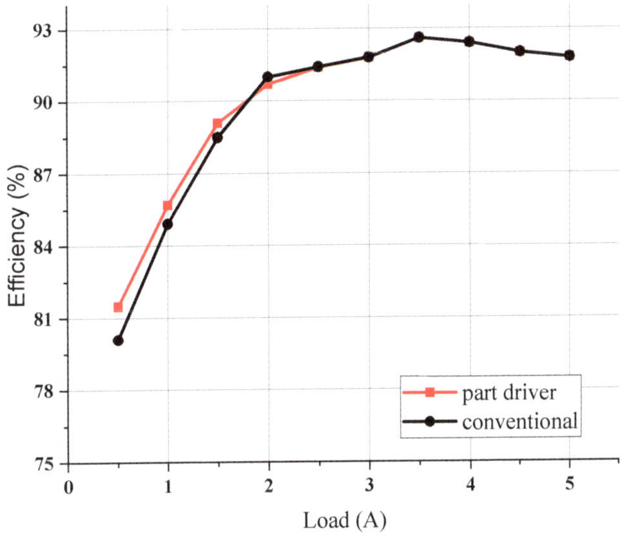

Figure 10. The efficiency results.

The real-world challenges of manufacturing variations and environmental factors must be considered. To reduce the power loss of the sense resistor, the gain of the current sense amplifier needed to be large enough. Manufacturing deviations in the sampling resistor and temperature coefficients of the current sense amplifier can cause a shift in the switching point of the hierarchical driving control strategy, which has an impact on the maximum gain of efficiency with the hierarchical driving control method. The efficiency improvement of this strategy was tested at different temperatures, and the test results are shown in Figure 11. This strategy worked effectively in the range from −25 °C to 75 °C. There were small fluctuations in the efficiency improvement at different operating temperatures, which were affected by the temperature characteristics of the key devices, such as the SiC MOSFET, the current sense amplifier, and the sense resistor.

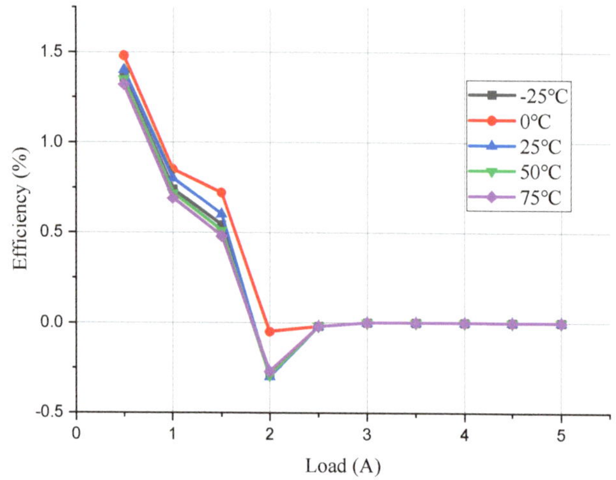

Figure 11. The efficiency improvements at different temperatures.

A comparison was made between state-of-the-art approaches and the approach proposed in this work; the results are shown in Table 5. The approaches proposed in [3,17] only verified the effectiveness of their modulation strategies, and efficiency was not mentioned. The approach proposed in [5] caused a 3% efficiency improvement at most with the optimal switching point design, but the highest efficiency was 85%. The authors of [4] introduced SiC MOSFETs to widen the ZVS range, and a 1.1% efficiency gain was realized with this strategy. All of the strategies focused on the number of converter cells, which was different from this work. Compared with the state-of-the-art approaches, this work had an improvement of 1.4% in efficiency with a negligible impact on the function under heavy loads.

Table 5. Comparison among state-of-the-art approaches.

State-of-the-Art Approach	Working Mode	Efficiency Gains
This work	Hierarchical driving control	1.4%
[3]	Multi-level state control scheme	Only verified the modulation strategy
[5]	Phase-shedding control scheme	3%
[4]	Phase-shedding control for QSW-ZVS	1.1%
[17]	Modulation strategy for 3P-DAB	Only verified the modulation strategy

5. Conclusions

In this study, for the first time, we analyzed the problem regarding the relationship of driving and conduction in high-frequency parallel applications with SiC MOSFETs. Based on the problems found in the loss analysis, a hierarchical driving strategy was proposed. As verified with simulations and experiments, the strategy was able to improve the efficiency by up to 1.4% under light loads and had almost no effects on the efficiency under heavy loads. This was the first time that SiC MOSFETs' drive and conduction losses were quantitatively analyzed, and a relative strategy was proposed to complete the optimization of efficiency under light loads. The application of high-frequency SiC MOSFETs can effectively improve the efficiency of converters under light loads, which is of great significance for giving full play to the performance advantages of SiC MOSFETs under high-frequency and high-current conditions. The application range of this hierarchical driving strategy is mainly in applications with large load ranges, such as in electric drive systems for electric vehicles, battery systems for the generation of new energy, and control processor power supply systems.

Author Contributions: Conceptualization, J.L. contributes to visualization, and writing—review & editing. Y.S. contributes to visualization, writing—original draft preparation. S.S. and X.T.; methodology, Y.L.; validation, Y.L., S.S. and X.T.; formal analysis, S.S.; investigation, X.T.; resources, X.T.; data curation, W.Y.; writing—original draft preparation, Y.L.; writing—review and editing, Y.L.; visualization, S.S.; supervision, X.T.; project administration, X.T.; funding acquisition, X.T. All authors have read and agreed to the published version of the manuscript.

Funding: This research was funded by the Key R&D Program of Shaanxi Province (grant number: 2023-YBGY-005).

Data Availability Statement: Data are contained within the article.

Conflicts of Interest: The authors declare no conflict of interest.

References

1. Tavassoli, M.; Kamran-Pirzaman, A. Comparison of effective greenhouse gases and global warming. In Proceedings of the 2023 8th International Conference on Technology and Energy Management (ICTEM), Babol, Iran, 8–9 February 2023; pp. 1–5. [CrossRef]
2. Alekseev, V.V.; Orlova, N.V.; Sarantseva, V.A.; Bryzgalo, V.S. The Analysis of Greenhouse Gas Emissions Rates for the Environmental Exposure Control System. In Proceedings of the 2023 XXVI International Conference on Soft Computing and Measurements (SCM), St. Petersburg, Russia, 24–26 May 2023; pp. 271–273. [CrossRef]
3. Wang, Z.; Lin, H.; Ma, Y.; Wang, X.; Wang, T.; Ze, Z. Analysis and Control Strategy of Modular Multilevel Converter with Integrated Battery Energy Storages System Based on Voltage Source Mode. In Proceedings of the 2018 20th European Conference on Power Electronics and Applications (EPE'18 ECCE Europe), Riga, Latvia, 17–21 September 2018; pp. P.1–P.9.
4. Vazquez, A.; Rodriguez, A.; Rogina, M.R.; Lamar, D.G. Different Modular Techniques Applied in a Synchronous Boost Converter with SiC MOSFETs to Obtain High Efficiency at Light Load and Low Current Ripple. *IEEE Trans. Ind. Electron.* **2017**, *64*, 8373. [CrossRef]
5. Su, J.T.; Liu, C.W. A Novel Phase-Shedding Control Scheme for Improved Light Load Efficiency of Multiphase Interleaved DC–DC Converters. *IEEE Trans. Power Electron.* **2013**, *28*, 4742–4752. [CrossRef]
6. Bashar, E.; Wu, R.; Agbo, N.; Mendy, S.; Jahdi, S.; Gonzalez, J.O.; Alatise, O. Comparison of Short Circuit Failure Modes in SiC Planar MOSFETs, SiC Trench MOSFETs and SiC Cascode JFETs. In Proceedings of the 2021 IEEE 8th Workshop on Wide Bandgap Power Devices and Applications (WiPDA), Redondo Beach, CA, USA, 7–11 November 2021; pp. 384–388. [CrossRef]
7. Karout, M.A.; Taha, M.; Fisher, C.A.; Deb, A.; Mawby, P.; Alatise, O. Impact of Diode Characteristics on 1.2 kV SiC MOSFET and Cascode JFET Efficiency: Body Diodes Vs SiC Schottky Barrier Diodes. In Proceedings of the 2023 IEEE Applied Power Electronics Conference and Exposition (APEC), Orlando, FL, USA, 19–23 March 2023; pp. 202–208.
8. Wirths, S.; Alfieri, G.; Romano, G.; Ceccarelli, E.; Arango, Y.; Mihaila, A.; Knoll, L. Gate Stress Study on SiN-Based SiC Power MOSFETs. In Proceedings of the 2022 IEEE 34th International Symposium on Power Semiconductor Devices and ICs (ISPSD), Vancouver, BC, Canada, 22–25 May 2022; pp. 245–248. [CrossRef]
9. Zhou, Y.; Yang, T.; Liu, H.; Wang, B. Failure Models and Comparison on Short-circuit Performances for SiC JFET and SiC MOSFET. In Proceedings of the 2018 1st Workshop on Wide Bandgap Power Devices and Applications in Asia (WiPDA Asia), Xi'an, China, 16–18 May 2018; pp. 123–129. [CrossRef]
10. She, X.; Huang, A.Q.; Lucia, O.; Ozpineci, B. Review of Silicon Carbide Power Devices and Their Applications. *IEEE Trans. Ind. Electron.* **2017**, *64*, 8193–8205. [CrossRef]
11. Zhang, Y.; Rong, G.; Qu, S.; Song, Q.; Tang, X.; Zhang, Y. A High-Power LED Driver Based on Single Inductor-Multiple Output DC-DC Converter with High Dimming Frequency and Wide Dimming Range. *IEEE Trans. Power Electron.* **2020**, *35*, 8501–8511. [CrossRef]
12. Peftitsis, D.; Rabkowski, J. Gate and Base Drivers for Silicon Carbide Power Transistors: An Overview. *IEEE Trans. Power Electron.* **2016**, *31*, 7194–7213. [CrossRef]
13. Zhou, M.; Lv, H.; Zhang, Y.; Xu, S.; Zhang, Y. High-speed gate driver circuit of SiC-MOSFET for high temperature application. *IET Power Electron.* **2020**, *13*, 3851–3860. [CrossRef]
14. Zhang, Y.; Song, Q.; Tang, X.; Zhang, Y. Gate driver for parallel connection SiC MOSFETs with over-current protection and dynamic current balancing scheme. *J. Power Electron.* **2020**, *20*, 319–328. [CrossRef]
15. Castellazzi, A.; Irace, A. *SiC Power Module Design: Performance, Robustness and Reliability*; The Institution of Engineering and Technology: London, UK, 2021; pp. 1–343.

16. Agarwal, A.; Kanale, A.; Han, K.; Baliga, B.J. Switching and Short-Circuit Performance of 27 nm Gate Oxide, 650 V SiC Planar-Gate MOSFETs with 10 to 15 V Gate Drive Voltage. In Proceedings of the 2020 32nd International Symposium on Power Semiconductor Devices and ICs (ISPSD), Vienna, Austria, 13–18 September 2020.
17. Choi, H.J.; Park, H.P.; Kim, M.A.; Sang-Gyu, C.; Lee, C.U.; Jung, J.H. Modulation Strategy of Three-Phase Dual-Active-Bridge Converter Using SiC-MOSFET for Improving Light Load Condition. In Proceedings of the 2019 IEEE Workshop on Wide Bandgap Power Devices and Applications in Asia (WiPDA Asia), Taipei, Taiwan, 23–25 May 2019; pp. 1–5.
18. Zhang, X.; Sheh, G.; Ji, I.H.; Banerjee, S. In Depth Analysis of Driving Loss and Driving Power Supply Structure for SiC MOSFETs. In Proceedings of the 2019 IEEE Applied Power Electronics Conference and Exposition (APEC), Anaheim, CA, USA, 17–21 March 2019.
19. Gonzalez, J.O.; Wu, R.; Wu, H.; Wang, X.; Pickert, V.; Mawby, P.; Alatise, O. Optimisation of the gate voltage in sic mosfets: Efficiency vs. reliability. In Proceedings of the The 10th International Conference on Power Electronics, Machines and Drives (PEMD 2020), Online, 15–17 December 2020; pp. 820–825.
20. Semicondutor, T. Half-Bridge Gate Driver. Technical Report TF2184. Rev.1.0. Available online: https://atta.szlcsc.com/upload/public/pdf/source/20210115/C2683757_99BA058D531FD039AC93C1414388FE01.pdf (accessed on 17 August 2020).

electronics

Article

Design of High-Performance Driving Power Supply for Semiconductor Laser

Bin Feng [1,*], Junfeng Zhao [1], Haofei Zhang [2], Tao Li [3] and Jianjun Mi [3]

1 School of Optoelectronic Engineering, Xi'an Technological University, Xi'an 710021, China; f3214449565@gmail.com
2 No. 208 Research Institute of China Ordnance Industries, Beijing 102202, China; jeamszhf@163.com
3 Xi'an Institute of Applied Optics, Xi'an 710065, China; litao6d7n@163.com (T.L.); hfgd2016@126.com (J.M.)
* Correspondence: fengbin@xatu.edu.cn

Abstract: High power semiconductor laser is a kind of photoelectric device with high efficiency and high stability, the performance of its drive system directly affects its output characteristics and service life. In order to solve the problems of stability and robustness of the output power of the semiconductor laser, a semiconductor laser driving power supply with high efficiency, low ripple and strong anti-interference ability was developed. In this paper, the topology of the LCC resonant converter is adopted (LCC refers to the type of resonant converter, because its resonator is composed of an inductor L and two capacitors C, it is called LCC resonant converter). The power supply adopts full-bridge LCC resonant power topology. Firstly, a mathematical model is established to analyze the relationship between LCC resonator parameters and output current gain. Secondly, an LCC resonator parameter design method is proposed to reduce the current stress of components, and the variable frequency phase shift (PFM-PWM) composite control strategy and linear active disturbance rejection control (LADRC) algorithm are proposed, which not only ensures the zero voltage (ZVS) conduction of MOS (Metal-Oxide-Semiconductor) tube, but also reduces the on-off loss of MOS tube. The PFM-PWM composite control strategy and LADRC algorithm not only improve the power efficiency of the drive power supply, suppress the output current ripple, but also ensure that the output current of the drive power supply is stable when the input voltage, load and parasitic parameters of the circuit change. Finally, the simulation and experimental results show that the power supply can be continuously adjustable in the output current range of 0–40 A, the current ripple is less than 0.8%, and the working efficiency is up to 92%. It has the characteristics of high stability, small ripple, high efficiency, low cost and good robustness.

Keywords: high power semiconductor laser; constant current driving; LCC-type resonant-circuit; low ripple; laser drive power circuit

Citation: Feng, B.; Zhao, J.; Zhang, H.; Li, T.; Mi, J. Design of High-Performance Driving Power Supply for Semiconductor Laser. *Electronics* **2023**, *12*, 4758. https://doi.org/10.3390/electronics12234758

Academic Editor: Nakkeeran Kaliyaperumal

Received: 8 November 2023
Revised: 19 November 2023
Accepted: 20 November 2023
Published: 23 November 2023

1. Introduction

High power semiconductor lasers have many advantages such as light weight, high efficiency, small size, high reliability and long service life, and are widely used in production, medical, aerospace, national defense and scientific research and other fields [1,2]. The extensive utilization of the high-power semiconductor laser is closely linked to the ongoing enhancement and advancement of its driving power supply [3,4]. The laser diode in the core device of the semiconductor laser pump source exhibits low resistance to electrical shock, making it susceptible to significant deviations in output optical power due to even slight current fluctuations. Additionally, transient current spikes in the drive power circuit can lead to damage of the laser diode and negatively impact its operational lifespan [5,6]. The laser diode in the core device of the semiconductor laser pump source exhibits limited resistance to electrical shock, and the performance of the laser power supply directly determines the overall performance of the entire laser system as the output power of the semiconductor laser continues to increase. Therefore, it is essential for the semiconductor laser drive power supply to possess high stability, high power density, low ripple, robustness, and adaptability to complex environmental applications [7].

Based on the load characteristics of semiconductor lasers, constant current mode is the preferred operating mode for their driving power supply. Both domestic and foreign scholars have conducted extensive research on designing driving powers for semiconductor lasers to meet various application requirements [8–10].

International research on semiconductor lasers commenced earlier, exemplified by the QCL (the code for a series of power supplies) driver power supply from Wavelength company and the LDC202C power supply developed by Thorlabs company [11,12]. These devices offer a precision of up to 0.05 mA, along with flexible adjustability and high stability; however, their cost is relatively high, making widespread adoption challenging. The researchers at Kassel University in Germany, Zhang et al. [13], employed an oscillator circuit to generate high-frequency pulses for precise control of semiconductor lasers, thereby introducing a novel approach to drive semiconductor lasers. Lu Yi et al., from the University of Electronic Science and Technology of China [14], proposed a double closed-loop constant current drive strategy. They utilized the LM25117 chip, which is based on a synchronous rectifier Buck power supply, to effectively regulate the steady-state loss of MOSFET in the constant current circuit. As a result, the steady-state operating efficiency of the power supply reached 88%, thereby significantly enhancing its overall power efficiency. The utilization of the LM2517 control chip exhibits limited flexibility and inadequate anti-interference capability. However, the LM2517 control chip lacks flexibility and has poor anti-interference ability. The researchers at Huazhong University of Science and Technology, led by Wang Chao et al. [15], developed a single-stage conversion switching DC drive power supply with a capacity of up to 2 kW using the parallel interleaving technology of single-tube positive cataclastic switching. This design achieved an impressive power efficiency of 85% and enabled dynamic regulation of output current from 0–100 A. However, precise control of the current balance for each individual tube converter is essential due to the large volume of the positive shock converter and the implementation of parallel interleaving technology, resulting in a complex control circuit and significant limitations in its application. Zhong Xulang et al. [16] from Dazu Laser Technology proposed a power supply structure that combines MOS tubes and operational amplifiers, enabling real-time acquisition of power data, multi-channel current output, and laser power error control within ±2%. But the use of operational amplifier MOS tube will cause the system power loss to be too large, the control flexibility to be poor, and the overall efficiency of the power supply to be low. The researchers at Yanshan University, led by Zhao Qinglin et al. [17,18], have developed a structure that combines the LCC resonant converter capacitor charging circuit with a pulse current circuit. This innovative design enables the pulse current output to reach an amplitude of 80 A, allowing for wide voltage output range and ensuring high flexibility and stability.

In order to address the limitations of the aforementioned semiconductor laser drive power supply, this paper adopts a topology consisting of full-bridge LCC resonant transformation, variable frequency phase shift (PFM-PWM) composite control strategy, and a linear active disturbance rejection control (LADRC) algorithm. This approach not only enhances the dynamic response speed, stability, and efficiency of the semiconductor laser drive power supply but also enables a wide range of input/output capabilities.

2. The Composition of a Power System

The proposed structure of the semiconductor laser power supply system is illustrated in Figure 1, comprising primarily of an LCC resonant converter module, a controller module, a current acquisition module, and an auxiliary power supply module.

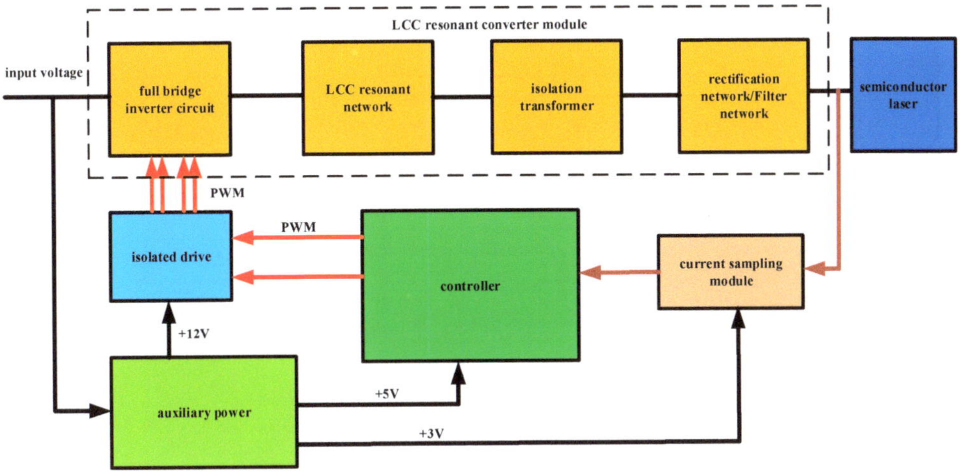

Figure 1. The power supply structure of semiconductor laser.

The LCC resonant converter module, depicted in Figure 1, serves as the cornerstone of the power supply system by delivering a constant current. It encompasses a full-bridge inverter network, an LCC resonant network, a full-bridge rectifier network, and an LC filter network. The current acquisition module primarily samples the output current into the controller module, establishing a closed-loop feedback control mechanism. The controller module utilizes the appropriate algorithm to calculate the sampling current, thereby generating the signal that drives the MOS tube. It also assists in supplying power to the controller, drive circuit, and current sampling circuit.

2.1. Mathematical Model Analysis of LCC Resonant Converter

The topology of the LCC resonant converter is illustrated in Figure 2. From a topological perspective, the LCC resonant converter exhibits not only the characteristics of a series resonant converter for DC component isolation and transformer protection but also possesses wide-range input/output regulation capabilities akin to those of a parallel resonant converter. Consequently, it demonstrates excellent constant current source characteristics and robust output short-circuit resistance [19,20]. In order to optimize the volume of resonant converter, reduce the circuit current and circuit equivalent capacitance, increase the characteristic impedance of resonant network, and reduce the current stress borne by the converter, the parallel resonant capacitor (C_P) is designed on the secondary side of the transformer, so that the parallel resonant capacitor (C_P) and series resonant inductor (L_r) become parasitic parameters that cannot be ignored by the transformer under high frequency operation. It has a good internal short-circuit protection function.

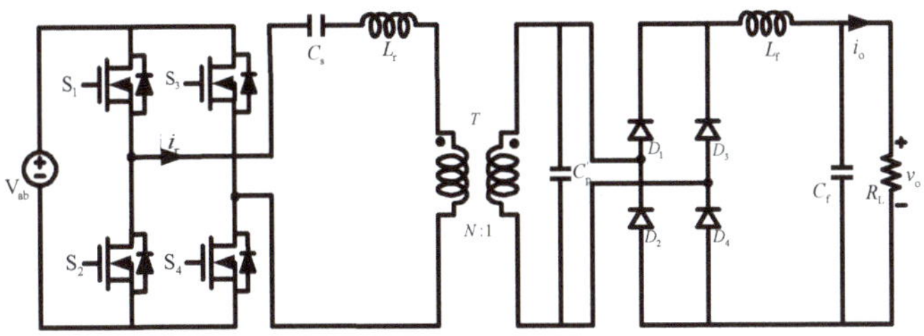

Figure 2. The Topology of LCC resonant converter.

Because the mathematical model of the LCC resonant converter is nonlinear, it is not conducive to the analysis of its working state, so it is necessary to linearize its nonlinear terms. When the frequency of the small-amplitude disturbed signal is much lower than the switching frequency of the system, the whole resonant converter can be considered as a quasi-steady-state system. In order to simplify the analysis, extended description function analysis [21,22] and linear differential equation theory are used to approximate the nonlinear terms, and the mathematical model of the steady-state system is obtained:

$$L_r(\frac{di_{rs}}{dt} - \omega_s i_{rc}) + v_{C_s s} + n v_{C'_p s} = \frac{4 v_g}{\pi} \sin(\frac{d}{2}) \tag{1}$$

$$L_r(\frac{di_{rc}}{dt} + \omega_s i_{rs}) + v_{C_s c} + n v_{C'_p c} = 0 \tag{2}$$

$$i_{rs} = C_s(\frac{dv_{C_s s}}{dt} - \omega_s v_{C_s c}) \tag{3}$$

$$i_{rc} = C_s(\frac{dv_{C_s c}}{dt} + \omega_s v_{C_s s}) \tag{4}$$

$$C'_p(\frac{dv_{C'_p s}}{dt} - \omega_s v_{C'_p c}) + \frac{4 i_{L_f}}{\pi A_p} v_{C'_p s} = n i_{rs} \tag{5}$$

$$C'_p(\frac{dv_{C'_p c}}{dt} + \omega_s v_{C'_p s}) + \frac{4 i_{L_f}}{\pi A_p} v_{C'_p c} = n i_{rc} \tag{6}$$

$$L_f \frac{di_{L_f}}{dt} + i_{L_f} r'_c + (1 - \frac{r'_c}{R_L}) v_{C_f} = \frac{2}{\pi} \sqrt{v^2_{C'_p s} + v^2_{C'_p c}} \tag{7}$$

$$\frac{R_L}{r_c + R_L} C_f \frac{dv_{C_f}}{dt} + \frac{1}{R_L} v_{C_f} = i_{L_f} + i_o \tag{8}$$

The output voltage v_o is as follows:

$$v_0 = R_L i_o = i_{L_f} r'_c + (1 - \frac{r'_c}{R_L}) v_{C_f} \tag{9}$$

i_o is the output current; v_o the output voltage; i_{rs} and i_{rc} are amplitude of sinusoidal component and cosine component of resonant current i_r, respectively; $v_{C_s s}, v_{C_s c}$ are the amplitude of the sinusoidal and cosine components of the series resonant capacitor voltage v_{C_s}, respectively; $v_{C'_p s}, v_{C'_p c}$ are the amplitude of the sinusoidal and cosine components of the series resonant capacitor voltage $v_{C'_p}$, respectively; ω_s is the operating angular frequency; n indicates the ratio of turns of the transformer; v_g indicates the input voltage; d represents duty cycle; i_{Lf} represents the current through the filtered inductor L_f; v_{Cf} represents the voltage of filter capacitor C_f. r_c represents the ESR of the filter capacitance; R_L indicates the resistance of the load.

Solve Equations (1) through (9), The voltage gain of the open loop DC point of the LCC resonant converter is obtained:

$$G_V = \frac{V_o}{V_g} = \frac{\omega_s R_L C_s \sin(\frac{d}{2})}{n\sqrt{(1 - \omega_s^2 L_r C_e)^2 + (\omega_s R_e (C_p + C_s)(1 - \omega_s^2 L_r C_e))^2}} \tag{10}$$

$C_e = C_s C_p/(C_p + C_s)$ represents the equivalent capacitance of the resonant network.

According to the relationship between voltage gain and current gain, the current gain can be obtained by normalizing Equation (10):

$$G_I = \frac{I_o}{I_i} = G_V Q_L = \frac{Q_L \sin(\frac{d}{2})}{n\sqrt{(1+A)^2(1-\omega)^2 + Q_L^2(\omega - \omega^{-1}\frac{A}{1+A})^2}} \tag{11}$$

$\omega = \omega_s/\omega_r$, ω_r is the angular frequency of the resonant network, $\omega_r = \sqrt{1/L_rC_e}$, $Q_L = Z_r/R_e$, $Z_r = \sqrt{L_r/C_e}$, Z_r indicates the characteristic impedance.

It can be seen from Equation (11) that when $\omega = 1$, the LCC resonant converter exhibits constant current characteristics, and the output current is independent of the load, but only related to the input voltage and resonance parameters. When the LCC resonant converter works in this mode, the output current can remain unchanged even if the semiconductor laser load changes. In order to study the influence of current gain G_I on Q_L, A, d and n, assuming $A = n = 1$ and $d = \pi$, the change curve of current gain with different values of Q_L and A is drawn, as shown in Figure 3.

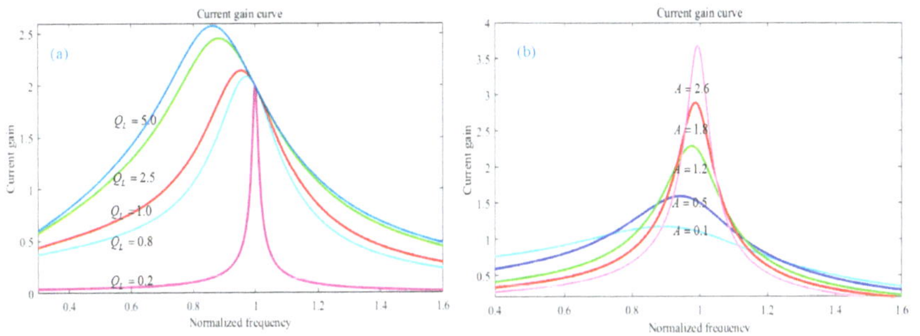

Figure 3. Parameter influence curve of current gain: (**a**) Q_L value current gain influence curve; (**b**) *A* value current gain influence curve.

As can be seen from Figure 3a, when the QF value increases continuously, the value of the current gain also increases. When the load is open ($\lim Q_L \rightarrow \infty$), the voltage gain G_I will increase rapidly and tend to infinity, which will damage the power system in serious cases. When $\omega = 1$, the current gain is independent of the load, showing constant current characteristics; As can be seen from Figure 3b, when the capacitance ratio (*A* value) increases, the current gain will change in the same direction as the *A* value, that is, with the increase in the *A* value, and when $\omega > 1$, the smaller *A*, the more gradual the gain change which was mentioned above.

The values of duty cycle (*d*) and coil turns (*n*) of the transformer also affect the LCC resonant converter, and the gain influence curve of its output current and voltage is shown in Figure 4. The above two graphs are the relationship curves of duty cycle (*d*) and frequency ratio (ω) with voltage gain and current gain. The *X*-axis is the duty cycle, and the *Y*-axis is the voltage gain and the current gain, respectively. The following two graphs are the relationship curves of the transformer turn ratio (*n*) and frequency ratio (ω) with voltage gain and current gain. The *X*-axis is the ratio of turns, and the *Y*-axis is the voltage gain and the current gain, respectively.

According to the influence curves of *d* and *n* values on current and voltage gain, it can be seen that the voltage/current gain is affected by the duty cycle *d*, the number of turns of the transformer *n*, and the frequency ratio ω. When the duty cycle and the number of turns of the transformer are determined, the output current and voltage gain can be adjusted by adjusting the frequency ratio ω. At the same time, under the same duty cycle and transformer turns, when $\omega > 1$, the voltage gain/current gain change is small. When $\omega < 1$, the voltage/current gain change difference is large.

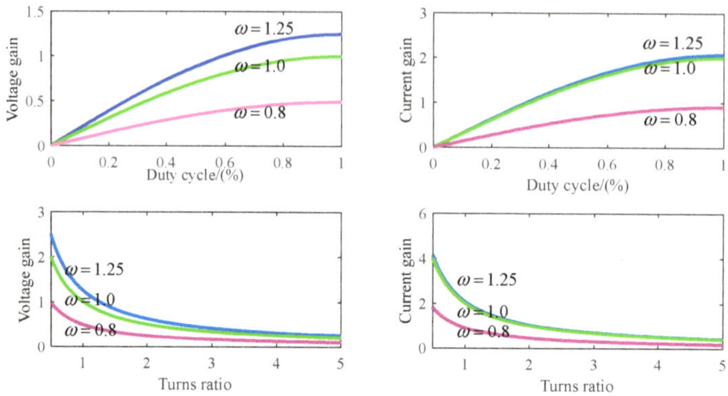

Figure 4. Effect curve of d and n values on current-voltage gain.

2.2. Design of LCC Resonator Parameters

Ignoring the loss in the power transmission of the LCC resonant converter, the resonant current peak I_r is obtained from the power conservation, which is expressed as follows:

$$I_r = \frac{2\pi V_i G_I^2}{Q_L^2 R_L \cos\varphi} \quad (12)$$

where, φ represents the impedance Angle.

The characteristic curve of resonant current peak I_r affected by capacitance ratio A, quality factor Q_L, impedance Angle φ and input voltage V_i is shown in Figure 5. It can be seen from the characteristic curve that when the converter current gain, load and input voltage are constant, the smaller the φ is, the smaller the value of I_r is, and the smaller the current stress of the switching tube is. Only by ensuring $\varphi > 0$ can the ZVS conduction of the MOS tube be ensured, and a sufficient margin is often left in the actual selection, generally within the range of $15^\circ < \varphi < 30^\circ$. At the same time, too small a quality factor and capacitance ratio will cause the resonant current peak to be too large, and the increase in input voltage will also cause the resonant current to increase.

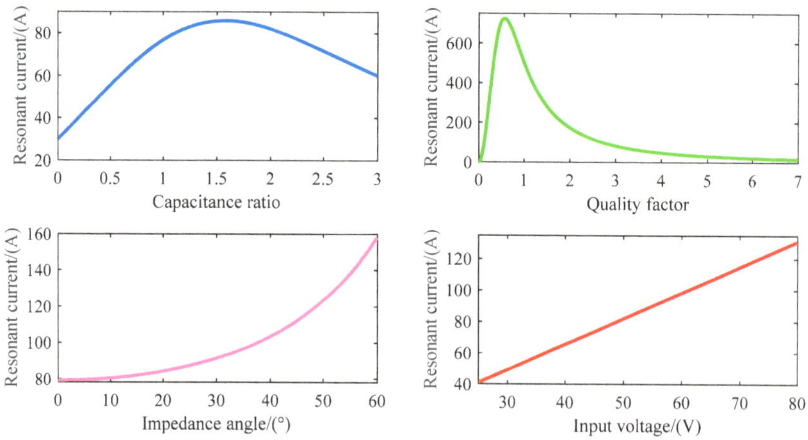

Figure 5. Resonant current peak characteristics.

In order to obtain more accurate parameters of the resonant network, reduce the current stress of the resonant components, and ensure that the resonant converter can achieve ZVS conduction, the following parameter design steps are given:

(1) The peak range of resonant current is determined according to input voltage, output power and output current;

(2) Select the appropriate resonant frequency and operating frequency range;

(3) The quality factor and capacitance ratio are determined according to the resonant current peak range;

(4) The transformer ratio n is calculated according to Formula (13):

$$n = \frac{G_v \sin^2(\frac{\theta}{2})}{2\cos\varphi} \tag{13}$$

In Formula (13), θ represents the conduction Angle of the rectifier tube;

(5) Finally, the LCC resonator parameters are calculated by Formulas (14)–(16).

Series resonant capacitance:

$$C_s = \frac{\sqrt{1+A}}{2\pi n^2 f_r Q_L R_e} \tag{14}$$

Parallel resonant capacitance:

$$C'_p = n^2 C_s / A \tag{15}$$

Resonant inductance:

$$L_r = \frac{1}{(2\pi f_r)^2}\frac{A+1}{C_p} \tag{16}$$

3. Power System Control Strategy and Algorithm Analysis

In order to overcome the disadvantages of the traditional control strategy, the variable frequency phase-shift composite control strategy is adopted, and its structure is shown in Figure 6.

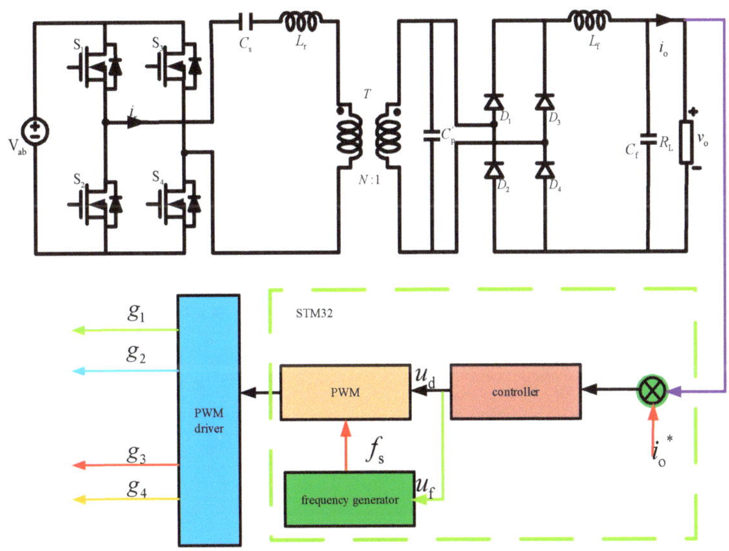

Figure 6. Variable frequency phase shifting compound control strategy structure.

In Figure 6, i_o* represents the given value of output current, u_d and u_f represent the control quantity of the phase and frequency of the controller output, respectively. f_s represents the carrier signal frequency, and g_1~g_4 represents the MOSFET drive signal, respectively.

This control strategy enables the controller to keep the output current stable by adjusting the switching frequency and the on-angle at the same time when the input/output changes, and still realize the soft switching under the condition of narrow switching frequency and wide input voltage/wide output range, thus improving the working efficiency and the service life of the device.

When the output current is increased, the increase in switching frequency f_s will lead to the decrease in the on-angle φ. Therefore, under the combined action of f_s and φ, the converter can achieve a wide input voltage in a narrow switching frequency range and achieve steady current output and soft switching under a wide output power condition.

The traditional PID controller relies too much on the system model when designing parameters. In view of the complexity of the semiconductor laser load characteristics, the PID controller is easy to produce integral saturation and insufficient anti-interference ability, which makes the PID controller difficult to meet the design requirements. Therefore, the LADRC control algorithm is adopted, and its current loop structure is shown in Figure 7, where $G_s(s)$ represents the control object transfer function, LESO represents the linear expansion observer, k_p represents the scale coefficient of the controller, k_d represents the differential coefficient, and b_0 represents the compensation factor. The design of this algorithm does not depend on the exact model and can observe and track the disturbance of the system in real time. It can overcome the shortcomings of the PID controller and restrain the current instability caused by multi-factor disturbance.

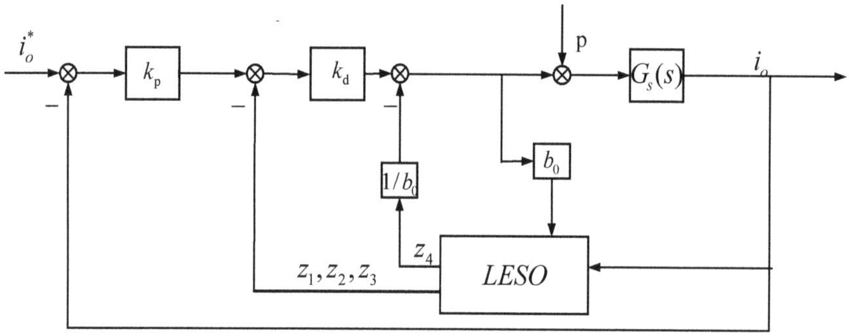

Figure 7. Linear active disturbance rejection current loop control algorithm.

The total disturbance p of the system can be estimated by the linear extended observer (LESO):

$$\begin{aligned} \dot{z} &= Az + Bu + L(y - \hat{y}) \\ \hat{y} &= Cz \end{aligned} \quad (17)$$

$$\begin{aligned} A &= \begin{bmatrix} 0 & 1 & 0 \\ 0 & 0 & 1 \\ 0 & 0 & 0 \end{bmatrix} \\ B &= \begin{bmatrix} b_0 & 0 & 0 \end{bmatrix}^T \\ E &= \begin{bmatrix} 0 & 0 & 0 \end{bmatrix}^T \\ C &= \begin{bmatrix} 1 & 0 & 0 \end{bmatrix} \\ L &= \begin{bmatrix} \beta_1 & \beta_2 & \beta_3 \end{bmatrix}^T \end{aligned} \quad (18)$$

where, u and y represent the input and output of the system, respectively. L is LESO gain matrix; z is LESO state variable; $\hat{y}$ is the output value estimated by the system.

The transfer function of the LCC resonant converter can be approximated as a second-order system, so it is necessary to design a third-order extended observer to observe the system state and the total disturbance.

$$\begin{aligned} \dot{z}_1 &= z_2 + l_1(y - z_1) \\ \dot{z}_2 &= z_3 + l_2(y - z_1) + b_0 u \\ \dot{z}_3 &= l_3(y - z_1) \end{aligned} \quad (19)$$

According to the bandwidth method [23], observer bandwidth ω_o, all poles of the extended observer are assigned in the left half plane, and the parameter tuning problem is simplified to the selection of observer bandwidth, and the corresponding observer gain is as follows:

$$\begin{aligned} l_1 &= 3\omega_o \\ l_2 &= 3\omega_o^2 \\ l_3 &= \omega_o^3 \end{aligned} \quad (20)$$

The reconstructed system after disturbance compensation can be approximated as an integral series system. When the disturbance is observable, the second-order system can be reconstructed as a second-order integral series system by disturbance compensation. After disturbance compensation, the series integral system can obtain good control performance through simple linear error feedback rate control. The controller design is as follows:

$$u_o = k_p(r - z_1) - k_d(\dot{r} - z_2) \quad (21)$$

where, r represents the given input; $\dot{r}$ represents the first derivative of a given input; u_0 indicates the controller output.

4. Simulation and Test Results

In order to verify the correctness of the semiconductor laser power supply scheme, Matlab/Simulink is used to build a semiconductor laser drive power supply simulation model. The prototype test platform with the LCC resonant converter and STM32 controller as the core is shown in Figure 8. By testing the constant current output characteristic, anti-interference characteristic, working efficiency and ripple characteristic of the semiconductor laser driving power supply, it is proved that the design is reasonable and meets the design requirements.

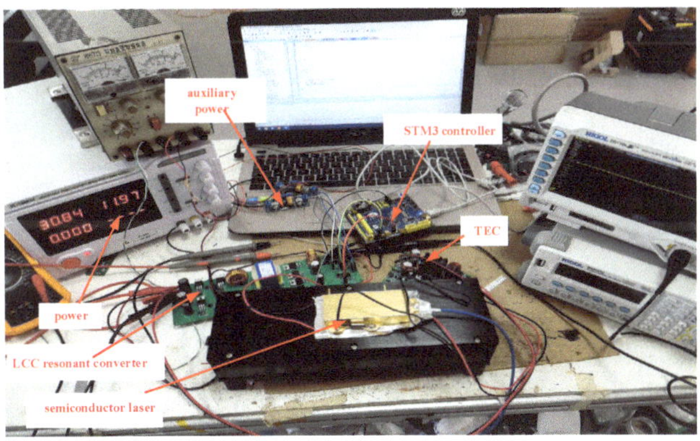

Figure 8. Prototype test platform.

The output voltage of the test prototype is 10~15 V, the output current is 0~40 A adjustable, the operating frequency is 51.5 kHz, the resonant capacitor C_s is 354 nF, the resonant capacitor C'_p is 9.03 μF, the resonant inductor L_r is 42.9 μH, and the transformer turns ratio n is 3.5.

4.1. Constant Current Characteristic Test of Semiconductor Laser Power Supply

The simulation and test results of the constant current (40 A) characteristics of the semiconductor laser power supply are shown in Figure 9, where V_{gs} and V_{ds} are, respectively, the drive voltage and the drain-source voltage of the MOS tube, V_{ab} is the input voltage of the resonant network, I_r is the resonant current, and V_{cp} is the shunt capacitor voltage. It can be seen from the simulation and test results that the variable frequency phase-shift composite control method extends the input voltage/output power range of the LCC resonant converter by synchronously adjusting the switching frequency and duty ratio, narifies the switching frequency range, and effectively reduces the current and voltage stress of the switching tube. The output current ripple coefficient is <0.8%, and the output voltage ripple coefficient is <0.45%. It can realize ZVS conduction under any power input voltage, effectively reduce the loss of power tube, and improve the performance of LCC resonant converter.

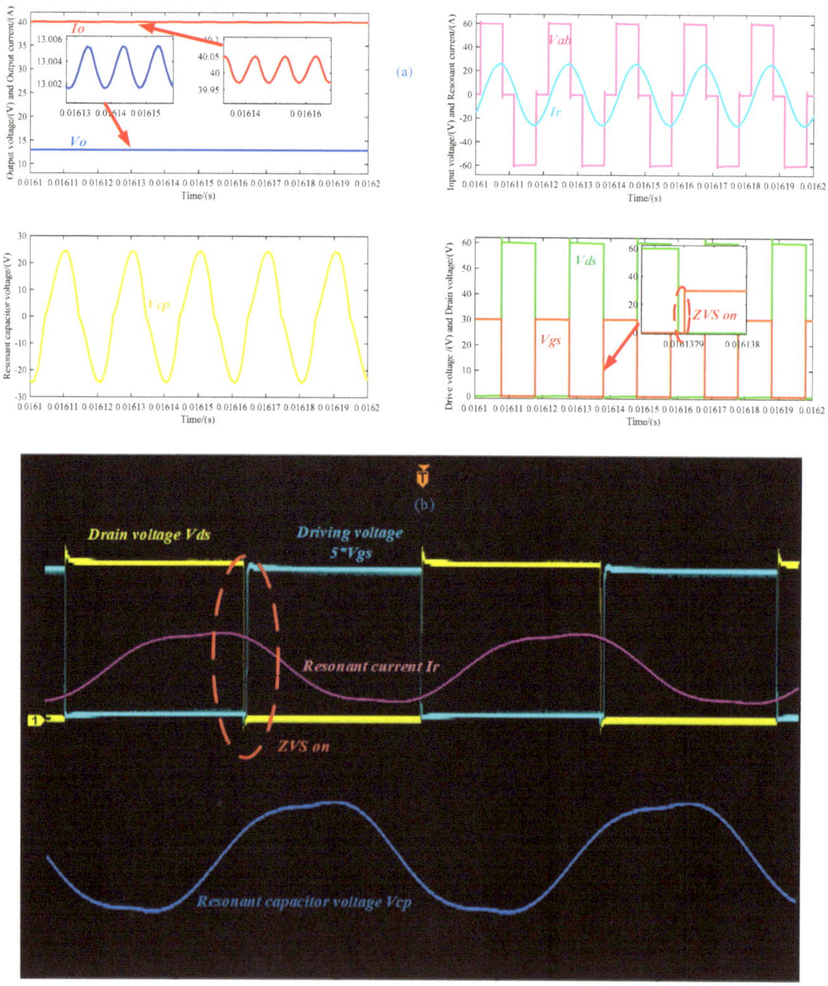

Figure 9. *Cont.*

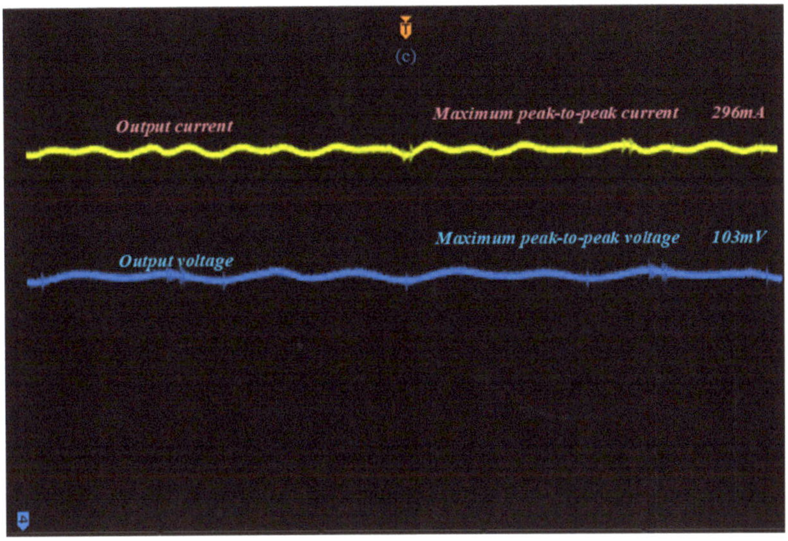

Figure 9. Simulation and experimental results of constant current characteristics of semiconductor laser power supply: (**a**) simulation waveform; (**b**) resonant network waveform; (**c**) output characteristic waveform.

4.2. Experiment on Anti-Interference Characteristics of Semiconductor Laser Power Supply

The simulation and test results of anti-interference characteristics of semiconductor laser power supply are shown in Figure 10. When t = 0.02~0.03 s, the input voltage V_{in} = 60 V changes to V_{in} = 80 V, and when t = 0.04~ 0.05 s, the load switches from full load to light load. According to the simulation and test results, the output current of the PID algorithm has a fast response speed, a large overjump, a large error and a large current ripple. When the input voltage or load is transformed, the ripple of the current and voltage increase, and the current and voltage remain at the set value. Therefore, the LADRC algorithm has a good performance in anti-interference.

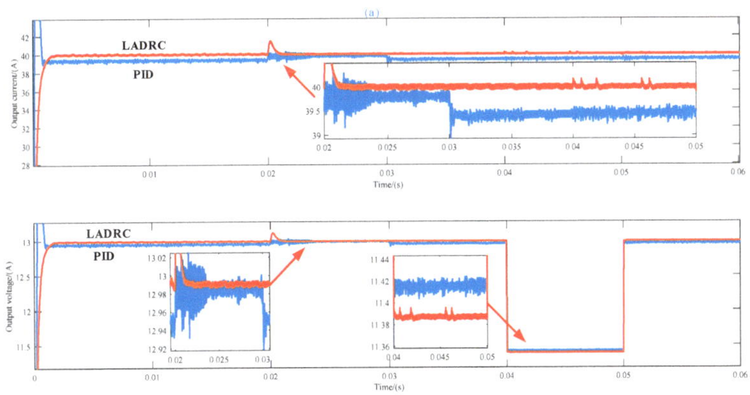

Figure 10. *Cont.*

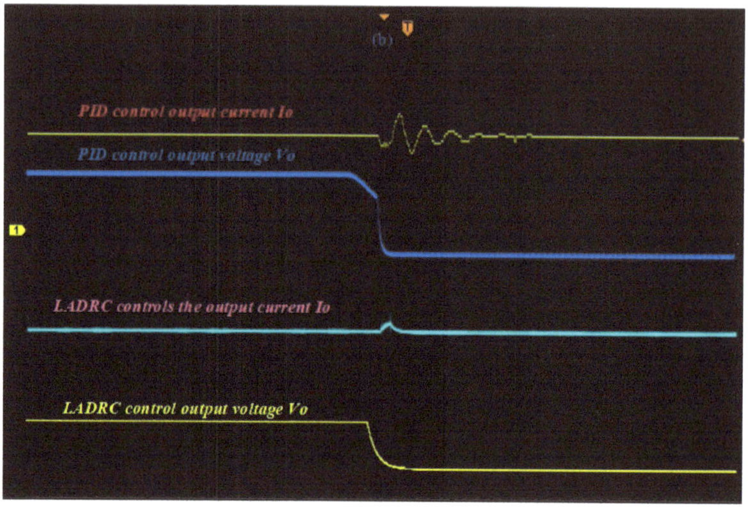

Figure 10. Simulation and experimental results of anti-jamming characteristics of semiconductor laser power supply: (**a**) simulated waveform; (**b**) experimental waveform.

4.3. Power Efficiency Characteristic Test of Semiconductor Laser

The simulation and test outcomes of the semiconductor laser power supply efficiency, as illustrated in Figure 11, reveal that the conventional frequency conversion control exhibits limited output adjustment capability and stability, leading to an overall low power supply efficiency. Consequently, only traditional phase shift control and composite control with frequency conversion phase shift are subjected to testing and analysis. Based on the simulation and test results, various control strategies yield differing power efficiencies; specifically, the phase-shifting control strategy induces hard switching states in the converter switch, thereby increasing switching losses and reactive power losses of the resonant converter. By implementing the inverter phase-shifting compound control under full load conditions, a maximum power supply efficiency of 92% can be achieved concurrently with superior constant current characteristics.

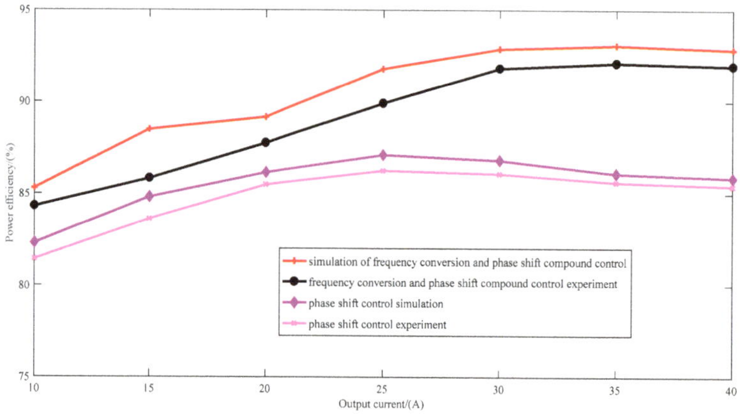

Figure 11. Simulation and experimental results of power efficiency of semiconductor laser.

5. Conclusions

In this paper, a high-performance semiconductor laser power supply system is designed. By establishing the mathematical model of the LCC resonant converter, the parameters of the LCC resonant device are designed, the disadvantages of traditional control strategy algorithm are analyzed, and the inverter phase-shift compound control strategy and LADRC algorithm are proposed, which not only improves the power efficiency of semiconductor laser, but also realizes the ZVS on-switching of switching tube. It reduces the on-off loss, reduces the overall power consumption by about 3% compared with the same type of power supply, and also has a strong anti-interference ability, fundamentally avoids the limitations of PID control algorithm applied in the control of the LCC resonant converter, reduces the current and voltage stress of the LCC resonant converter components, and improves the life and reliability of semiconductor laser drive power supply. The above characteristics show the unique characteristics of the LCC resonant converter application, which is suitable for high performance semiconductor laser drive power supply.

Author Contributions: Conceptualization, B.F.; Methodology, B.F.; Investigation, J.Z. and T.L.; Data curation, H.Z. and J.M. All authors have read and agreed to the published version of the manuscript.

Funding: This research was funded by Shaanxi Science and Technology Department (Program No.2020GY-158) and 2022 "Insight Action" Achievement Transformation and Application Project (Program No.628020320).

Data Availability Statement: Data are contained within the article.

Acknowledgments: The authors thank all the anonymous reviewers for their very helpful comments to improve the paper.

Conflicts of Interest: The authors declare no conflict of interest.

References

1. Yang, Y.F. Research on high power semiconductor laser. *Electron. Compon. Inf. Technol.* **2022**, *6*, 41–43.
2. Ning, Y.Q.; Chen, Y.Y.; Zhang, J.; Song, Y.; Lei, Y.; Qiu, C.; Liang, L.; Jia, P.; Qin, L.; Wang, L. Brief review of development and techniques for high power semiconductor lasers. *Acta Opt. Sin.* **2021**, *41*, 191–200.
3. Liux, X.; Han, J.H.; Cai, H.; Yang, F.; Rong, K.; An, G.; Wang, Y. Review of high repetition-rate mid-infrared lasers for photoelectric countermeasures. *Laser Technol.* **2021**, *45*, 271–279.
4. Xu, Y.F.; Shi, Y.J.; Shao, J.Z.; Lin, Y.; Shen, Q.; Liang, X. Design of high precision pulse power supply for high power semiconductor laser. *Laser Technol.* **2023**, *47*, 108–114.
5. Tian, Y.L.; Li, C.S.; Zhang, Z.Y. High-accuracy and high-stability constant current power for semiconductor lasers. *J. Xi'an Jiaotong Univ.* **2019**, *53*, 1–5.
6. Yang, C.L.; Jia, H.Z.; Xia, G.Z. Design for the circuit to eliminate surge of the laser diode power supply. *Appl. Laser* **2008**, *28*, 310–313.
7. Wu, J.W.; Qi, Y.; Wu, X.K. High-efficiency laser diode driver for aerospace applications. *Aerosp. Shanghai* **2020**, *37*, 81–87.
8. Thompson, M.T.; Schlecht, M.F. High power laser diode driver based on power converter technology. *IEEE Trans. Power Electron.* **1997**, *12*, 46–52. [CrossRef]
9. Yang, J.T.; Wang, J.A.; Wang, Y.; Hu, X. Design and research of a new high performance power supply for semiconductor lasers. *Chin. J. Electron. Devices* **2021**, *44*, 817–823.
10. Zhao, Q.L.; Li, S.; Cao, R.R.; Wang, D.; Yuan, J. Design of pulse power supply for high-power semiconductor laser diode arrays. *IEEE Access* **2019**, *7*, 92805–92812. [CrossRef]
11. Zhang, L.; Chen, J.S.; Gao, J.; Tan, H.; Wu, X. Design of driving power and temperature control system for high power semiconductor laser. *Infrared Laser Eng.* **2018**, *47*, 102–108.
12. Liu, J.N.; Zheng, H.; Sun, Y.H.; Fang, Y.; Tie, J.; Fan, C.Y.; Zhang, C.B. Research on a multi-branch micro-second laser drive power and control technology for laser cleaning. *Chin. J. Electron. Devices* **2021**, *44*, 322–327.
13. Zhang, Z.F.; Zhai, Y.S.; Zhan, S.U.; Su, Y.-L.; Wang, X.-J.; Du, Y.-X. A novel method to drive semiconductor laser for optical frequency standard based on constant voltage source. *Optik* **2015**, *126*, 3924–3926. [CrossRef]
14. Lü, Y.; Gao, J.Y.; Jiang, Y.J. Design of efficient power supply for high power infrared laser diode. *Laser Infrared* **2019**, *49*, 309–314.
15. Wang, C.; Ma, X.M.; Yu, P.P.; Guo, F.; Wang, Z.Q. Research of laser diode driver with interleaved single ended forward conversion. *Laser Technol.* **2016**, *40*, 665–669.
16. Zhong, X.L.; Wang, J.; Luo, Y.H.; Deng, J.H.; Li, X.T. Research on power control system of semiconductor laser with active temperature. *Appl. Laser* **2021**, *41*, 608–613.
17. Zhao, Q.L.; Cao, R.R.; Wang, D.Y.; Yuan, Y. Design of pulse power supply for diode pumped solid state laser. *High Power Laser Part. Beams* **2018**, *30*, 129–134.
18. Cao, R.R.; Wang, D.Y.; Zhao, Q.L.; Li, S. Wide range voltage of diode laser driver. *High Power Laser Part. Beams* **2018**, *30*, 12–15.

19. Yuan, Y.S.; Yi, C.Y.; Peng, N. T-type half-bridge LCC resonant converter with L-R composite modulation. *Trans. China Electrotech. Soc.* **2022**, *37*, 892–904.
20. Liao, H.F.; Shuai, D.X.; Long, T.Y. Research on Li-ion battery charger based on LCC resonant converter. *Chin. J. Power Sources* **2021**, *45*, 669–672.
21. Zhan, L.N.; Wang, C.F.; Ma, C. Modeling of LLC resonant converter based on extended describing function method. *Syst. Simul. Technol.* **2014**, *10*, 211–216.
22. Hu, F.Z.; Xiong, J.F.; Qi, C.C. Modeling and analysis of bidirectional full-bridge CLLLC resonant converter. *J. Air Space Early Warn. Res.* **2019**, *33*, 446–450.
23. Ma, M.; Liao, P.; Cai, Y.X.; Lei, R.T.; He, Y. First-order active disturbance rejection control and parameter tuning method based on particle swarm optimization for LCL grid-connected inverter. *Electr. Power Autom. Equip.* **2021**, *41*, 174–182.

MDPI AG
Grosspeteranlage 5
4052 Basel
Switzerland
Tel.: +41 61 683 77 34

Electronics Editorial Office
E-mail: electronics@mdpi.com
www.mdpi.com/journal/electronics

www.ingramcontent.com/pod-product-compliance
Lightning Source LLC
Chambersburg PA
CBHW041358210926
51850CB00013B/3113
* 9 7 8 3 7 2 5 8 3 2 1 6 3 *